Gong Harmonic Analysis on Classical Groups

Gong Sheng

Harmonic Analysis on Classical Groups

Springer-Verlag Berlin Heidelberg GmbH

Gong Sheng
University of Science and Technology of China
Hefei, Anhui
The People's Republic of China

Revised edition of the original Chinese edition published by Science Press Beijing 1983 as the 12th volume in the Series in Pure and Applied Mathematics.

Mathematics Subject Classification (1980): 20-XX, 43-XX

ISBN 978-3-540-17652-7 ISBN 978-3-642-58189-2 (eBook)
DOI 10.1007/978-3-642-58189-2

Originally published by Springer-Verlag Berlin Heidelberg and Science Press Beijing in 1991

Typesetting: Science Press, Beijing. The People's Republic of China

41/3140 –553210

Preface to the English Version

The study of harmonic analysis on classical groups in China was began by my advisor, Professor Hua Luogeng. In the development of theory for harmonic analysis on classical domains in several complex variables, Professor Hua determined the complete orthonormal system of continuous functions on unitary groups and proved the Abel summability theorem for the Fourier series of the continuous functions on unitary groups. Under his influence, a systematical study of harmonic analysis on the most important classical groups—unitary, rotation and unitary symplectic groups was undertaken. Professor Hua guided and encouraged us to pursue research in this direction. He also inspired me to compile the results in the area into this monograph and wrote a preface for it. It is unfortunate that Professor Hua did not live to see the subsequent fruitful development of the harmonic analysis on groups, including the compact Lie groups.

I am very glad to see the appearence of the English edition and hope that it is useful for those who want to study harmonic analysis on groups.

Sheng Gong

June 20, 1991

Preface

H.Weyl studied harmonic analysis on compact groups of finite dimension. He proved that an orthonormal system exists and that any continuous function on these groups can be approximated by some finite linear combination of functions in this system. His research, however, seems to be too abstract to yield an explicit expression for the orthonormal system. Thus, we cannot talk about the form of the approximation, nor about its convergence.

The simplest example of compact groups is $\{e^{i\theta}\}$, on which there exists an orthonormal system

$$\{e^{in\theta}\}, \ n = 0, \pm 1, \pm 2, \cdots,$$

namely

$$\frac{1}{2\pi}\int_0^{2\pi} e^{in\theta}e^{-im\theta}d\theta = \begin{cases} 1, & \text{for } n = m; \\ 0, & \text{for } n \neq m. \end{cases}$$

The harmonic analysis on this compact group refers to the whole Fourier analysis. So far, extensive literature has been available on this topic. Its remarkable progress is evidenced by the great monograph of seven-hundred pages in two volumes written by A. Zygmund in 1959.

An immediate extension for $\{e^{i\theta}\}$ is group U_n, which consists of all $n \times n$ square matrices U satisfying

$$U\overline{U}' = I,$$

where $\overline{U}'$ denotes the conjugate transpose matrix of U. As for construction, there is a close relation between the group U_2 and the group SO_3. Besides, the application of U_n has been found more and more important in physics.

So far in harmonic analysis [beyond abstract compact groups and the simplest case $U_1 = \{e^{i\theta}\}$], no further research on concrete compact groups has been conducted. In the study of harmonic analysis on classical domains of several complex variables, the preface's author has determined the complete orthonormal system on U_n and proved the Abel summability theorem. From this, no doubt, one might easily deduce a linear expression for approximation.

In the meantime any compact group can be "embedded" into U_n. Thus the object under discussion can have both concreteness and generality.

On the other hand, Abel summation is considered as a preliminary method for treating harmonic analysis. In the theory of Fourier analysis there exist a great number of fundamental problems. For example, $(C, 1)$ summation, contributed by Fejér and Lebesgue, is the most suitable method in summation.

To the problem of extending the Cesàro summability to the harmonic analysis on U_n, Professor Gong Sheng (Sheng Kung) has contributed, with some creative techniques, some results parallel to those obtained by Lebesgue and others. He has also obtained a series of results on the harmonic analysis on U_n. This theory could be beneficial to modern physics.Thus one would ask whether there are results in harmonic analysis on U_n corresponding to the classical theory of Fourier analysis.

This is a direction with brilliant prospects. So it is my great pleasure to preface this book.

Hua Luogeng (Hua Loo-Keng)

Contents

PART I

HARMONIC ANALYSIS ON UNITARY GROUPS

Chapter 0. Preliminary

§ 0.1 Introduction

In this book the representation theory of groups is used as a main tool to study harmonic analysis on classical groups. The whole book consists of three parts. The first part deals with harmonic analysis on unitary groups, the second part with rotation groups, and the last part with 'unitary symplectic groups. To some extent, the methods used in the three parts are similar. Besides, each part is self-contained.

Part I contains the author's systematic study of harmonic analysis on unitary groups under the guidance of Prof. Hua during 1959—1962. It is Prof. Hua who successfully applied representation theory of groups in studying harmonic analysis on classical domains of several complex variables. The related results are summed up in Hua's famous monograph [1]. His work has exerted a profound influence on many fields. For example, the following theorem can be deduced from Hua's results (see Hua[2]): Fourier series of any continuous function on unitary groups is Abel summable to itself. This is a pioneer work for harmonic analysis both on unitary and on compact Lie groups and is an interesting theorem from the point of view of the representation theory of groups. The well-known Peter-Weyl theorem ([1]) asserts that any continuous function on a compact group can be approximated by a linear combination of irreducible representations of the group. The Peter-Weyl theorem is an abstract one concerning approximation, while Hua's theorem concerns the convergence of the Fourier series of continuous functions on any finite-dimensional (>1) compact group. Evidently the latter result is superior to the former.

From group theory, the unit circle is equivalent to the unitary group of dimension 1 and the general unitary groups can be considered as the direct extension of the unit circle. As a consequence, harmonic analysis on unitary groups is referred to a direct extension of classical harmonic analysis. Since any compact topological group can be imbedded in some unitary group to become its subgroup, any compact homogeneous space can be regarded as a quotient of two compact groups. Besides, a continuous function on any compact subgroup of a unitary group can be extended to a continuous function on the unitary group. Therefore, the study of harmonic analysis on unitary groups is meaningful even for general compact groups and for compact homogeneous space. From the function theory of several complex variables, if the Fourier analysis of several variables is harmonic analysis on characteristic manifolds of a complex polydisc, then the Fourier analysis on unitary groups can be viewed as the harmonic analysis on characteristic manifolds for classical domains $\mathscr{R}_I(n, C)$ of the first class of several complex variables; thus harmonic analysis on rotation groups can be regarded as harmonic analysis on characteristic manifolds[1)] of real classical domains $\mathscr{R}_I(n,R)$ of the first class; and that on unitary symplectic groups can be considered as the harmonic analysis on characteristic manifolds of classical domains $\mathscr{R}_I(n, Q)$ of the first class on quaternion fields (for the details readers are referred to the last two parts of this book). In other words, what we shall discuss is the harmonic analysis on characteristic manifolds of classical domains of the first class on complex fields, real number fields and quaternion fields respectively.

The method used to deal with harmonic analysis on groups may have some other applications. By using this method, Zhong Jiaqing [3] obtained an explicit formula for Schubet calculation on the Grassmann manifold.

The manuscript of the first part of this book was completed in 1965 with a view to publication. For some reason, this did not come true. In this English version some minor revisions were made, some mistakes corrected and some new materials added.

§ 0.2 Harmonic Analysis on Unitary Groups

Let U_n be a unitary group of order n. If $U \in U_n$, write $A_{f_1 \cdots f_n}(U)$ for the unitary representation of U with the signature $(f_1, f_2, \cdots, f_n)$, where $f_1, f_2, \cdots, f_n$ are integers such that $f_1 \geqslant f_2 \geqslant \cdots \geqslant f_n$ and $(f_1, f_2, \cdots, f_n)$ is usually denoted by f. Set $N(f_1, f_2, \cdots, f_n)$ to be the order of $A_{f_1 \cdots f_n}(U)$. If

$$A_f(U) = (a_{ij}^f(U)), 1 \leqslant i, j \leqslant N(f),$$

then $\{\varphi_{ij}^f(U)\}$ stands for an orthonormal system on U_n, where

$$\varphi_{ij}^f(U) = \sqrt{\frac{N(f)}{C}}\, a_{ij}^f(U),$$

1) The characteristic manifold $\mathscr{C}$ of $\mathscr{R}$ is by definition the part of the boundary of $\mathscr{R}$ which possesses the following properties: i) Any harmonic function on $\mathscr{R}$ attains its maximum on $\mathscr{C}$; ii) For any point ξ on $\mathscr{C}$ there exists a harmonic function $f(x)$ on $\mathscr{R}$ which attains its maximal absolute value at the point.

C is the volume of U_n, i. e.

$$C=(2\pi)^{\frac{1}{2}n(n+1)}/((n-1)!(n-2)!\cdots 2!1!).$$

Besides, $\{\varphi_{ij}^{f}\}$ denotes a complete system of integrable functions. Let

$$\Phi_f(U)=(\varphi_{ij}^{f}(U))_{1\leqslant i,j\leqslant N(f)}=\sqrt{\frac{N(f)}{C}}\,A_f(U).$$

Then $u(U)$, if integrable, can be expanded into the Fourier series

$$\sum_{f_1\geqslant f_2\geqslant\cdots\geqslant f_n}\operatorname{tr}\,(c_{f_1\cdots f_n}\Phi'_{f_1\cdots f_n}(U)),\tag{0.2.1}$$

where

$$c_{f_1\cdots f_n}=\int_{U_n}u(V)\overline{\Phi_{f_1\cdots f_n}(V)}\dot{V},$$

$\operatorname{tr}B$ represents the trace of B, and B' the transpose of B. More clearly there exists

$$u(U)\sim\sum_{f_1\geqslant\cdots\geqslant f_n}\operatorname{tr}\left(\frac{N(f)}{C}\int_{U_n}u(V)\overline{A_{f_1\cdots f_n}(V\overline{U}')}\dot{V}\right)$$

$$=\frac{1}{C}\int_{U_n}u(V)\sum_{f_1\geqslant\cdots\geqslant f_n}N(f)\chi_{f_1\cdots f_n}(\overline{V}U')\dot{V},\tag{0.2.2}$$

where $\chi_{f_1\cdots f_n}(U)$ is the character of $A_{f_1\cdots f_n}(U)$. First, it can be easily proved that the Fourier series of $u(U)$ is equal to the real part if $u(U)$ belongs to a real function, i. e.

$$\sum_{f_1\geqslant\cdots\geqslant f_n}\operatorname{tr}(C_{f_1\cdots f_n}\Phi_{f_1\cdots f_n}(U))=\sum_{f_1\geqslant\cdots\geqslant f_n}\operatorname{Re}\operatorname{tr}\,(C_{f_1\cdots f_n}\Phi'_{f_1\cdots f_n}(U)).$$

If $U,V\in U_n$, then

$$u(VU)\sim\frac{1}{C}\int_{U_n}u(W)\sum_{f_1\geqslant\cdots\geqslant f_n}N(f)\chi_{f_1\cdots f_n}(\overline{W}U'V')\dot{W}$$

$$=\sum_{f_1\geqslant\cdots\geqslant f_n}\frac{N(f)}{C}\operatorname{tr}\left(\left(\int_{U_n}u(W)\overline{A_{f_1\cdots f_n}(W)}\ \overline{A_{f_1\cdots f_n}(U')}\right.\right.$$

$$\left.\overline{A_{f_1\cdots f_n}(\overline{V}')}\dot{W}\right)$$

$$=\sum_{f_1\geqslant\cdots\geqslant f_n}\sqrt{\frac{N(f)}{C}}\operatorname{tr}\,(C_{f_1\cdots f_n}A'_{f_1\cdots f_n}(U)A'_{f_1\cdots f_n}(V)).$$

However, we have

$$u(\bar{V}U) \sim \sum_{f_1 \geqslant \cdots \geqslant f_n} \frac{N(f)}{C} \operatorname{tr}\left(\int_{U_n} u(W) A_{f_1 \cdots f_n}(\bar{W}U'\bar{V}')\dot{W}\right)$$

$$= \sum_{f_1 \geqslant \cdots \geqslant f_n} \frac{N(f)}{C} \operatorname{tr}\left(\int_{U_n} u(W) A_{-f_n \cdots -f_1}(W\bar{U}'V')\dot{W}\right)$$

$$= \sum_{f_1 \geqslant \cdots \geqslant f_n} \frac{N(f)}{C} \operatorname{tr}\left(\int_{U_n} u(W) A_{f_1 \cdots f_n}(W) \overline{A'_{f_1 \cdots f_n}(U)}\right.$$

$$\left. \cdot A_{f_1 \cdots f_n}(V')\dot{W}\right)$$

$$= \sum_{f_1 \geqslant \cdots \geqslant f_n} \sqrt{\frac{N(f)}{C}} \operatorname{tr}\left(\bar{C}_{f_1 \cdots f_n} \overline{A'_{f_1 \cdots f_n}(U)} A_{f_1 \cdots f_n}(V')\right).$$

Since

$$A_{f_1 \cdots f_n}(U) = \overline{A'_{-f_n, \cdots, -f_1}(U)}$$

and

$$N(f_1, \cdots, f_n) = N(-f_n, \cdots, -f_1).$$

Hence

$$\frac{u(VU) + u(\bar{V}U)}{2} \sim \frac{1}{2} \sum_{f_1 \geqslant \cdots \geqslant f_n} \sqrt{\frac{N(f)}{C}} \operatorname{tr}\left[C_{f_1 \cdots f_n} A'_{f_1 \cdots f_n}(U)\right.$$

$$\left. \cdot A'_{f_1 \cdots f_n}(V) + \overline{C_{f_1 \cdots f_n} A'_{f_1 \cdots f_n}(U) A_{f_1 \cdots f_n}(V')}\right].$$

Taking $V = I$, it follows that

$$u(U) \sim \sum_{f_1 \geqslant \cdots \geqslant f_n} \operatorname{Re} \operatorname{tr}\left(C_{f_1 \cdots f_n} \Phi'_{f_1 \cdots f_n}(U)\right).$$

§ 0.3 Harmonic Functions

Suppose $\mathscr{M}$ is a Riemann manifold of dimension m, the fundamental tensor $g_{ij}(i, j = 1, 2, \cdots, m)$ have continuous partial derivatives of order 1 and $x = (x^1, \cdots, x^m)$ denotes any local coordinate system. Then we have the Beltrami operator

$$\Delta = \sum_{i,j=1}^{m} g^{ij}\left(\frac{\partial^2}{\partial x^i \partial x^j} - \sum_{k=1}^{m} \begin{Bmatrix} k \\ ij \end{Bmatrix} \frac{\partial}{\partial x^k}\right),$$

where g^{ij} signifies the contravariant tensor of g_{ij} and

$$\begin{Bmatrix} k \\ ij \end{Bmatrix} = \frac{1}{2} \sum_{l=1}^{m} g^{kl}\left(\frac{\partial g_{il}}{\partial x^j} + \frac{\partial g_{jl}}{\partial x^i} - \frac{\partial g_{ij}}{\partial x^l}\right)$$

denotes the Christoffel symbol.

Let $u(x)$ be a real function defined on $\mathscr{M}$ with continuous partial derivatives of order 2. Then $u(x)$ is said to be harmonic if

$$\Delta u(x) = 0. \tag{0.3.1}$$

For classical domains $\mathscr{R}_I: I^{(n)} - Z\bar{Z}' > 0$, where $I^{(n)}$ denotes the unit matrix of order n, $Z = (z_{ij})_{1\leqslant i,j\leqslant n}$, the equation (0.3.1) becomes

$$\sum_{\alpha,\beta=1}^{n}\sum_{j,k=1}^{n}\left(\delta_{\alpha\beta} - \sum_{l=1}^{n}\bar{z}_{l\alpha}z_{l\beta}\right)\cdot\left(\delta_{jk} - \sum_{r=1}^{n}\bar{z}_{jr}z_{kr}\right)\frac{\partial^2 u(Z)}{\partial\bar{z}_{j\alpha}\partial z_{k\beta}} = 0. \quad (0.3.2)$$

This is known as the Beltrami-Hua's equation on R_I and the operator Δ is called the Beltrami-Hua operator.

Hua Luogeng [3] solved the following Dirichlet problem. Given a continuous function on U_n, there exists a unique solution of (0.3.2) which takes the same values on U_n as the given function does. Precisely speaking, the solution is the so-called Poisson integral

$$\frac{1}{C}\int_{U_n} u(U)\frac{\det^n(I - Z\bar{Z}')}{|\det(I - Z\bar{U}')|^{2n}}\dot{U},$$

where $u(U)$ is a given function on U_n and C the volume of U_n. In view of these facts, Hua defined the Abel summation of the Fourier series on U_n. From the theory of harmonic functions we observe

$$\lim_{Z\to U}\frac{1}{C}\int_{U_n} u(V)\frac{\det^n(I - Z\bar{Z}')}{|\det(I - Z\bar{V}')|^{2n}}\dot{V} = u(U), \quad (0.3.3)$$

where Z tends to U along a path out of contact with the boundary. For the related results and their further development, readers are referred to Hua's [1] and Lu Qikeng's[1].

Particularly, taking $Z = rU (0 < r < 1)$ in (0.3.3), we have

$$\lim \frac{1}{C}\int_{U_n} u(V)\frac{(1 - r^2)^{n^2}}{|\det(I - rU\bar{V}')|^{2n}}\dot{V} = u(U). \quad (0.3.4)$$

Thus the Abel mean of the Fourier series of $u(U)$

$$\sum_{f_1\geqslant\cdots\geqslant f_n} \rho^f(r)\mathrm{tr}(C_{f_1\cdots f_n}\Phi'_{f_1\cdots f_n}(U)), \quad (0.3.5)$$

tends to $u(U)$ as $r\to 1$, where

$$\rho^f(r) = \frac{1}{N(f)}\cdot\frac{1}{C}\int_{U_n}\frac{(1 - r^2)^{n^2}\chi_f(U)\dot{U}}{|\det(I - rU)|^{2n}}. \quad (0.3.6)$$

§0.4 Summation of the Fourier Series

In Chapter 1, two different proofs of (0.3.4) are given. The simpler one was proposed by Hua Luogeng. The other proof is deduced by directly using techniques in harmonic analysis, instead of applying the theory on harmonic functions of several complex variables. It is representative and is applicable to those cases where suitable conditions are satisfied. For example, the method can prove: If $u(U)$ is continuous on U_n, then the Fejer mean of its Fourier series converges to itself.

For the value of $\rho^f(r)$, we have

(1) $\rho^f(r) \to 1$, as $r \to 1$;

(2) $$\rho^f(r) = \begin{cases} r^{f_1+\cdots+f_n}, & \text{if } f_1 \geqslant \cdots \geqslant f_n \geqslant 0; \\ r^{-f_1-\cdots-f_n}, & \text{if } 0 \geqslant f_1 \geqslant \cdots \geqslant f_n. \end{cases}$$

In the same chapter, the explicit expression of $\rho^f(r)$ is given. For calculating it there are two different methods. One is due to the present author and the other was given by Zhong Jiaqing in 1974.

In Chapter 2, the Cesàro mean of the Fourier series (0.2.1) is well defined and the following Riesz-type theorem is proved: Assume that $u(U)$ is continuous on U_n. Then its Fourier series is (C, α)-summable to itself if $\alpha > \frac{n-1}{n}$. Whether the constant $\frac{n-1}{n}$ can be further improved is an open problem. As is well-known, it is the best possible constant for $n = 1$.

Besides, also in Chapter 2, the Fejér mean is under discussion, which is the most important and the most typical case of Cesàro mean.

It should be indicated that all (C, α) kernels obtained below can be simply expressed by matrices. For example, Fejér kernels read

$$\frac{1}{\mathrm{D}}\left|\frac{\det\,(I - V^{N+1})}{\det\,(I - V)}\right|^{2n},$$

where $V \in U_n$ and D is an absolute constant depending on n and N.

§ 0.5 Criteria of Convergence

The partial sum of the Fourier series (0.2.1) is

$$\sum_{N \geqslant l_1 > \cdots > l_n \geqslant -N} \operatorname{tr}\,(C_{f_1 \cdots f_n} \Phi'_{f_1 \cdots f_n}(U)), \tag{0.5.1}$$

which is just the so-called "cubical" partial sum and whose Dirichlet kernels are given in Chapter 3. There are also two methods for proving it. One is the algebraic method given by Hua. Dirichlet kernels of (0.5.1) can be written as

$$\frac{1}{(n-1)!\cdots 2!1!} \cdot \frac{D\left(\frac{\partial}{\partial \lambda_1}, \cdots, \frac{\partial}{\partial \lambda_n}\right)}{D(\bar{\lambda}_1, \cdots, \bar{\lambda}_n)} \cdot \frac{\det\,(\bar{V}'^N - V^{N+1})}{\det\,(I - V)},$$

where $\lambda_1, \cdots, \lambda_n$ are characteristic roots of V, and D is the Vandermonde determinant.

Moreover, even if $u(U)$ is continuous on U_n, (0.5.1) does not necessarily converge as $N \to \infty$. A natural problem arises: What conditions would guarantee its convergence? In Chapter 3, the following criterion of convergence is proved: (0.5.1) is convergent and converges to $u(U)$ if $u(U)$ belongs to the class $C^{\frac{n(n-1)}{2}+p}(U_n)$ $(0 < p < 1)$. In addition,other criteria of absolute convergence are given as well.

§ 0.6 Approximation Theory on Compact Topological Groups

The Peter-Weyl theorem only indicated the existence of approximation, but did not offer an explicit expression. As a compact topological group can be mapped into a unitary group isomorphically, the approximation theory on compact topological groups can be established by the results of the first three chapters. This will be discussed in Chapter 4.

In Chapter 4, first of all, a definition of continuous modular of a continuous function on any compact topological group is given. Here, not only a finite linear approximation formula of a continuous function but the estimation of possible errors are given. There it is proved that, for example, if $u(g)$ is continuous on a compact topological group G and belongs to Lip p, then there must exist a finite linear approximation $P_N(g)$ depending on N such that

$$|u(g) - P_N(g)| < C/N^p,$$

where C is an absolute constant and $0 < p < 1$.

§ 0.7 Spherical Summation

Chapter 5 discusses spherical summation. The so-called spherical summation means that (0.2.1) can be regarded as

$$\sum_{m=0}^{\infty} \sum_{\substack{f_1 \geqslant \cdots \geqslant f_n \\ l_1^2+\cdots+l_n^2=m}} \operatorname{tr}\left(C_{f_1\cdots f_n}\Phi'_{f_1\cdots f_n}(U)\right), \tag{0.7.1}$$

where $l_1 = f_1 + n - 1, l_2 = f_2 + n - 2, \cdots, l_n = f_n$.

Naturally, the meaning of the convergence and the summation of (0.7.1) is different from that given before.

In that chapter, a general theorem on the convergence of spherical means of (0.7.1) is given first. The important Riesz mean (in the sense of spherical summation) is included in the theorem, i. e., the following theorem is proved:

If $u(U)$ is continuous on U_n, then the Riesz mean of order δ of its Fourier series (0.7.1)

$$\left(1-\frac{(2n-1)n(n-1)}{6R^2}\right)^{-\delta}\sum_{\substack{f_1\geqslant\cdots\geqslant f_n\\ l_1^2+\cdots+l_n^2\leqslant R^2}}\left(1-\frac{l_1^2+\cdots+l_n^2}{R^2}\right)^{\delta}$$

$$\cdot\,\mathrm{tr}(C_{f_1\cdots f_n}\Phi'_{f_1\cdots f_n}(U)),\tag{0.7.2}$$

converges to $u(U)$ for $\delta>\frac{n^2-1}{2}$ as $R\to\infty$.

In addition, another Tauber-type convergence theorem is presented.

Chapter 1. Abel Summation of Fourier Series on Unitary Groups

§ 1.1 Poisson-Hua Kernels of Classical Domains

In this section, we introduce briefly the Poisson-Hua kernels of classical domains of Class I of several complex variables. Readers may refer to Hua's work [1] for further details.

E. Cartan [1] proved that there are only six possible classes of transitive and irreducible bounded symmetric domains. Two of them are exceptional, of which one is of 16 dimensions and the other 27 dimensions. According to Hua [1], the rest four classes are called classical domains and are completely expressed in terms of matrices. As a consequence they are included in the category of geometry of matrices studied by Hua himself, in which Class I means the so-called hyperbolic space of $m \times n$ matrices and is denoted by $\mathscr{R}_1$. More precisely, Class I consists of $m \times n$ complex matrices Z such that

$$I^{(m)} - Z\bar{Z}' > 0,$$

where $I^{(m)}$ is the unit matrix of order m, $\bar{Z}'$ is the conjugate transpose of Z (thus $\bar{Z}'$ is a nxm matrix) and $H > 0$ denotes a positively definite H for the Hermite square matrix H.

The characteristic manifold $\mathscr{C}$ is such a part of the boundary of $\mathscr{R}$ that

(1) Any holomorphic function on $\mathscr{R}$ takes its maximum modulus on $\mathscr{C}$.

(2) For a given point on $\mathscr{C}$, there exists an holomorphic function on $\mathscr{R}$ of which the maximum modulus is reached at that point.

Those matrices Z on the boundary of $\mathscr{R}_I$ satisfying $Z\bar{Z}' = I$ constitute the characteristic manifold $\mathscr{C}_I$.

Now consider the case when $m = n$. Here the two conditions satisfied by the elements of $\mathscr{R}_I$ and $\mathscr{C}_I$ become

$$I^{(n)} - Z\bar{Z}' > 0$$

and

$$Z\bar{Z}' = I^{(n)}$$

respectively (i. e. $\mathscr{C}_I$ becomes the unitary group U_n of order n). The volume of $\mathscr{C}_I$ is written as

$$C = (2\pi)^{\frac{1}{2}n(n-1)}/((n-1)!\,(n-2)!\cdots 2!1!).$$

The group Γ_I of all holomorphic automorphisms of $\mathscr{R}_I$ consists of all the transformations

$$W = (AZ + B)\,(CZ + D)^{-1}, \tag{1.1.1}$$

where

$$\bar{A}A' - \bar{B}B' = I, \quad \bar{A}C' = \bar{B}D', \quad \bar{C}C' - \bar{D}D' = -I$$

and

$$\det\begin{pmatrix} A & B \\ C & D \end{pmatrix} = 1.$$

It is easily shown that A, B, C and D also satisfy

$$A'\bar{A} - C'\bar{C} = I, \quad A'\bar{B} = C'\bar{D}, \quad B'\bar{B} - D'\bar{D} = -I. \tag{1.1.2}$$

From the identity

$$(Z\bar{B}' + \bar{A}')(AZ + B) = (Z\bar{D}' + \bar{C}')(CZ + D)$$

we get another expression of (1.1.1)

$$W = (Z\bar{B}' + \bar{A}')^{-1}(Z\bar{D} + \bar{C}'). \tag{1.1.3}$$

(1.1.2) and (1.1.3) lead to

$$I - \bar{W}W' = (\bar{Z}B' + A')^{-1}\,(I - \bar{Z}Z')(\bar{B}Z' + \bar{A})^{-1},$$

i. e. (1.1.1) changes $\mathscr{R}_I$ and $\mathscr{C}_I$ into $\mathscr{R}_I$ and $\mathscr{C}_I$ respectively.

For $\mathscr{C}_I$ (i. e. U_n), define (1.1.1) by

$$V = (AU + B)\,(CU + D)^{-1}. \tag{1.1.4}$$

This would change a unitary matrix U into a unitary matrix V. Differentiating (1.1.4) and using (1.1.3), we obtain

$$dV = (U\bar{B}' + \bar{A}')^{-1}dU(CU + D)^{-1}.$$

Let $\delta V = V^{-1}dV$ and $\delta U = U^{-1}dU$. It follows from (1.1.3) that

$$\delta V = \overline{(CU + D)}'^{-1}\delta U(CU + D)^{-1}.$$

Write $\dot{U}$ to be the unitary volume element spanned by the differential vectors, then

$$\dot{V} = |\det(CU + D)|^{-2n}\dot{U}. \tag{1.1.5}$$

The reason is: If matrix $X = AYB$ is regarded as a linear relation between independent parameters, then the determinant of the linear relation is $\det (AB)^n$.

Choose P to be an arbitrary point in $\mathscr{R}_I$ (i. e. $I - \bar{P}P' > 0$). Evidently it is equivalent to $I - P'\bar{P} > 0$. Hence, there exist two nonsingular matrices Q and R such that

$$\bar{Q}(I - \bar{P}P')Q' = I, \quad \bar{R}(I - P'\bar{P})R' = I.$$

Let $A = Q$, $B = -QP$, $C = -R\bar{P}'$ and $D = R$. They obviously satisfy (1.1.2). Therefore

$$W = Q(Z - P)\ (-\bar{P}'Z + I)^{-1}R^{-1} \tag{1.1.6}$$

belongs to Γ_I and maps $Z = P$ onto $W = 0$.

By (1.1.5), (1.1.6) changes the volume element into

$$\begin{aligned}\dot{V} &= |\det\ (-R\bar{P}'U + R)|^{-2n}\dot{U} \\ &= |\det\ R|^{-2n}|\det\ (\bar{U}'P - I)|^{-2n}\dot{U}.\end{aligned}$$

Considering the relation $|\det\ R|^2 = |\det\ (I - P'\bar{P})|^{-1}$, we have

$$\dot{V} = \frac{|\det\ (I - P'\bar{P})|^n}{|\det\ (P - U)|^{2n}}\dot{U}.$$

For any $Z \in \mathscr{R}_I$ and $U \in \mathscr{C}_I$, we define the function

$$P(Z,\ U) = \frac{\det^n(I - Z\bar{Z}')}{|\det(Z - U)|^{2n}}$$

to be the Poisson-Hua kernel of $\mathscr{R}_I$. Starting from this definition we can prove the following

Theorem 1.1.1 *If $u(U)$ is continuous on U_n, then*

$$u(U) = \lim_{r\to 1}\frac{1}{C}\int_{U_n} P(rU, V)u(V)\dot{V}. \tag{1.1.7}$$

Proof. Without loss of generality, we may suppose $U = I$. For if we let $\phi(U) = u(UU_0)$ for any $U_0 \in U_n$ and assume that (1.1.7) is valid for $U = I$, then

$$\phi(I) = \lim_{r\to 1}\frac{1}{C}\int_{U_n} P(rI,\ V)\ \phi\ (V)\dot{V},$$

which is equivalent to

$$u(U_0) = \lim_{r\to 1}\frac{1}{C}\int_{U_n} P(rI,\ V)\ u\ (VU_0)\dot{V}.$$

By substituting V by $r\ U_0^{-1}$ in the formula

$$P(rU_0, V) = \frac{(1-r^2)^{n^2}}{|\det\ (I - rU_0\bar{V}')|^{2n}} = P(rI, V\bar{U}_0'),$$

(1.1.7) follows. Thus we only need to prove

$$u(I) = \lim_{r\to 1} \frac{1}{C} \int_{U_n} P(rI, V)\, u\,(V)\dot{V}. \tag{1.1.8}$$

The relation between volume elements of the transformation

$$W = -(V - rI)\ (-rV + I)^{-1}$$

is

$$\dot{W} = (1 - r^2)^{n^2}\ |\det\,(I - rV)|^{-2n}\, \dot{V} = P(rI, V)\dot{V}.$$

Therefore

$$\int_{U_n} P(rI, V)\, u\ (V)\dot{V} = \int_{U_n} u((rI - W)\ (I - rW)^{-1})\dot{W}.$$

For proving (1.1.8) it suffices to prove

$$\lim_{r\to 1} \int_{U_n} |u(I) - u\,(rI - W)/(I - rW)\ |\dot{W} = 0,$$

which immediately follows from the Lebesgue Theorem.

The original proof of this theorem is very complicated. The theorem together with its succinct proof given here was due to Hua Luogeng. The method used can also be applied to general transitive domains. Besides, the important idea that the Poisson kernels are regarded as coefficients between volume elements corresponding to elements of holomorphic automorphisms was also advanced by Hua. It can be seen that this theorem only belongs to "local properties". Thus it can be easily extended to the case where $u(U)$ is Lebesgue integrable.

§ 1.2 Expansion of Poisson-Hua Kernels

To begin with, we give some algebraic identities.

Let $f_1, f_2, \cdots, f_n$ be a set of integers such that $f_1 \geqslant f_2 \geqslant \cdots \geqslant f_n$. Again let $l_1 = f_1 + n - 1, l_2 = f_2 + n - 2, \cdots, l_n = f_n$, and

$$M_{f_1, f_2, \cdots, f_n}(x_1, \cdots, x_n) = \begin{vmatrix} x_1^{l_1}, & \cdots, & x_n^{l_1} \\ x_1^{l_2}, & \cdots, & x_n^{l_2} \\ \cdots & \cdots & \cdots \\ x_1^{l_n}, & \cdots, & x_n^{l_n} \end{vmatrix}.$$

Particularly,

$$M_{0,0,\cdots 0}(x_1, \cdots, x_n) = D(x_1, \cdots, x_n)$$

is the Vandermonde determinant.

Set

$$N(f_1, \cdots, f_n) = \lim_{\substack{x_1 \to 1 \\ \cdots \\ x_n \to 1}} \frac{M_{f_1}, \cdots f_n(x_1, \cdots, x_n)}{D(x_1, \cdots, x_n)},$$

and denote $(f_1, f_2, \cdots, f_n)$ by f hereafter.

Theorem 1.2.1 *Let*

$$f(z) = \sum_{-M \leqslant \nu \leqslant N} a_\nu z^\nu ,$$

where M and N are either finite or infinite, and, in the latter case, the series should be convergent. Then we have the following identity

$$\frac{(-1)^{\frac{1}{2}n(n-1)}}{1!2!\cdots(n-1)!} \begin{vmatrix} f(x_1), & \cdots, & f(x_n) \\ x_1 f'(x_1), & \cdots, & x_n f'(x_n) \\ \cdots & \cdots & \cdots \\ x_1^{n-1} f^{(n-1)}(x_1), & \cdots, & x_n^{n-1} f^{(n-1)}(x_n) \end{vmatrix}$$

$$= \sum_{N \geqslant l_1 > l_2 > \cdots > l_n \geqslant -M} a_{l_1} \cdots a_{l_n} N(f_1, \cdots, f_n) M_f(x_1, \cdots, x_n).$$

For the proof see Hua [1].

Taking $M = 0$, $N = \infty$ and $f(x) = \dfrac{1}{1-x}$, particularly, we obtain

Theorem 1.2.2 *If* $|x_\nu| < 1 (\nu = 1, \cdots, n)$, *then*

$$\frac{1}{\left(\prod_{\nu=1}^{n} (1 - x_\nu)\right)^n} = \sum_{f_1 \geqslant f_2 \geqslant \cdots \geqslant f_n \geqslant 0} N(f_1, \cdots, f_n) \frac{M_{f_1, \cdots, f_n}(x_1, \cdots, x_n)}{D(x_1, \cdots, x_n)}.$$

Taking $z = xe^{i\theta}$ and setting $f(e^{i\theta}) = g(\theta)$ in Theorem 1.2.1, we have

$$g'(\theta) = if'(e^{i\theta})e^{i\theta} = i \frac{\partial}{\partial x} f(xe^{i\theta})|_{x=1};$$

$$g''(\theta) = i^2 f''(e^{i\theta})e^{2i\theta} + i^2 f'(e^{i\theta})e^{i\theta}$$

$$= i^2 \frac{\partial^2}{\partial x^2} f(xe^{i\theta})|_{x=1} + i^2 \frac{\partial}{\partial x} f(xe^{i\theta})|_{x=1};$$

$$g'''(\theta) = i^3 f'''(e^{i\theta})e^{3i\theta} + 3i^3 f''(e^{i\theta})e^{2i\theta} + i^3 f'(e^{i\theta})e^{i\theta}$$

$$= i^3 \frac{\partial^3}{\partial x^3} f(xe^{i\theta})|_{x=1} + 3i^3 \frac{\partial^2}{\partial x^2} f(xe^{i\theta})|_{x=1}$$

$$+ i^3 \frac{\partial}{\partial x} f(xe^{i\theta})|_{x=1};$$

$$\cdots\cdots\cdots\cdots\cdots\cdots$$

Thus $g^{(l)}(\theta)$ is a linear combination of

$$\left.\frac{\partial^l}{\partial x^l} f(xe^{i\theta})\right|_{x=1}, \cdots, \left.\frac{\partial}{\partial x} f(xe^{i\theta})\right|_{x=1},$$

where the coefficient of $\left.\frac{\partial^l}{\partial x^l} f(xe^{i\theta})\right|_{x=1}$ is i^l. Hence we have

$$\left.\begin{vmatrix} f(x_1e^{i\theta_1}), & \cdots, & f(x_ne^{i\theta_n}) \\ \frac{\partial f(x_1e^{i\theta_1})}{\partial x_1}, & \cdots, & \frac{\partial}{\partial x_n} f(x_ne^{i\theta_n}) \\ \cdots & \cdots & \cdots \\ \frac{\partial^{n-1}}{\partial x_1^{n-1}} f(x_1e^{i\theta_1}), & \cdots, & \frac{\partial^{n-1}}{\partial x_n^{n-1}} f(x_ne^{i\theta_n}) \end{vmatrix}\right|_{x_1=1,\cdots,x_n=1}$$

$$= (-i)^{\frac{1}{2}n(n-1)} \begin{vmatrix} g(\theta_1), & \cdots, & g(\theta_n) \\ g'(\theta_1), & \cdots, & g'(\theta_n) \\ \cdots & \cdots & \cdots \\ g^{(n-1)}(\theta_1), & \cdots, & g^{(n-1)}(\theta_n) \end{vmatrix}.$$

Thus we obtain

Theorem 1.2.3 *Let*

$$g(\theta) = \sum_{-M\leqslant \nu \leqslant N} a_\nu e^{i\nu\theta},$$

where M and N are either finite or infinite and, in the latter case, the series should be convergent. Then

$$\frac{i^{1/2n(n-1)}}{1!2!\cdots(n-1)!} \begin{vmatrix} g(\theta_1), & \cdots, & g(\theta_n) \\ g'(\theta_1), & \cdots, & g'(\theta_n) \\ \cdots & \cdots & \cdots \\ g^{(n-1)}(\theta_1), & \cdots, & g^{(n-1)}(\theta_n) \end{vmatrix}$$

$$= \sum_{N\geqslant l_1>l_2>\cdots>l_n\geqslant -M} a_{l_1}\cdots a_{l_n} N(f_1,\cdots,f_n) M_f(e^{i\theta_1},\cdots,e^{i\theta_n}).$$

Now we are ready to give the expansion of the Poisson-Hua kernels.

Let U_n be a unitary group of order n and $A_{f_1,\cdots,f_n}(U)$ be an irreducible unitary representation with the signature $f=(f_1,\cdots,f_n)$ $(f_1 \geqslant f_2 \geqslant \cdots \geqslant f_n \geqslant 0)$. The representation is an $N(f)\times N(f)$ unitary matrix written as

$$A_f(U) = (a^f_{ij}(U))_{1\leqslant i,j\leqslant N(f)}.$$

For any $N(f)\times N(f)$ matrix $\mathscr{X}$, we have

$$\int_{U_n} A_f(U)\mathscr{X}\overline{A_g(U)}'\dot{U} = \mathscr{Y},$$

which has the following property

$$A_f(V)\mathscr{Y}\overline{A_g(V)}' = \int_{U_n} A_f(V)A_f(U)\mathscr{X}\overline{A_g(U)}'\overline{A_g(V)}'\dot{U}$$
$$= \int_{U_n} A_f(VU)\mathscr{X}\overline{A_g(VU)}'\dot{U} = \mathscr{Y}.$$

Besides, by Schur's lemma, $\mathscr{Y} = 0$ for $f \neq g$ and $\mathscr{Y} = \lambda_f I^{(N_f)}$ for $f = g$, i. e.

$$\{a^f_{ij}(U)\},\ f_1 \geqslant f_2 \geqslant \cdots \geqslant f_n \geqslant 0,\ i,\ j = 1,\ 2,\ \cdots,\ N(f)$$

form an orthogonal system such that

$$\int_{U_n} |a^f_{ij}(U)|^2\dot{U} = \lambda_f,$$

where λ_f is dependent only on f but independent of i and j. Since $A_f(U)$ is a unitary representation, i. e.

$$A_f(U)\overline{A_f(U)}' = I^{(N(f))},$$

we have

$$\int_{U_n} |a^f_{ij}(U)|^2\dot{U} = \frac{1}{(N(f))^2}\int_{U_n}\sum_i\sum_j |a^f_{ij}(U)|^2\dot{U}$$
$$= \frac{1}{(N(f))^2}\int_{U_n} t_r(A_f(U)\overline{A_f(U)}')\dot{U}$$
$$= \frac{1}{N(f)}\int_{U_n}\dot{U} = C/N(f_1,\ \cdots,\ f_n),$$

where C is the volume of U_n. Therefore

$$\varphi^f_{ij}(U) = \sqrt{\frac{N(f_1,\ \cdots,\ f_n)}{C}}\, a^f_{ij}(U),$$
$$f_1 \geqslant f_2 \geqslant \cdots \geqslant f_n \geqslant 0,\ i,\ j = 1,\ \cdots,\ N(f)$$

form an orthonormal system.

Resorting to

$$(\det U)^l A_{f_1,\cdots,f_n}(U) = A_{f_1+l,\cdots,f_n+l}(U),$$

we extend the definition of A_f for $f_n \neq 0$. Hence

$$\varphi^f_{ij}(U),\ f_1 \geqslant f_2 \geqslant \cdots \geqslant f_n,\ i,\ j = 1,\ \cdots,\ N(f) \tag{1.2.1}$$

still form an orthonormal system.

Assume that $e^{i\theta_1},\ \cdots,\ e^{i\theta_n}$ are characteristic roots of U, then

$$\mathrm{tr}A_f(U) = \chi_f(U) = \frac{M_f(e^{i\theta_1},\ \cdots,\ e^{i\theta_n})}{D(e^{i\theta_1},\ \cdots,\ e^{i\theta_n})}.$$

From Theorem 1.2.2 we have

$$(\det(I - r[e^{i\theta_1}, \cdots, e^{i\theta_n}]))^{-n}$$

$$= \sum_{f_1 \geqslant \cdots \geqslant f_n \geqslant 0} N(f_1, \cdots, f_n) \frac{M_f(re^{i\theta_1}, \cdots, re^{i\theta_n})}{D(re^{i\theta_1}, \cdots, re^{i\theta_n})}$$

$$= \sum_{f_1 \geqslant \cdots \geqslant f_n \geqslant 0} N(f_1, \cdots, f_n) \chi_f([e^{i\theta_1}, \cdots, e^{i\theta_n}]) r^{f_1 + \cdots + f_n},$$

i.e.

$$\frac{1}{C}(\det(I - rU\bar{V}'))^{-n}$$

$$= \frac{1}{C} \sum_{f_1 \geqslant \cdots \geqslant f_n \geqslant 0} N(f) \chi_f(U\bar{V}') r^{f_1 + \cdots + f_n}$$

$$= \frac{1}{C} \sum_{f_1 \geqslant \cdots \geqslant f_n \geqslant 0} N(f) \operatorname{tr}(A_f(U)\overline{A_f(V)'}) r^{f_1 + \cdots + f_n}$$

$$= \frac{1}{C} \sum_{f_1 \geqslant \cdots \geqslant f_n \geqslant 0} N(f) \sum_{i,j} a^f_{ij}(U)\overline{a^f_{ij}(V)'} r^{f_1 + \cdots + f_n}$$

$$= \sum_{f_1 \geqslant \cdots \geqslant f_n \geqslant 0} \sum_{i,j} \varphi^f_{ij}(U)\overline{\varphi^f_{ij}(V)} r^{f_1 + \cdots + f_n}.$$

Therefore the Poisson-Hua kernel can be written as

$$\frac{1}{C} P(rU, V) = \frac{1}{C} \frac{(1 - r^2)^{n^2}}{|\det(I - rU\bar{V}')|^{2n}}$$

$$= (1 - r^2)^{n^2} \sum_{f,i,j} \varphi^f_{ij}(U)\overline{\varphi^f_{ij}(V)} r^{f_1 + \cdots + f_n}$$

$$\cdot \sum_{g,s,t} \overline{\varphi^g_{st}(U)} \varphi^g_{st}(V) r^{g_1 + \cdots + g_n}.$$

$\varphi^f_{ij}(U)\overline{\varphi^g_{st}(U)}$ can be expressed by a direct sum of A_h, since it appears in the representation $A_f \times A_g$. So, $\varphi^f_{ij}(U)\varphi^g_{st}(U)$ can be expressed by a linear combination of functions in (1.2.1), i.e.

$$\frac{1}{C} P(rU, V) = \sum_f \sum_{i,j} \Psi^f_{ij}(rU)\overline{\varphi^f_{ij}(V)},$$

where f runs over all such integers that $f_1 \geqslant f_2 \geqslant \cdots \geqslant f_n$. And, the preceding series is both absolutely and uniformly convergent for $r < 1$.

Taking $u(U) = \varphi^f_{ij}(U)$ in Theorem 1.1.1, we obtain by orthogonality of φ^f_{ij} that

$$\lim_{r \to 1} \Psi^f_{ij}(rV) = \varphi^f_{ij}(V).$$

Let

$$B_f(rV) = \sqrt{\frac{C}{N(f)}} (\Psi^f_{ij}(rV))_{1 \leqslant i,j \leqslant N(f)}.$$

From Theorem 1.1.1, we deduce

$$B_f(rV) = \frac{1}{C}\int_{U_n} \frac{(1-r^2)^{n^2}}{|\det(I - rU\bar{V}')|^{2n}} A_f(U)\dot{U}. \tag{1.2.2}$$

If $W \in U_n$, U and V in (1.2.2) are replaced by WU and WV respectively, then

$$\begin{aligned} B_f(rWV) &= \frac{1}{C}\int_{U_n} \frac{(1-r^2)^{n^2}}{|\det(I - rU\bar{V}')|^{2n}} A_f(WU)\dot{U} \\ &= A_f(W)B_f(rV). \end{aligned} \tag{1.2.3}$$

Similarly, we are able to justify that

$$B_f(rVW) = B_f(rV)A_f(W).$$

Taking $V = I$, we obtain

$$A_f(W)B_f(rI) = B_f(rI)A_f(W).$$

By Schur's lemma, it yields

$$B_f(rI) = \rho^f(r)I^{(N(f))}, \tag{1.2.4}$$

and

$$\lim_{r\to 1} \rho^f(r) = 1. \tag{1.2.5}$$

Inserting (1.2.4) into (1.2.3), we find

$$B_f(rV) = \rho^f(r)A_f(V). \tag{1.2.6}$$

Thus the Poisson-Hua kernel has the following expansion

$$\frac{1}{C}P(rU,V) = \sum_f \rho^f(r) \sum_{i,j} \varphi^f_{ij}(U)\overline{\varphi^f_{ij}(V)}.$$

Choose

$$\Phi_{f_1,\dots,f_n}(U) = (\varphi^f_{ij}(U))_{1\leqslant i,j\leqslant N(f)},$$

then the above expansion can be expressed by

$$\frac{1}{C}P(rU,\ V) = \sum_f \rho^f(r)\operatorname{tr}(\Phi_f(U)\overline{\Phi_f(V)'}).$$

By (1.2.2) and (1.2.6), we have

$$\rho^f(r)A_f(U) = \frac{1}{C}\int_{U_n} \frac{(1-r^2)^{n^2}}{|\det(I - rU\bar{V}')|^{2n}} A_f(V)\dot{V}.$$

Putting $U = I$ and taking trace, we obtain

$$\rho^f(r)N(f) = \frac{1}{C}\int_{U_n} \frac{(1-r^2)^{n^2}}{|\det(I - rV)|^{2n}} \chi_f(V)\dot{V}, \tag{1.2.7}$$

which can be rewritten as

$$\rho^f(r)N(f) = \frac{1}{(2\pi)^n} \int \cdots \int_{\pi \geqslant \theta_1 \geqslant \cdots \geqslant \theta_n \geqslant -\pi} \frac{(1-r^2)^{n^2}}{\prod_{\nu=1}^{n} |1 - re^{i\theta_\nu}|^{2n}}$$

$$\cdot \begin{vmatrix} e^{il_1\theta_1}, & \cdots, & e^{il_1\theta_n} \\ \cdots & \cdots & \cdots \\ e^{il_n\theta_1}, & \cdots, & e^{il_n\theta_n} \end{vmatrix} \prod_{\nu<\mu} (e^{-i\theta_\nu} - e^{-i\theta_\mu}) d\theta_1 \cdots d\theta_n. \quad (1.2.8)$$

It can be shown directly that

$$\rho^f(r) = \begin{cases} r^{f_1+\cdots+f_n}, & \text{if } f_1 \geqslant \cdots \geqslant f_n \geqslant 0, \\ r^{-f_1-\cdots-f_n}, & \text{if } 0 \geqslant f_1 \geqslant \cdots \geqslant f_n. \end{cases}$$

§ 1.3 The Abel Summation

Having expanded Poisson-Hua kernels, Hua Luogeng [2] defined the Abel mean of the Fourier series

$$\sum_{f_1 \geqslant \cdots \geqslant f_n} \operatorname{tr}(C_{f_1\cdots f_n} \Phi'_{f_1\cdots f_n}(U)) \quad (1.3.1)$$

of an integrable function $u(U)(U \in U_n)$ on the unitary group U_n of order n as

$$\sum_{f_1 \geqslant \cdots \geqslant f_n} \rho^f(r) \operatorname{tr}(C_{f_1\cdots f_n} \Phi'_{f_1\cdots f_n}(U)), \quad (1.3.2)$$

where $\rho^f(r)$ is the function defined by (1.2.7). As we know, (1.3.2) can be expressed

$$\frac{1}{C} \int_{U_n} u(VU) \frac{(1-r^2)n^2}{|\det(I - r\bar{V}')|^{2n}} \dot{V}. \quad (1.3.3)$$

Thus Theorem 1.1.1 can also be put into the following form (Hua [2])

Theorem 1.3.1 *If $u(U)$ is continuous on U_n, then its Fourier series is Abel-summable to itself.*

Theorem 1.1.1 is already proved in § 1.1, and so is Theorem 1.3.1. In what follows we shall give a direct proof of Theorem 1.3.1. This means that another proof of Theorem 1.1.1 is provided.

For the value of $\rho^f(r)$, it is known that

(1) $\rho^f(r) \to 1$, as $r \to 1$;

(2) $\rho^f(r) = \begin{cases} r^{f_1+\cdots+f_n}, & \text{if } f_1 \geqslant \cdots \geqslant f_n \geqslant 0, \\ r^{-f_1-\cdots-f_n}, & \text{if } 0 \geqslant f_1 \geqslant \cdots \geqslant f_n. \end{cases}$

When f are not all nonnegative (or nonpositive), an explicit expression of $\rho^f(r)$ invokes a complicated calculation. In the following, three forms of the explicit expressions of $\rho^f(r)$ will be presented. Obviously, each of them will exactly be the above formula (2) provided all $f_i (1 \leqslant i \leqslant n)$ are nonnegative or nonpositive (cf. Gong Sheng (Kung Sheng) [1]).

Theorem 1.3.2 *If* $l_1 > l_2 > \cdots > l_s \geqslant 0 > l_{s+1} > \cdots > l_n$ $(n \geqslant s \geqslant 0)$, then

$$\rho^f(r) = r^{f_1+\cdots+f_s-f_{s+1}-\cdots-f_n} \cdot \sum_{s \geqslant g_{s+1} \geqslant \cdots \geqslant g_n \geqslant 0} \frac{N_s(f,g)N_s(g,f)}{N(f)N(g)} r^{2(g_{s+1}+\cdots+g_n)}, \tag{1.3.4}$$

where $N_s(a, b)$ *denotes* $N(a_1, \cdots, a_s, b_{s+1}, \cdots, b_n)$, $g = (g_1, \cdots, g_n)$, $(g_1 + n - 1, g_2 + n - 2, \cdots, g_n)$ *is a permutation of* $(0, 1, \cdots, n-1)$, *and* $l_1 = f_1 + n - 1$, $l_2 = f_2 + n - 2$, $\cdots$, $l_n = f_n$, $0 \geqslant g_1 \geqslant g_2 \geqslant \cdots \geqslant g_s \geqslant s - n$.

$\rho^f(r)$ can also be put into the following forms

Theorem 1.3.3 *Let* $(\nu_1, \nu_2, \cdots, \nu_n)$ *be a permutation of* $(0, 1, \cdots, n-1)$ *and* $\nu_1 > \cdots > \nu_s$. *Then*

$$\rho^f(r) = r^{|l_1|+\cdots+|l_n| - \frac{n(n-1)}{2}} \cdot \sum_{n-1 \geqslant \nu_{s+1} > \cdots > \nu_n \geqslant 0} \prod_{j=1}^{s} \prod_{k=s+1}^{n} \frac{(l_j - \nu_k)(\nu_j - l_k)}{(\nu_j - \nu_k)(l_j - l_k)} r^{2(\nu_{s+1}+\cdots+\nu_n)}. \tag{1.3.5}$$

Theorem 1.3.4 *Under the same assumptions of the preceding theorem we have*

$$\rho^f(r) = r^{f_1+\cdots+f_s-f_{s+1}-\cdots-f_n} \sum_{\nu=0}^{s(n-s)} b_\nu r^{2\nu}, \tag{1.3.6}$$

where b_ν *is equal to*

$$\sum_{\substack{n-1 \geqslant \nu_{s+1} > \cdots > \nu_n \geqslant 0 \\ \nu_{s+1}+\cdots+\nu_n = \nu + \frac{(n-s)(n-s-1)}{2}}} \prod_{j=1}^{s} \prod_{k=s+1}^{n} \frac{(l_j - \nu_k)(\nu_j - l_k)}{(\nu_j - \nu_k)(l_j - l_k)},$$

i.e.

$$\sum_{\substack{s \geqslant g_{s+1} \geqslant \cdots \geqslant g_n \geqslant 0 \\ g_{s+1}+\cdots+g_n = \nu}} \frac{N_s(f, g)N_s(g, f)}{N(f)N(g)}.$$

From those expressions, it can be seen clearly that $\rho^f(r)$ is a polynomal of order $f_1 + \cdots + f_s - f_{s+1} - \cdots - f_n + 2s(n-s)$ and $\rho^f(r)$ is just (2) for $s = 0$ or $s = n$. Besides,

$$\rho^f(r) = r^{f_1+\cdots+f_s-f_{s+1}-\cdots-f_n} + O(1-r). \tag{1.3.7}$$

Thus the coefficients of Abel summation are already given explicitly. The proof will be given in subsequent sections.

§ 1.4 Poisson Integral

As is well known, the Abel mean (1.3.2) of the Fourier series (1.3.1) can be expressed by the Poisson integral (1.3.3). To begin with, let us prove

$$\frac{1}{C}\int_{U_n} \frac{(1-r^2)^{n^2}}{|\det(I-r\bar{V}')|^{2n}} \dot{V} = 1. \tag{1.4.1}$$

Let $e^{i\theta_1}, \cdots, e^{i\theta_n}$ be characteristic roots of $\bar{V}'$. Then the left side of (1.4.1) can be written as

$$\frac{1}{(2\pi)^n}\int_{\pi\geqslant\theta_1\geqslant\cdots\geqslant\theta_n\geqslant-\pi}\cdots\cdots\int \frac{(1-r^2)^{n^2}\prod_{j<k}|e^{i\theta_j}-e^{i\theta_k}|^2}{|1-re^{i\theta_1}|^{2n}\cdots|1-re^{i\theta_n}|^{2n}}\cdot d\theta_1\cdots d\theta_n \tag{1.4.2}$$

(see [1]). Since the integrand of (1.4.2) is a symmetric function in $\theta_1, \cdots, \theta_n$, (1.4.2) can be put into

$$\frac{1}{n!}\cdot\frac{1}{(2\pi)^n}\int_{-\pi}^{\pi}\cdots\int_{-\pi}^{\pi}\frac{(1-r^2)^{n^2}\prod_{j<k}|e^{i\theta_j}-e^{i\theta_k}|^2}{|1-re^{i\theta_1}|^{2n}\cdots|1-re^{i\theta_n}|^{2n}}$$

$$\cdot\, d\theta_1\cdots d\theta_n. \tag{1.4.3}$$

Notice that

$$\prod_{j<k}|e^{i\theta_j}-e^{i\theta_k}|^2 = \prod_{j<k}(e^{i\theta_j}-e^{i\theta_k})\prod_{j<k}(e^{-i\theta_j}-e^{-i\theta_k})$$

$$= \prod_{j\neq k}(e^{i\theta_j}-e^{i\theta_k})\cdot \det{}^{n-1}V$$

$$= \prod_{j\neq k}((r-e^{i\theta_k})-(r-e^{i\theta_j}))\det{}^{n-1}V \tag{1.4.4}$$

and let $re^{i\theta_j}=p_j$. It is easily seen that any term in the expression of $\prod_{j\neq k}(p_k-p_j)$ other than $n!p_1^{n-1}\cdots p_n^{n-1}$ has at least one factor, say p_k, of which the index is greater than or equal to n. If $p_k=n$, then there is a factor equal to zero in (1.4.3)

$$\int_{-\pi}^{\pi}\frac{e^{-i(n-1)\theta_k}(r-e^{i\theta_k})^n d\theta_k}{|(1-re^{i\theta_k})|^{2n}} = \int_{-\pi}^{\pi}\frac{e^{i\theta_k}d\theta_k}{(1-re^{i\theta_k})^n} = 0.$$

If $p_k>n$, there is also a factor equal to zero. Thus (1.4.3) is equal to

$$\frac{1}{(2\pi)^n}\int_{-\pi}^{\pi}\cdots\int_{-\pi}^{\pi}\frac{(1-r^2)^{n^2}e^{i(\theta_1+\cdots+\theta_n)}d\theta_1\cdots d\theta_n}{(1-re^{i\theta_1})^n\cdots(1-re^{i\theta_n})^n(e^{i\theta_1}-r)\cdots(e^{i\theta_n}-r)}.$$

This is just

$$\left(\frac{1}{2\pi i}\int_{|\zeta_1|=1}\frac{(1-r^2)^n d\zeta_1}{(1-r\zeta_1)^n(\zeta_1-r)}\right)$$
$$\cdot\left(\frac{1}{2\pi i}\int_{|\zeta_2|=1}\frac{(1-r^2)^n d\zeta_2}{(1-r\zeta_2)^n(\zeta_2-r)}\right)\cdots$$
$$\cdots\left(\frac{1}{2\pi i}\int_{|\zeta_n|=1}\frac{(1-r^2)^n d\zeta_n}{(1-r\zeta_n)^n(\zeta_n-r)}\right),$$

where $\zeta_1 = e^{i\theta_1}, \cdots, \zeta_n = e^{i\theta_n}$. By Cauchy integral theorem, (1.4.1) is immediate.

§ 1.5 Proof of Theorem 1.3.1

According to § 1.4, in order to prove Theorem 1.3.1, it suffices to prove that

$$\frac{1}{C}\int_{U_n}(u(VU)-u(U))\frac{(1-r^2)^{n^2}}{|\det(I-r\bar{V}')|^{2n}}\dot{V} \tag{1.5.1}$$

tends to 0 as $r\to 1$. Let $e^{i\theta_1}, \cdots, e^{i\theta_n}$ be characteristic roots of $\bar{V}'$ and

$$\varphi(\theta_1, \cdots, \theta_n) = \frac{1}{\omega'}\int_{[U_n]}(u(VU)-u(U))[\dot{V}], \tag{1.5.2}$$

where ω' is the volume of the coset of unitary group of order n and is equal to

$$(2\pi)^{\frac{1}{2}n(n-1)}/(n-1)!\cdots 2!1!.$$

Thus (1.5.1) is equal to

$$\frac{1}{(2\pi)^n}\int\cdots\int_{\pi\geqslant\theta_1\geqslant\cdots\geqslant\theta_n\geqslant-\pi}\varphi(\theta_1, \cdots, \theta_n)$$
$$\cdot\frac{(1-r^2)^{n^2}\prod_{j<k}|e^{i\theta_j}-e^{i\theta_k}|^2}{|1-re^{i\theta_1}|^{2n}\cdots|1-re^{i\theta_n}|^{2n}}d\theta_1\cdots d\theta_n. \tag{1.5.3}$$

Partition the integral domain into the following sub-domains

$$R_1:\ \delta\geqslant\theta_1\geqslant\theta_2\geqslant\cdots\geqslant\theta_n\geqslant-\pi,$$
$$R_2:\pi\geqslant\theta_1\geqslant\delta\geqslant\theta_2\geqslant\cdots\geqslant\theta_n\geqslant-\pi,$$
$$R_3:\ \pi\geqslant\theta_1\geqslant\theta_2\geqslant\delta\geqslant\theta_3\geqslant\cdots\geqslant\theta_n\geqslant-\pi,$$
$$\cdots\cdots\cdots\cdots\cdots\cdots\cdots\cdots\cdots$$
$$R_n:\ \pi\geqslant\theta_1\geqslant\cdots\geqslant\theta_{n-1}\geqslant\delta\geqslant\theta_n\geqslant-\pi,$$
$$R_{n+1}:\ \pi\geqslant\theta_1\geqslant\cdots\geqslant\theta_n\geqslant\delta.$$

I_j denotes the part of (1.5.3) which corresponds to the subdomain R_j. Then I_2 is equal to

$$\frac{1}{(2\pi)^n}\int\!\!\cdots\!\!\int_{\pi\geqslant\theta_1\geqslant\delta\geqslant\theta_2\geqslant\cdots\geqslant\theta_n\geqslant-\pi}\varphi(\theta_1,\cdots,\theta_n)$$
$$\cdot\frac{(1-r^2)^{n^2}\prod_{j<k}|e^{i\theta_j}-e^{i\theta_k}|^2}{|1-re^{i\theta_1}|^{2n}\cdots|1-re^{i\theta_n}|^{2n}}d\theta_1\cdots d\theta_n. \tag{1.5.4}$$

As $u(U)$ is continuous on U_n, $|u(U)|$ is bounded on U_n. Suppose that $|u(U)|\leqslant M$ holds on U_n, and then we have

$$|\varphi(\theta_1,\cdots,\theta_n)|\leqslant 2M.$$

Thus $|I_2|$ is less than or equal to

$$\frac{2M}{(2\pi)^n}\int\!\!\cdots\!\!\int_{\pi\geqslant\theta_1\geqslant\delta\geqslant\theta_2\geqslant\cdots\geqslant\theta_n\geqslant-\pi}\frac{(1-r^2)^{n^2}\prod_{j<k}|(e^{i\theta_j}-e^{i\theta_k}|}{|1-re^{i\theta_1}|^{2n}\cdots|1-re^{i\theta_n}|^{2n}}d\theta_1\cdots d\theta_n, \tag{1.5.5}$$

which does not exceed

$$\frac{2M(1-r^2)^{2n+1}}{(2\pi)^n|1-re^{i\delta}|^{2n}}\frac{1}{(1-r)^{2(n-1)}}$$
$$\cdot\int\!\!\cdots\!\!\int_{\delta\geqslant\theta_2\geqslant\cdots\geqslant\theta_n\geqslant-\pi}\frac{(1-r^2)^{(n-1)^2}\prod_{2\leqslant j<k}|e^{i\theta_j}-e^{i\theta_k}|^2}{|1-re^{i\theta_2}|^{2(n-1)}\cdots|1-re^{i\theta_n}|^{2(n-1)}}d\theta_2\cdots d\theta_n$$
$$\cdot\int_\delta^\pi\prod_{j=2}^n|e^{i\theta_1}-e^{i\theta_j}|^2d\theta_1.$$

By (1.4.1), this is just

$$O\left(\frac{1-r}{|1-re^{i\delta}|^{2n}}\right).$$

Similarly

$$|I_3|=O\left(\frac{(1-r)^{2n}}{|1-re^{i\delta}|^{4n}}\right),$$

$$\cdots\cdots\cdots\cdots\cdots\cdots$$

$$|I_p|=O\left(\frac{(1-r)^{2np-4n-p^2+4p-3}}{|1-re^{i\delta}|^{2n(p-1)}}\right),\quad p=2,3,\cdots,n.$$

Obviously, $2np-4n-p^2+4p-3\geqslant 0$. Let us consider I_1 and again partition R_1 into

$$s_1:\ \delta\geqslant\theta_1\geqslant\theta_2\geqslant\cdots\geqslant\theta_n\geqslant-\delta,$$
$$s_2:\ \delta\geqslant\theta_1\geqslant\cdots\geqslant\theta_{n-1}\geqslant-\delta\geqslant\theta_n\geqslant-\pi,$$
$$\cdots\cdots\cdots\cdots\cdots\cdots\cdots\cdots$$
$$s_n:\ \delta\geqslant\theta_1\geqslant-\delta\geqslant\theta_2\geqslant\cdots\geqslant\theta_n\geqslant-\pi,$$
$$s_{n+1}:\ -\delta\geqslant\theta_1\geqslant\theta_2\geqslant\cdots\geqslant\theta_n\geqslant-\pi,$$

J_i denotes the part of (1.3.3) which corresponds to the subdomain s_i. In the same way we get

$$|J_p| = O\left(\frac{(1-r)^{(p-1)^2}}{|1-re^{is}|^{2n(p-1)}}\right), \quad p = 2, 3, \cdots, n.$$

As to J_1, in virtue of the continuity of $u(U)$, for any given $\varepsilon > 0$, there must exist δ such that

$$|\varphi(\theta_1, \cdots, \theta_n)| < \frac{\varepsilon}{2},$$

whenever $\delta \geqslant \theta_1 \geqslant \cdots \geqslant \theta_n \geqslant -\delta$. Choose r fully approaching 1 such that

$$|I_2| + \cdots + |I_{n+1}| + |J_2| + \cdots + |J_{n+1}| < \varepsilon/2.$$

Thus, for any given $\varepsilon > 0$, there exists r fully approaching 1 such that the absolute value of (1.5.1) is less than ε.

This proves Theorem 1.3.1. By the transitivity of classical domains, Theorem 1.1.1 is proved as well.

§ 1.6 Calculation of Coefficients

In the following section, we shall prove Theorem 1.3.2 and other conclusions.

Let $e^{i\theta_1}, \cdots, e^{i\theta_n}$ be characteristic roots of U. Then (1.3.3) becomes

$$\rho^f(r)N(f) = \frac{1}{(2\pi)^n} \int \cdots \int_{\pi \geqslant \theta_1 \geqslant \cdots \geqslant \theta_n \geqslant -\pi} \frac{(1-r^2)^{n^2}}{\prod_{\nu=1}^{n} |1-re^{i\theta_\nu}|^{2n}}$$

$$\cdot \begin{vmatrix} e^{il_1\theta_1}, & \cdots, & e^{il_1\theta_n} \\ e^{il_2\theta_1}, & \cdots, & e^{il_2\theta_n} \\ \cdots & \cdots & \cdots \\ e^{il_n\theta_1}, & \cdots, & e^{il_n\theta_n} \end{vmatrix} \prod_{\nu<\mu} (e^{-i\theta_\nu} - e^{-i\theta_\mu}) d\theta_1 \cdots d\theta_n. \tag{1.6.1}$$

Expanding the determinant of the integrand in (1.6.1), we obtain

$$\rho^f(r)N(f) = \sum \delta^{1,2,\cdots,n}_{\alpha_1,\alpha_2,\cdots,\alpha_n} \frac{1}{(2\pi)^n} \int \cdots \int_{\pi \geqslant \theta_1 \geqslant \cdots \geqslant \theta_n \geqslant -\pi}$$

$$\cdot \frac{(1-r^2)^{n^2}}{\prod_{\nu=1}^{n} |1-re^{i\theta_\nu}|^{2n}} \times e^{i(l_{\alpha_1}\theta_1 + \cdots + l_{\alpha_n}\theta_n)}$$

$$\cdot D(e^{-i\theta_1}, \cdots, e^{-i\theta_n}) d\theta_1 \cdots d\theta_n. \tag{1.6.2}$$

In the above integral, we change variables so that $l_{\alpha_1}\theta_1 + \cdots + l_{\alpha_n}\theta_n$ is replaced by $l_1\theta_1 + \cdots + l_n\theta_n$. Then we observe

$$\rho^f(r)N(f) = \frac{1}{(2\pi)^n}\int_0^{2\pi}\cdots\int_0^{2\pi} \frac{(1-r^2)^{n^2}}{\prod_{\nu=1}^{n}|1-re^{i\theta_\nu}|^{2n}} \cdot e^{i(l_1\theta_1+\cdots+l_n\theta_n)}D(e^{-i\theta_1}, \cdots, e^{-i\theta_n})d\theta_1\cdots d\theta_n. \tag{1.6.3}$$

Expanding the determinant of the integrand in (1.6.3) again, we get

$$\rho^f(r)N(f) = (-1)^{\frac{1}{2}n(n-1)}\sum \delta^{0,1,\cdots,n-1}_{k_0,k_1,\cdots,k_{n-1}} \frac{1}{(2\pi)^n}\int_0^{2\pi}\cdots\int_0^{2\pi} \frac{(1-r^2)^{n^2}}{\prod_{\nu=1}^{n}|1-re^{i\theta_\nu}|^{2n}} \cdot e^{i[(l_1-k_0)\theta_1+\cdots+(l_n-k_{n-1})\theta_n]}d\theta_1\cdots d\theta_n,$$

which is just

$$\rho^f(r)N(f) = (-1)^{\frac{1}{2}n(n-1)}(1-r^2)^{n^2}\begin{vmatrix} b_{11}, & b_{12}, & \cdots, & b_{1n} \\ b_{21}, & b_{22}, & \cdots, & b_{2n} \\ \cdots & \cdots & \cdots & \cdots \\ b_{n1}, & b_{n2}, & \cdots, & b_{nn}\end{vmatrix}, \tag{1.6.3'}$$

where $b_{\alpha\beta} = \frac{1}{2\pi}\int_0^{2\pi}\frac{e^{i(l_\beta-\alpha+1)\theta}}{|1-re^{i\theta}|^{2n}}d\theta$. In the preceding determinant, multiplying the first row by $C_0^{k-1}r^0(-1)^0$, the second row by $C_1^{k-1}r^1(-1)^1$, $\cdots$, the k-th row by $C_{k-1}^{k-1}r^{k-1}(-1)^{k-1}$ and then adding the first $(k-1)$ rows to the k-th row, $k = 1, 2, \cdots, n$, we deduce

$$\rho^f(r)N(f) = (1-r^2)^{n^2}r^{-\frac{n(n-1)}{2}} \cdot \begin{vmatrix} c_{11}, & c_{12}, & \cdots, & c_{1n} \\ c_{21}, & c_{22}, & \cdots, & c_{2n} \\ \cdots & \cdots & \cdots & \cdots \\ c_{n1}, & c_{n2}, & \cdots, & c_{nn}\end{vmatrix}, \tag{1.6.4}$$

where

$$c_{\alpha\beta} = \frac{1}{2\pi}\int_0^{2\pi}\frac{e^{il_\beta\theta}d\theta}{(1-re^{i\theta})^n(1-re^{-i\theta})^{n-\alpha+1}}.$$

In fact, (1.6.4) can be verified directly from (1.6.3). By virtue of the properties of $D(x_1, x_2, \cdots, x_n)$, we note

$$\begin{aligned} D(e^{-i\theta_1}, \cdots, e^{-i\theta_n}) &= r^{-\frac{1}{2}n(n-1)}D(re^{-i\theta_1}, \cdots, re^{-i\theta_n}) \\ &= r^{-\frac{1}{2}n(n-1)}D(re^{-i\theta_1}-1, \cdots, re^{-i\theta_n}-1) \\ &= (-r)^{-\frac{1}{2}n(n-1)}D(1-re^{-i\theta_1}, \cdots, 1-re^{-i\theta_n}) \\ &= r^{-\frac{1}{2}n(n-1)}\prod_{j=1}^{n}(1-re^{-i\theta_j})^{n-1} \\ &\quad \cdot D\left(\frac{1}{1-re^{-i\theta_1}}, \cdots, \frac{1}{1-re^{-i\theta_n}}\right). \end{aligned}$$

Inserting the above equality into (1.6.3), we obtain

$$\rho^f(r)N(f) = r^{-\frac{1}{2}n(n-1)} \frac{1}{(2\pi)^n}$$

$$\cdot \int_0^{2\pi} \cdots \int_0^{2\pi} \frac{(1-r^2)^{n^2} e^{i(l_1\theta_1+\cdots+l_n\theta_n)}}{\prod_{\nu=1}^{n}(1-re^{i\theta_\nu})^n \prod_{\nu=1}^{n}(1-re^{-i\theta_\nu})}$$

$$\cdot D\left(\frac{1}{1-re^{-i\theta_1}}, \cdots, \frac{1}{1-re^{-i\theta_n}}\right) d\theta_1 \cdots d\theta_n.$$

Expanding the determinant in the integrand, (1.6.4) follows.

Introduce the following notation

$$A^{p_1,\cdots,p_n}_{q_1,\cdots,q_n} = \begin{vmatrix} d_{11}, & d_{12}, & \cdots, & d_{1n} \\ d_{21}, & d_{22}, & \cdots, & d_{2n} \\ \cdots & \cdots & \cdots & \cdots \\ d_{n1}, & d_{n2}, & \cdots, & d_{nn} \end{vmatrix}, \tag{1.6.5}$$

where

$$d_{\alpha\beta} = \frac{1}{2\pi}\int_0^{2\pi} \frac{e^{ip_\beta\theta} d\theta}{(1-re^{i\theta})^{q_\beta}(1-re^{-i\theta})^{n-\alpha+1}}.$$

Thus (1.6.4) leads to

$$\rho^f(r)N(f) = (1-r^2)^{n^2} r^{-\frac{n(n-1)}{2}} A^{l_1,\cdots,l_n}_{n,\cdots,n}. \tag{1.6.6}$$

§ 1.7 Several Algebraic Identities

In this section, we suppose

$$l_1 > l_2 > \cdots \geqslant l_s \geqslant 0 > l_{s+1} > \cdots > l_n (n \geqslant s \geqslant 0).$$

By the definition (1.4.5) of A, it follows immediately that

$$A^{l_1,\cdots,l_n}_{n,\cdots,n} = A^{l_1,\cdots,l_{n-1},l_n}_{n,\cdots,n,n-1} + rA^{l_1,\cdots,l_{n-1},l_n+1}_{n,\cdots,n,n}.$$

However

$$A^{l_1,\cdots,l_{n-1},l_n}_{n,\cdots,n,n-1} = A^{l_1,\cdots,l_{n-1},l_n}_{n,\cdots,n,n-2} + rA^{l_1,\cdots,l_{n-1},l_n+1}_{n,\cdots,n,n-1},$$

$$\cdots\cdots\cdots\cdots\cdots\cdots\cdots\cdots$$

$$A^{l_1,\cdots,l_{n-1},l_n}_{n,\cdots,n,1} = A^{l_1,\cdots,l_{n-1},l_n}_{n,\cdots,n,0} + rA^{l_1,\cdots,l_{n-1},l_n+1}_{n,\cdots,n,1}.$$

Sum up those identities given above. As

$$A^{l_1,\cdots,l_{n-1},l_n}_{n,\cdots,n,0} = 0, \text{ for } l_n < 0,$$

we get the following identity

$$A^{l_1,\cdots,l_n}_{n,\cdots,n} = r\sum_{j=0}^{n-1} A^{l_1,\cdots,l_{n-1},l_n+1}_{n,\cdots,n,n-j}. \tag{1.7.1}$$

However

$$A^{l_1,\cdots,l_{n-1},l_n+1}_{n,\cdots,n,n-j} = A^{l_1,\cdots,l_{n-1},l_n+1}_{n,\cdots,n,n-j-1} + rA^{l_1,\cdots,l_{n-1},l_n+2}_{n,\cdots,n,n-j},$$

$$A^{l_1,\cdots,l_{n-1},l_n+1}_{n,\cdots,n,n-j-1} = A^{l_1,\cdots,l_{n-1},l_n+1}_{n,\cdots,n,n-j-2} + rA^{l_1,\cdots,l_{n-1},l_n+2}_{n,\cdots,n,n-j-1},$$

$$\cdots\cdots\cdots\cdots\cdots\cdots\cdots\cdots$$

$$A^{l_1,\cdots,l_{n-1},l_n+1}_{n,\cdots,n,1} = A^{l_1,\cdots,l_{n-1},l_n+1}_{n,\cdots,n,0} + rA^{l_1,\cdots,l_{n-1},l_n+2}_{n,\cdots,n,1}.$$

Summing up these identities, we get the identity

$$A^{l_1,\cdots,l_{n-1},l_n+1}_{n,\cdots,n,n-j} = r\sum_{k=1}^{n-1} A^{l_1,\cdots,l_{n-1},l_n+2}_{n,\cdots,n,n-k},$$

since

$$A^{l_1,\cdots,l_{n-1},l_n+1}_{n,\cdots,n,0} = 0.$$

Inserting the above identity into (1.7.1), we obtain

$$A^{l_1,\cdots,l_n}_{n,\cdots,n} = r^2\sum_{j=0}^{n-1}\sum_{k=j}^{n-1} A^{l_1,\cdots,l_{n-1},l_n+2}_{n,\cdots,n,n-k} = r^2\sum_{k=0}^{n-1}(k+1)A^{l_1,\cdots,l_{n-1},l_n+2}_{n,\cdots,n,n-k}.$$

Repeating the procedure mentioned above, finally we obtain

$$A^{l_1,\cdots,l_n}_{n,\cdots,n} = r^{-l_n}\sum_{k=0}^{n-1}\frac{(k+1)\cdots(k-l_n-1)}{(-l_n-1)!}A^{l_1,\cdots,l_{n-1},0}_{n,\cdots,n,n-k}$$

$$= r^{-l_n}\sum_{k=0}^{n-1}\frac{(k-l_n-1)!}{k!(-l_n-1)!}A^{l_1,\cdots,l_{n-1},0}_{n,\cdots,n,n-k}. \tag{1.7.2}$$

Considering that $l_1 > \cdots > l_s \geqslant 0 > l_{s+1} > \cdots > l_n$ and carrying the foregoing procedure on $l_{n-1}, l_{n-2}, \cdots, l_{s+1}$ successively on the right side of (1.7.2) until each of them increasingly becomes 0, finally, we arrive at

$$A^{l_1,\cdots,l_n}_{n,\cdots,n} = r^{-l_{s+1}-\cdots-l_n}\sum_{k_{s+1}=0}^{n-1}\cdots\sum_{k_n=0}^{n-1}\frac{(k_{s+1}-l_{s+1}-1)!}{k_{s+1}!(-l_{s+1}-1)!}\cdot\cdots$$

$$\cdot\frac{(k_n-l_n-1)!}{k_n!(-l_n-1)!}\cdot A^{l_1,\cdots,l_s,0,\cdots,0}_{n,\cdots,n,n-k_{s+1},\cdots,n-k_n}. \tag{1.7.3}$$

§ 1.8 The Value of Λ

For calculating the value of $\rho^f(r)$, by (1.6.6) it is sufficient to calculate $\Lambda^{l_1,\cdots,l_n}_{n,\cdots,n}$, which is reduced to calculating $\Lambda^{l_1,\cdots,l_s,0,\cdots,0}_{n,\cdots,n,n-k_{s+1},\cdots,n-k_n}$ in the light of (1.7.3). But $\Lambda^{l_1,\cdots,l_s,0,\cdots,0}_{n,\cdots,n,n-k_{s+1},\cdots,n-k_n}$ is equal to

$$\det(h_{fg})_{1\leqslant f,g\leqslant n},$$

where

$$h_{fg}=\frac{1}{2\pi}\int_0^{2\pi}\frac{e^{il_g\theta}d\theta}{(1-re^{i\theta})^n(1-re^{-i\theta})^{(n-f+1)}}$$

for $g=1,2,\cdots,s$, and

$$h_{fg}=\frac{1}{2\pi}\int_0^{2\pi}\frac{d\theta}{(1-re^{i\theta})^{n-k_g+1}(1-e^{-i\theta})^{(n-f+1)}}\tag{1.8.1}$$

for $g=s+1,\cdots,n$. For calculating the preceding determinants we first prove

Lemma 1.8.1 *If $p\geqslant 0$, $q>0$, $m>0$ and $1>r>0$, then*

$$\frac{1}{2\pi}\int_0^{2\pi}\frac{e^{ip\theta}d\theta}{(1-re^{i\theta})^q(1-re^{-i\theta})^m}=\frac{r^{-p}}{(m-1)!}$$

$$\cdot\left[\frac{d^{m-1}}{dt^{m-1}}\left(t^{m+p-1}(1-t)^{-q}\right)\right]_{t=r^2}\tag{1.8.2}$$

holds.

Proof. As $1>r>0$, we have

$$(1-re^{i\theta})^{-q}=\sum_{k=0}^{\infty}\frac{(q+k-1)!r^ke^{ik\theta}}{(q-1)!k!}$$

and

$$(1-re^{-i\theta})^{-m}=\sum_{l=0}^{\infty}\frac{(m+l-1)!r^le^{-il\theta}}{(m-1)!l!}.$$

Therefore

$$\frac{1}{2\pi}\int_0^{2\pi}\frac{e^{ip\theta}d\theta}{(1-re^{i\theta})^q(1-re^{-i\theta})^m}$$

$$=\sum_{\substack{l-k=p\\ l\geqslant 0,k\geqslant 0}}\frac{(q+k-1)!(m+l-1)!}{(m-1)!l!(q-1)!k!}r^{k+2}$$

$$=r^p\sum_{k=0}^{\infty}\frac{(q+k-1)!(m+k+p-1)!}{(m-1)!(k+p)!(q-1)!k!}r^{2k}.$$

By the expansion of $(1-t)^{-q}$, it is readily shown that

$$\sum_{k=0}^{\infty} \frac{(q+k-1)!(m+k+p-1)!}{(m-1)!(k+p)!(q-1)!k!} t^k = \frac{t^{-p}}{(m-1)!} \frac{d^{m-1}}{dt^{m-1}} (t^{m+p-1}(1-t)^{-q}),$$

which is just (1.8.2).

Inserting the value of (1.8.2) into (1.8.1), we can show that

$$A^{l_1,\cdots,l_s,0,\cdots,0}_{n,\cdots,n,n-k_{s+1},\cdots,n-k_n} = \frac{r^{-l_1-\cdots-l_s}}{(n-1)!\cdots 2!1!} \begin{vmatrix} e_{11}, e_{12}, \cdots, e_{1n} \\ e_{21}, e_{22}, \cdots, e_{2n} \\ \cdots\cdots\cdots\cdots \\ e_{n1}, e_{n2}, \cdots, e_{nn} \end{vmatrix}, \quad (1.8.3)$$

where

$$e_{ij} = \frac{d^{n-i}}{dt^{n-i}} (t^{n+l_j-i}(1-t)^{-n}),$$

if $j \leqslant s$;

$$e_{ij} = \frac{d^{n-i}}{dt^{n-i}} (t^{n-i}(1-t)^{-n+k_j}),$$

if $s+1 \leqslant j \leqslant n$.

However

$$\frac{d^{n-i}}{dt^{n-i}}(t^{n+l-i}(1-t)^{-n}) = \sum_{k=0}^{n-i} C_k^{n-i}(t^{n-i})^{(k)}(t^l(1-t)^{-n})^{(n-j-k)},$$

$$\frac{d^{n-i}}{dt^{n-i}}(t^{n-i}(1-t)^{-n+\nu}) = \sum_{k=0}^{n-j} C_k^{n-i}(t^{n-i})^{(k)}((1-t)^{-n+\nu})^{(n-j-k)}.$$

Inserting these into (1.8.3) and taking $t = r^2$, we immediately obtain

$$A^{l_1,\cdots,l_s,0,\cdots,0}_{n,\cdots,n,n-k_{s+1},\cdots,n-k_n} = \frac{r^{-l_1-\cdots-l_s+n(n-1)}}{(n-1)!\cdots 2!1!} \begin{vmatrix} f_{11}, f_{12}, \cdots, f_{1n} \\ f_{21}, f_{22}, \cdots, f_{2n} \\ \cdots\cdots\cdots\cdots \\ f_{n1}, f_{n2}, \cdots, f_{nn} \end{vmatrix},$$

where

$$f_{ij} = \frac{d^{n-i}}{dt^{n-i}} (t^{l_j}(1-t)^{-n}),$$

if $j \leqslant s$;

$$f_{ij} = \frac{d^{n-i}}{dt^{n-i}} ((1-t)^{-n+k_j}),$$

if $s+1 \leqslant j \leqslant n$.

By Theorem 1.2.4 of Hua [1], it follows that the foregoing expression is equal to

$$r^{-l_1-\cdots-l_s+n(n-1)}$$

$$\cdot \lim_{t_1\to t,\cdots,t_n\to t}\begin{vmatrix} t_1^{l_1}(1-t_1)^{-n}, & \cdots, & t_1^{l_s}(1-t_1)^{-n}, & (1-t_1)^{-n+k_{s+1}}, & \cdots, & (1-t_1)^{-n+k_n} \\ t_2^{l_1}(1-t_2)^{-n}, & \cdots, & t_2^{l_s}(1-t_2)^{-n}, & (1-t_2)^{-n+k_{s+1}}, & \cdots, & (1-t_2)^{-n+k_n} \\ \cdots & \cdots & \cdots & \cdots & \cdots & \cdots \\ t_n^{l_1}(1-t_n)^{-n}, & \cdots, & t_n^{l_s}(1-t_n)^{-n}, & (1-t_n)^{-n+k_{s+1}}, & \cdots, & (1-t_n)^{-n+k_n} \end{vmatrix}$$

$$\cdot \frac{1}{D(t_1, \cdots, t_n)} = r^{-l_1-\cdots-l_s+n(n-1)}(1-t)^{-n^2}$$

$$\cdot \lim_{t_1\to t,\cdots,t_n\to t}\begin{vmatrix} t_1^{l_1}, & \cdots, & t_1^{l_s}, & (1-t_1)^{k_{s+1}}, & \cdots, & (1-t_1)^{k_n} \\ t_2^{l_1}, & \cdots, & t_2^{l_s}, & (1-t_2)^{k_{s+1}}, & \cdots, & (1-t_2)^{k_n} \\ \cdots & \cdots & \cdots & \cdots & \cdots & \cdots \\ t_n^{l_1}, & \cdots, & t_n^{l_s}, & (1-t_n)^{k_{s+1}}, & \cdots, & (1-t_n)^{k_n} \end{vmatrix}$$

$$\cdot \frac{1}{D(t_1, \cdots, t_n)}. \tag{1.8.4}$$

Inserting (1.8.4) into (1.7.3) and noticing that

$$g_p(t) = \frac{1}{(-p-1)!}\sum_{k=0}^{n-1}\frac{(k-p-1)!}{k!}(1-t)^k = \sum_{\nu=0}^{n-1}A_\nu^p t^\nu,$$

and

$$p<0,\ A_\nu^p = \frac{(-1)^\nu(n-p-1)!}{(\nu-p)\nu!(-p-1)!(n-\nu-1)!}, \tag{1.8.5}$$

we get

$$A_{n,\cdots,n}^{l_1,\cdots,l_n} = r^{-l_1-\cdots-l_n+n(n-1)}(1-r^2)^{-n^2}$$

$$\cdot \lim_{t_1\to t,\cdots,t_n\to t}\begin{vmatrix} t_1^{l_1}, & \cdots, & t_1^{l_s}, & g_{l_{s+1}}(t_1), & \cdots, & g_{l_n}(t_1) \\ t_2^{l_1}, & \cdots, & t_2^{l_s}, & g_{l_{s+1}}(t_2), & \cdots, & g_{l_n}(t_2) \\ \cdots & \cdots & \cdots & \cdots & \cdots & \cdots \\ t_n^{l_1}, & \cdots, & t_n^{l_s}, & g_{l_{s+1}}(t_n), & \cdots, & g_{l_n}(t_n) \end{vmatrix}$$

$$\cdot \frac{1}{D(t_1, \cdots, t_n)}. \tag{1.8.6}$$

Now we turn to prove (1.8.5). To this end we first notice an algebraic identity

$$\sum_{r=0}^{s}\frac{(r+l)!}{r!} = \frac{(s+l+1)!}{(l+1)\cdot s!}, \tag{1.8.7}$$

where both s and l are integers.

Proof. It is clear that

$$\begin{aligned}\sum_{r=0}^{s}\frac{(r+l+1)!}{r!} &= \sum_{r=1}^{s}\frac{(r+l)!}{(r-1)!}+(l+1)\cdot\sum_{r=0}^{s}\frac{(r+l)!}{r!}\\ &= \sum_{r=0}^{s-1}\frac{(r+l+1)!}{r!}+(l+1)\cdot\sum_{r=0}^{s}\frac{(r+l)!}{r!},\end{aligned}$$

from which (1.8.7) is immediate.

Thus

$$\begin{aligned}g_p(t) &= \frac{1}{(-p-1)!}\sum_{k=0}^{n-1}\frac{(k-p-1)!}{k!}(1-t)^k\\ &= \sum_{k=0}^{n-1}\frac{(k-p-1)!}{k!(-p-1)!}\sum_{\nu=0}^{k}\frac{k!(-t)^\nu}{(k-\nu)!\nu!}\\ &= \sum_{k=0}^{n-1}\sum_{\nu=0}^{k}\frac{(k-p-1)!(-t)^\nu}{(-p-1)!(k-\nu)!\nu!}\\ &= \sum_{\nu=0}^{n-1}\sum_{k=\nu}^{n-1}\frac{(k-p-1)!(-t)^\nu}{(-p-1)!(k-\nu)!\nu!},\end{aligned}$$

$$\begin{aligned}A_\nu^p &= \sum_{k=\nu}^{n-1}\frac{(k-p-1)!(-1)^\nu}{(-p-1)!(k-\nu)!\nu!}\\ &= \frac{(-1)^\nu}{\nu!(-p-1)!}\sum_{r=0}^{n-\nu-1}\frac{(r+\nu-p-1)!}{r!}.\end{aligned}$$

Taking $s=n-\nu-1$ and $l=\nu-p-1$ in (1.8.7), we derive

$$A_\nu^p=\frac{(-1)^\nu(n-p-1)!}{\nu!(-p-1)!(n-\nu-1)!(\nu-p)},$$

which is just (1.8.5). By a theorem of Hua [1] we find out that (1.8.6) turns out exactly:

$$A_{n,\dots,n}^{l_1,\dots,l_n}=\frac{r^{-l_1-\dots-l_n+n(n-1)}(1-r^2)^{-n^2}}{(n-1)!\cdots 2!1!}\cdot\begin{vmatrix}\frac{d^{n-1}t^{l_1}}{dt^{n-1}}, & \dots, & \frac{d^{n-1}t^{l_s}}{dt^{n-1}}, & \frac{d^{n-1}g_{l_{s+1}}(t)}{dt^{n-1}}, & \dots, & \frac{d^{n-1}g_{l_n}(t)}{dt^{n-1}}\\ \frac{d^{n-2}t^{l_1}}{dt^{n-2}}, & \dots, & \frac{d^{n-2}t^{l_s}}{dt^{n-2}}, & \frac{d^{n-2}g_{l_{s+1}}(t)}{dt^{n-2}}, & \dots, & \frac{d^{n-2}g_{l_n}(t)}{dt^{n-2}}\\ \cdots & \cdots & \cdots & \cdots & \cdots & \cdots\\ t^{l_1}, & \dots, & t^{l_s}, & g_{l_{s+1}}(t), & \dots, & g_{l_n}(t)\end{vmatrix}. \quad (1.8.8)$$

§ 1.9 Proof of Theorems in § 1.3

Inserting the value of (1.8.8) into (1.6.6) leads to

$$\rho^j(r)N(f) = \frac{r^{-l_1-\cdots-l_n+\frac{n(n-1)}{2}}}{(n-1)!\cdots 2!1!}$$

$$\cdot \sum_{\nu_{s+1}=0}^{n-1} \cdots \sum_{\nu_n=0}^{n-1} A^{l_{s+1}}_{\nu_{s+1}} \cdots A^{l_n}_{\nu_n} \cdot \begin{vmatrix} g_{11}, g_{12}, \cdots, g_{1n} \\ g_{21}, g_{22}, \cdots, g_{2n} \\ \cdots\cdots\cdots\cdots \\ g_{n1}, g_{n2}, \cdots, g_{nn} \end{vmatrix}, \tag{1.9.1}$$

where

$$g_{ij} = \begin{cases} \dfrac{d^{n-i}t^{l_j}}{dt^{n-i}}, & \text{if } j \leqslant s; \\ \dfrac{d^{n-i}t^{\nu_j}}{dt^{n-i}}, & \text{if } s+1 \leqslant j \leqslant n. \end{cases}$$

From (1.9.1) it is readily seen that

$$\rho^j(r)N(f) = \frac{r^{-l_1-\cdots-l_n+\frac{n(n-1)}{2}}}{(n-1)!\cdots 2!1!}$$

$$\cdot \sum_{n-1\geqslant \nu_{s+1}>\cdots>\nu_n\geqslant 0} B^{l_{s+1},\cdots,l_n}_{\nu_{s+1},\cdots,\nu_n} \begin{vmatrix} g_{11}, \cdots, g_{1n} \\ g_{21}, \cdots, g_{2n} \\ \cdots\cdots\cdots \\ g_{n1}, \cdots, g_{nn} \end{vmatrix}, \tag{1.9.2}$$

where

$$\begin{aligned} B^{l_{s+1},\cdots,l_n}_{\nu_{s+1},\cdots,\nu_n} &= \det(A^{l_{s+i}}_{\nu_{s+k}})_{1\leqslant j,k\leqslant n-s} \\ &= \frac{(-1)^{\nu_{s+1}+\cdots+\nu_n}}{\nu_{s+1}!\cdots\nu_n!} \cdot \frac{(n-l_{s+1}-1)!}{(-l_{s+1}-1)!(n-\nu_{s+1}-1)!} \cdots \\ &\quad \cdot \frac{(n-l_n-1)!}{(-l_n-1)!(n-\nu_n-1)!} \det\left(\frac{1}{\nu_{s+k}-l_{s+j}}\right), \\ &\quad 1 \leqslant j,k \leqslant n-s. \end{aligned}$$

By Theorem 1.1.3 of Hua [1], it follows that

$$\begin{aligned} B^{l_{s+1},\cdots,l_n}_{\nu_{s+1},\cdots,\nu_n} &= \frac{(-1)^{\nu_{s+1}+\cdots+\nu_n}(n-l_{s+1}-1)!\cdots(n-l_n-1)!}{\nu_{s+1}!\cdots\nu_n!(-l_{s+1}-1)!\cdots(-l_n-1)!} \\ &\quad \cdot \frac{D(\nu_{s+1},\cdots,\nu_n)\cdot D(-l_{s+1},\cdots,-l_n)}{(n-\nu_{s+1}-1)!\cdots(n-\nu_n-1)!\prod_{i=1}^{n-s}\prod_{j=1}^{n-s}(\nu_{s+i}-l_{s+j})} \\ &= (-1)^{\nu_{s+1}+\cdots+\nu_n} \cdot \frac{(n-l_{s+1}-1)!}{(-l_{s+1}-1)!n!} n! \end{aligned}$$

$$\cdot \frac{(n-1)!}{\nu_{s+1}!(n-\nu_{s+1}-1)!} \cdot \frac{1}{(n-1)!} \cdots \frac{(n-l_n-1)!}{(-l_n-1)!n!}$$
$$\cdot\, n!\, \frac{(n-1)!}{\nu_n!(n-\nu_n-1)!}$$
$$\cdot \frac{1}{(n-1)!} \frac{D(\nu_{s+1},\cdots,\nu_n)D(-l_{s+1},\cdots,-l_n)}{\prod_{j=s+1}^{n}\prod_{k=s+1}^{n}(\nu_j-l_k)}$$

$$= (-1)^{\nu_{s+1}+\cdots+\nu_n} \cdot n^{n-s} \cdot \prod_{j=s+1}^{n} C_n^{n-l_j-1} C_{\nu_j}^{n-1}$$
$$\cdot \frac{D(\nu_{s+1},\cdots,\nu_n)D(-l_{s+1},\cdots,-l_n)}{\prod_{j=s+1}^{n}\prod_{k=s+1}^{n}(\nu_j-l_k)}. \tag{1.9.3}$$

On the other hand, owing to

$$\begin{vmatrix} g_{11}, g_{12},\cdots, g_{1n} \\ g_{21}, g_{22},\cdots, g_{2n} \\ \cdots\cdots\cdots \\ g_{n1}, g_{n2},\cdots, g_{nn} \end{vmatrix} = \lim_{t_1\to t,\cdots,t_n\to t} \begin{vmatrix} t_1^{l_1}, & \cdots, & t_n^{l_1} \\ \cdots & \cdots & \cdots \\ t_1^{l_s}, & \cdots, & t_n^{l_s} \\ t_1^{\nu_{s+1}}, & \cdots, & t_n^{\nu_{s+1}} \\ \cdots & \cdots & \cdots \\ t_1^{\nu_n}, & \cdots, & t_n^{\nu_n} \end{vmatrix} \Big/ D(t_1,\cdots,t_n)$$

$$= t^{l_1+\cdots+l_s+\nu_{s+1}+\cdots+\nu_n-\frac{1}{2}n(n-1)}$$
$$\cdot\, N(f_1,\cdots,f_s,\nu_{s+1}+s+1-n,\cdots,\nu_n), \tag{1.9.4}$$

we obtain

$$\rho^f(r)N(f) = r^{|l_1|+\cdots+|l_n|-\frac{n(n-1)}{2}} \sum_{n-1\geqslant\nu_{s+1}>\cdots>\nu_n\geqslant 0} B^{l_{s+1},\cdots,l_n}_{\nu_{s+1},\cdots,\nu_n}$$
$$\cdot\, N(f_1,\cdots,f_s,\nu_{s+1}+s+1-n,\cdots,\nu_n)$$
$$\cdot\, r^{2(\nu_{s+1}+\cdots+\nu_n)}, \tag{1.9.5}$$

by inserting (1.9.4) into (1.9.2).

Applying the following two important and obvious identities

$$C_n^{n-l_k-1} = \frac{1}{n!}\prod_{j=1}^{n}(\nu_j-l_k)$$

and

$$\nu_k!(n-\nu_k-1)! = \prod_{\substack{j=1\\ j\neq k}}^{n}(\nu_j-\nu_k)(-1)^{\nu_k}$$

to (1.9.3), we obtain that

$$B^{l_{s+1},\cdots,l_n}_{\nu_{s+1},\cdots,\nu_n} = (-1)^{\nu_{s+1}+\cdots+\nu_n}$$

$$\cdot \prod_{k=s+1}^{n}\left[\frac{1}{n!}\prod_{j=1}^{n}(\nu_j - l_k)(n-1)!\Big/\prod_{\substack{j=1\\ j\neq k}}^{n}(\nu_j-\nu_k)(-1)^{\nu_k}\right]^{n}$$

$$\cdot \frac{D(-l_{s+1},\cdots,-l_n)\cdot D(\nu_{s+1},\cdots,\nu_n)}{\prod_{j=s+1}^{n}\prod_{k=s+1}^{n}(\nu_j-l_k)}$$

$$= \frac{D(-l_{s+1},\cdots,-l_n)\cdot D(\nu_{s+1},\cdots,\nu_n)\prod_{k=s+1}^{n}\prod_{j=1}^{n}(\nu_j-l_k)}{\prod_{k=s+1}^{n}\prod_{j=1}^{n}(\nu_j-\nu_k)D(\nu_{s+1},\cdots,\nu_n)}$$

$$= \frac{D(\nu_1,\cdots,\nu_s,l_{s+1},\cdots,l_n)}{D(\nu_1,\cdots,\nu_s,\nu_{s+1},\cdots,\nu_n)}, \tag{1.9.6}$$

where $\nu_1>\nu_2>\cdots>\nu_s$ and $(\nu_1,\cdots,\nu_n)$ is a permutation of $(0,1,\cdots,n-1)$. From (1.9.6) and

$$N(\mu_1,\cdots,\mu_n) = \frac{D(\mu_1+n-1,\mu_2+n-2,\cdots,\mu_{n-1}+1,\mu_n)}{D(n-1,n-2,\cdots,1,0)},$$

Theorem 1.3.2 immediately follows.By the definition of $N(\mu_1,\cdots\mu_n)$ and

$$D(x_1,\cdots,x_n) = \prod_{j<k}(x_j-x_k)$$

via simplification, Theorem 1.3.3 follows. Collecting terms in Theorem 1.3.2 and 1.3.3 according to the powers of r, we complete the proof of Theorem 1.3.4.

As consequences of Theorems 1.3.1 and 1.3.2, we have the following two interesting algebraic identities.

Let $n\geqslant s\geqslant 0$, $f_1\geqslant f_2\geqslant\cdots\geqslant f_n$ and $l_j=f_j+n-j(j=1,2,\cdots,n)$. Then

(1) $$\sum_{s\geqslant g_{s+1}\geqslant\cdots\geqslant g_n\geqslant 0}\frac{N_s(f,g)N_s(g,f)}{N(f)N(g)} = 1, \tag{1.9.7}$$

where $g=(g_1,g_2,\cdots,g_n)$, $(g_1+n-1,g_2+n-2,\cdots,g_n)$ is a permutation of $(0,1,\cdots,n-1)$ and $0\geqslant g_1\geqslant g_2\geqslant\cdots\geqslant g_s\geqslant s-n$;

(2) $$\sum_{n-1\geqslant\nu_{s+1}>\cdots>\nu_n\geqslant 0}\prod_{j=1}^{s}\prod_{k=s+1}^{n}\frac{(l_j-\nu_k)(\nu_j-l_k)}{(\nu_j-\nu_k)(l_j-l_k)} = 1, \tag{1.9.8}$$

where $(\nu_1,\cdots,\nu_n)$ is a permutation of $(0,1,\cdots,n-1)$ and $\nu_1>\nu_2>\cdots>\nu_s$.

It can easily be checked that the property (2) of $\rho^j(r)$ in § 1.3 is satisfied. Thus, from (1.9.7) and (1.9.8), (1.3.7) immediately follows.

From (1.3.7) a Tauber-type theorem on convergence can be easily deduced. The details and the proof are omitted here.

§ 1.10 A Class of Integral Determinants

In 1959 the auther found out $\rho^j(r)$. In 1973 Zhong Jiaqing gave another simpler proof by using a class of integral determinants. With the aid of this class of integral determinants, he also obtained some results on representation theory of groups. For the details readers are referred to Zhong Jiaqing [1].

Let Γ be a simple closed curve on the complex plane, $f_i(z)$, $g_j(z)$ $(i, j = 1, 2, \cdots, n)$ be analytic functions both inside Γ and on Γ, and $A(z)$ be a polynomial in z of order $m (m \geqslant n)$, all zeros of $A(z)$ be located in the interior of Γ. Define the determinant

$$L(A; f_1, \cdots, f_n; g_1, \cdots, g_n) = \det\left(\frac{1}{2\pi i}\int_\Gamma \frac{f_i(z)g_j(z)}{A(z)}\,dz\right)_{1\leqslant i,j\leqslant n}. \quad (1.10.1)$$

Theorem 1.10.1 *Let* $A(z) = \prod_{k=1}^{n}(z - z_k)$, *all* z_k *be in the interior of* Γ *and be distinct. Then*

$$L(A; f_1, \cdots, f_n; g_1, \cdots, g_n) = \frac{(-1)^{\frac{n(n-1)}{2}}}{\prod_{i<j}(z_i - z_j)^2}$$

$$\cdot\begin{vmatrix} f_1(z_1), & \cdots, & f_1(z_n) \\ \cdots & \cdots & \cdots \\ f_n(z_1), & \cdots, & f_n(z_n)\end{vmatrix} \cdot \begin{vmatrix} g_1(z_1), & \cdots, & g_1(z_n) \\ \cdots & \cdots & \cdots \\ g_n(z_1), & \cdots, & g_n(z_n)\end{vmatrix}. \quad (1.10.2)$$

Proof. As all z_k are distinct, by the Cauchy formula we have

$$\frac{1}{2\pi i}\int_\Gamma \frac{f_i(z)g_j(z)}{\prod_{k=1}^{n}(z - z_k)}\,dz = \sum_{k=1}^{n}\frac{f_i(z_k)g_j(z_k)}{\prod_{l\neq k}(z_k - z_l)}.$$

Thus

$$\det\left(\frac{1}{2\pi i}\int_\Gamma \frac{f_i(z)g_j(z)}{A(z)}\,dz\right) = \det\left(\sum_{k=1}^{n}\frac{f_i(z_k)g_j(z_k)}{\prod_{l\neq k}(z_k - z_l)}\right)$$

$$= \begin{vmatrix} f_1(z_1), & \cdots, & f_1(z_n) \\ \cdots & \cdots & \cdots \\ f_n(z_1), & \cdots, & f_n(z_n) \end{vmatrix} \cdot \begin{vmatrix} \dfrac{g_1(z_1)}{\prod\limits_{l \neq 1} (z_1 - z_l)}, & \cdots, & \dfrac{g_n(z_1)}{\prod\limits_{l \neq 1} (z_1 - z_l)} \\ \cdots & \cdots & \cdots \\ \dfrac{g_1(z_n)}{\prod\limits_{l \neq n} (z_n - z_l)}, & \cdots, & \dfrac{g_n(z_n)}{\prod\limits_{l \neq n} (z_n - z_l)} \end{vmatrix}$$

$$= \left[\prod_{l \neq 1} (z_1 - z_l) \prod_{l \neq 2} (z_2 - z_l) \cdots \prod_{l \neq n} (z_n - z_l) \right]^{-1}$$

$$\cdot \begin{vmatrix} f_1(z_1), & \cdots, & f_1(z_n) \\ \cdots & \cdots & \cdots \\ f_n(z_1), & \cdots, & f_n(z_n) \end{vmatrix} \cdot \begin{vmatrix} g_1(z_1), & \cdots, & g_1(z_n) \\ \cdots & \cdots & \cdots \\ g_n(z_1), & \cdots, & g_n(z_n) \end{vmatrix}$$

$$= \frac{(-1)^{\frac{n(n-1)}{2}}}{\prod\limits_{i<j} (z_i - z_j)^2} \begin{vmatrix} f_1(z_1), & \cdots, & f_1(z_n) \\ \cdots & \cdots & \cdots \\ f_n(z_1), & \cdots, & f_n(z_n) \end{vmatrix} \cdot \begin{vmatrix} g_1(z_1), & \cdots, & g_1(z_n) \\ \cdots & \cdots & \cdots \\ g_n(z_1), & \cdots, & g_n(z_n) \end{vmatrix}.$$

Corollary 1.10.1 *Choose* $g_1(z) = 1,\ g_2(z) = z, \cdots,\ g_n(z) = z^{n-1}$ *in Theorem* 1.10.1. *Then* $\det(g_i(z_j))$ *differs from the Vandermonde determinant* $D(z_1, \cdots, z_n)$ *in the sign* $(-1)^{\frac{n(n-1)}{2}}$ *only. Thus*

$$L(A; f_1, \cdots, f_n; 1, z, \cdots, z^{n-1}) = \frac{1}{\prod\limits_{i<j} (z_i - z_j)} \begin{vmatrix} f_1(z_1), & \cdots, & f_1(z_n) \\ \cdots & \cdots & \cdots \\ f_n(z_1), & \cdots, & f_n(z_n) \end{vmatrix}. \tag{1.10.3}$$

If $f_1, f_2, \cdots, f_n$ are taken as polynomials, then $L(A; f_1, \cdots, f_n; 1, z, \cdots, z^{n-1})$ turns out to be the Lopatenski determinant defined by Reiko Sakamoto [1]. Thus (1.10.1) is an extension of the Lopetenski determinant.

Put $f_i(z) = z^{m_i+n-i} (i = 1, \cdots, n)$, where $m_1 \geqslant \cdots \geqslant m_n \geqslant 0$. Then

$$L(A; z^{m_1}, \cdots, z^{m_n}, 1, \cdots, z^{n-1}) = \left[\prod_{i<j} (z_i - z_j) \right]^{-1}$$

$$\cdot \begin{vmatrix} z_1^{m_1+n-1}, & \cdots, & z_n^{m_1+n-1} \\ z_1^{m_2+n-2}, & \cdots, & z_n^{m_2+n-2} \\ \cdots & \cdots & \cdots \\ z_1^{m_n}, & \cdots, & z_n^{m_n} \end{vmatrix} = \chi_{(m_1, \cdots, m_n)}(z_1, \cdots, z_n),$$

where $\chi_{(m_1, \cdots, m_n)}(z_1, \cdots, z_n)$ denotes the character of the irreducible representation of the full linear group with signature $(m_1, \cdots, m_n)$. Therefore (1.10.1) can be regarded as an extension of the character of irreducible representations of the full linear group.

Theorem 1.10.2 *Let $A(z)=(z-a)^n$ and a be inside Γ. Then*

$$L(A;f_1,\cdots,f_n;g_1,\cdots,g_n)=\frac{(-1)^{\frac{n(n-1)}{2}}}{(1!2!\cdots(n-1)!)^2}\cdot\begin{vmatrix}f_1(a), & \cdots, & f_n(a)\\ f_1'(a), & \cdots, & f_n'(a)\\ \cdots & \cdots & \cdots\\ f_1^{(n-1)}(a), & \cdots, & f_n^{(n-1)}(a)\end{vmatrix}\begin{vmatrix}g_1(a), & \cdots, & g_n(a)\\ g_1'(a), & \cdots, & g_n'(a)\\ \cdots & \cdots & \cdots\\ g_1^{(n-1)}(a), & \cdots, & g^{(n-1)}(a)\end{vmatrix}. \tag{1.10.4}$$

Proof. Resorting to the Cauchy and the Leibnitz formulas we have

$$\frac{1}{2\pi i}\int_\Gamma\frac{f_i(z)g_j(z)}{(z-a)^n}\,dz=\frac{1}{(n-1)!}[f_i(z)g_j(z)]_{z=a}^{(n-1)}$$

$$=\frac{1}{(n-1)!}\sum_{k=0}^{n-1}C_{n-1}^k f_i^{(k)}(a)g_j^{(n-1-k)}(a).$$

Thus

$$L(A;f_1,\cdots,f_n;g_1,\cdots,g_n)$$

$$=\det\left(\frac{1}{2\pi i}\int_\Gamma\frac{f_i(z)g_j(z)}{(z-a)^n}\,dz\right)$$

$$=\left[\frac{1}{(n-1)!}\right]^n\det\left(\sum_{k=0}^{n-1}C_k^{n-1}f_i^{(k)}(a)g_j^{(n-1-k)}(a)\right)$$

$$=(-1)^{\frac{n(n-1)}{2}}\left[\frac{1}{(n-1)!}\right]^n C_0^{n-1}\cdots C_{n-1}^{n-1}$$

$$\cdot\begin{vmatrix}f_1(a), & f_1'(a), & \cdots, & f_1^{(n-1)}(a)\\ \cdots & \cdots & \cdots & \cdots\\ f_n(a), & f_n'(a), & \cdots, & f_n^{(n-1)}(a)\end{vmatrix}\begin{vmatrix}g_1(a), & \cdots, & g_n(a)\\ \cdots & \cdots & \cdots\\ g_1^{(n-1)}(a), & \cdots, & g_n^{(n-1)}(a)\end{vmatrix},$$

which is (1.10.4).

Suppose $l_1>l_2>\cdots>l_k\geqslant 0>l_{k+1}>\cdots>l_n$ and let $l_{k+1}=-m_1,\cdots,l_n=-m_{n-k}$. Then $m_{n-k}>\cdots>m_1>0$. We now proceed to discuss such form of integral as follows:

$$\frac{1}{2\pi i}\int_{|\tau|=1}\frac{\tau^j d\tau}{\tau^m(1-r\tau)^n(\tau-r)^n},\ m>0,\ 0\leqslant j\leqslant n-1, \tag{1.10.5}$$

which is a special case of (1.8.2). Obviously we have

Lemma 1.10.1 *If an integer $P\leqslant n-2$, then*

$$\frac{1}{2\pi i}\int_{|\tau|=1}\frac{\tau^P}{(\tau-r)^n}\,d\tau=0.$$

Lemma 1.10.2 *For positive integers* $m>0$ *and* $n-1\geqslant j\geqslant 0$, *there exists a polynomial* $P_m(x)$ *of order* $\leqslant n-1$ *such that*

$$\frac{1}{2\pi i}\int_{|\tau|=1}\frac{\tau^j d\tau}{\tau^m(1-r\tau)^n(\tau-r)^n}$$
$$=\frac{r^m}{2\pi i}\int_{|\tau|=1}\frac{P_m(r\tau)\tau^j d\tau}{(1-r\tau)^n(\tau-r)^n}, \tag{1.10.6}$$

where $P_m(x)$ *is defined by*

$$P_m(x)=\frac{1-(1-x)^n(a_0+a_1x+\cdots+a_{m-1}x^{m-1})}{x^m}, \tag{1.10.7}$$

and

$$a_k=C_k^{n+k-1}=\frac{(n+k-1)\cdots n}{k!}.$$

Proof. To begin with, let us check that $P_m(x)$ should be a polynomial of order $\leqslant n-1$. By

$$(1-x)^{-n}=1+nx+\frac{n(n+1)}{2!}x^2+\cdots=a_0+a_1x+a_2x^2+\cdots,$$

we get

$$(1-x)^{-n}-(a_0+a_1x+\cdots+a_{m-1}x^{m-1})=a_mx^m+\cdots.$$

Therefore

$$1-(1-x)^n(a_0+a_1x+\cdots+a_{m-1}x^{m-1})=(1-x)^n(a_mx^m+\cdots).$$

The order on the left side is greater than or equal to m, thus it can be divided by x^m. Moreover, since the highest order on the left side is $n+m-1$, the result of division means that the order of $P_m(x)$ is no higher than $n-1$. Now let us prove (1.10.6). We have

$$\frac{1}{2\pi i}\int_{|\tau|=1}\frac{\tau^j d\tau}{\tau^m(\tau-r)^n(1-r\tau)^n}=\frac{1}{2\pi i}\int_{|\tau|=1}\frac{\tau^j}{(\tau-r)^n\tau^m}$$
$$\cdot\left[\sum_{i=0}^{m-1}a_ir^i\tau^i+\frac{1}{(1-r\tau)^n}-\sum_{i=0}^{m-1}a_ir^i\tau^i\right]d\tau$$
$$=\frac{1}{2\pi i}\int_{|\tau|=1}\frac{\tau^{j-m}}{(\tau-r)^n}\cdot\sum_{i=0}^{m-1}a_ir^i\tau^i d\tau+\frac{1}{2\pi i}\int_{|\tau|=1}\frac{\tau^j}{(\tau-r)^n(1-r\tau)^n}$$
$$\cdot\frac{1-(1-r\tau)^n\sum_{i=0}^{m-1}a_ir^i\tau^i}{\tau^m}d\tau.$$

Since $j-m+i\leqslant n-1+m-1-m=n-2$, the first integral of the right member, by Lemma 1.10.1, vanishes and the latter one is just

$$\frac{r^m}{2\pi i}\int_{|\tau|=1}\frac{P_m(r\tau)\tau^j}{(1-r\tau)^n(\tau-r)^n}\,d\tau.$$

Lemma 1.10.3 *The polynomial $P_m(x)$ in Lemma 1.10.2 is*

$$P_m(x)=\sum_{s=0}^{n-1} b_m^s x^s,$$

where

$$b_m^s=(-1)^s\frac{(n+m-1)\cdots(n-s)}{s!(m-1)!(m+s)!}. \tag{1.10.8}$$

Proof. From

$$P_{m+1}(x)=\frac{1-(1-x)^n(a_0+a_1x+\cdots+a_{m-1}x^{m-1}+a_mx^m)}{x^{m+1}},$$

it follows that

$$xP_{m+1}(x)=\frac{1-(1-x)^n(a_0+a_1x+\cdots+a_{m-1}x^{m-1})}{x^m}$$
$$-a_m(1-x)^n=P_m(x)-a_m(1-x)^n.$$

By this recurrence formula and $a_k=C_k^{n+k-1}$, (1.10.8) follows readily by induction.

Turning back to (1.6.4), we have by Lemma 1.10.2 that

$$\rho^f(r)N(f)=(1-r^2)^{n^2}$$
$$\cdot\begin{vmatrix}
\frac{1}{2\pi i}\int_{|\tau|=1}\frac{\tau^{l_1}d\tau}{(1-r\tau)^n(\tau-r)^n}, & \cdots, & \frac{1}{2\pi i}\int_{|\tau|=1}\frac{\tau^{l_1+n-1}d\tau}{(1-r\tau)^n(\tau-r)^n}\\
\cdots & \cdots & \cdots\\
\frac{1}{2\pi i}\int_{|\tau|=1}\frac{\tau^{l_k}d\tau}{(1-r\tau)^n(\tau-r)^n}, & \cdots, & \frac{1}{2\pi i}\int_{|\tau|=1}\frac{\tau^{l_1+n-1}d\tau}{(1-r\tau)^n(\tau-r)^n}\\
\frac{r^{m_1}}{2\pi i}\int_{|\tau|=1}\frac{P_{m_1}(r\tau)d\tau}{(1-r\tau)^n(\tau-r)^n}, & \cdots, & \frac{r^{m_1}}{2\pi i}\int_{|\tau|=1}\frac{P_{m_1}(r\tau)\tau^{n-1}d\tau}{(1-r\tau)^n(\tau-r)^n}\\
\cdots & \cdots & \cdots\\
\frac{r^{m_{n-k}}}{2\pi i}\int_{|\tau|=1}\frac{P_{m_{n-k}}(r\tau)d\tau}{(1-r\tau)^n(\tau-r)^n}, & \cdots, & \frac{r^{m_{n-k}}}{2\pi i}\int_{|\tau|=1}\frac{P^{m_{n-k}}(r\tau)\tau^{n-1}d\tau}{(1-r\tau)^n(\tau-r)^n}
\end{vmatrix}$$
$$=(1-r^2)^{n^2}r^{m_1+\cdots+m_{n-k}}L\Big((\tau-r)^n;\tau^{l_1},\cdots,\tau^{l_k},P_{m_1}(r\tau),\cdots,$$
$$P_{m_{n-k}}(r\tau);\frac{1}{(1-r\tau)^n},\cdots,\frac{\tau^{n-1}}{(1-r\tau)^n}\Big),$$

where $l_1>l_2>\cdots>l_k\geqslant 0>l_{k+1}>\cdots>l_n$. By. Theorem 1.10.2,

$$\rho^f(r)N(f)=(1-r^2)^{n^2}\cdot r^{-(l_{k+1}+\cdots+l_n)}\cdot\frac{(-1)^{\frac{n(n-1)}{2}}}{[1!2!\cdots(n-1)!]^2}$$

$$\cdot\begin{vmatrix} \tau^{l_1}, & \cdots, & \tau^{l_k}, & P_{m_1}(r\tau), & \cdots, & P_{m_{n-k}}(r\tau) \\ (\tau^{l_1})', & \cdots, & (\tau^{l_k})', & P'_{m_1}(r\tau), & \cdots, & P'_{m_{n-k}}(r\tau) \\ \cdots & \cdots & \cdots & \cdots & \cdots & \cdots \\ (\tau^{l_1})^{(n-1)}, & \cdots, & (\tau^{l_k})^{(n-1)}, & P^{(n-1)}_{m_1}(r\tau), & \cdots, & P^{(n-1)}_{m_{n-k}}(r\tau) \end{vmatrix}$$

$$\cdot\begin{vmatrix} \frac{1}{(1-r\tau)^n}, & \cdots, & \frac{\tau^{n-1}}{(1-r\tau)^n} \\ \cdots & \cdots & \cdots \\ \left[\frac{1}{(1-r\tau)^n}\right]^{(n-1)}, & \cdots, & \left[\frac{\tau^{n-1}}{(1-r\tau)^n}\right]^{(n-1)} \end{vmatrix}_{\tau=r}.$$

For calculating these two determinants, we need

Lemma 1.10.4

$$\begin{vmatrix} hg_1, & \cdots, & hg_n \\ (hg_1)', & \cdots, & (hg_n)' \\ \cdots & \cdots & \cdots \\ (hg_1)^{(n-1)}, & \cdots, & (hg_n)^{(n-1)} \end{vmatrix} = h^n \begin{vmatrix} g_1, & \cdots, & g_n \\ g'_1, & \cdots, & g'_n \\ \cdots & \cdots & \cdots \\ g_1^{(n-1)}, & \cdots, & g_n^{(n-1)} \end{vmatrix}.$$

Proof. The second row in the left determinant is

$$((hg_1)',\cdots,(hg_n)') = (\cdots, hg'_j + h'g_j, \cdots)$$
$$\equiv (hg'_1,\cdots,hg'_n) \text{ (mod the first row)}.$$

The other rows are all alike in conditions.

By this Lemma, we have

$$\begin{vmatrix} \frac{1}{(1-r\tau)^n}, & \cdots, & \frac{\tau^{n-1}}{(1-r\tau)^n} \\ \cdots & \cdots & \cdots \\ \left[\frac{1}{(1-r\tau)^n}\right]^{(n-1)}, & \cdots, & \left[\frac{\tau^{n-1}}{(1-r\tau)^n}\right]^{(n-1)} \end{vmatrix}_{\tau=r}$$

$$= (1-r^2)^{-n^2}\begin{vmatrix} 0! & & & * \\ & 1! & & \\ & & \ddots & \\ 0 & & & (n-1)! \end{vmatrix}$$

$$= (1-r^2)^{-n^2}\cdot 1!2!\cdots(n-1)!.$$

Hence

$$\rho^f(r)N(f) = r^{-(l_{k+1}+\cdots+l_n)}\frac{(-1)^{\frac{n(n-1)}{2}}}{1!2!\cdots(n-1)!}$$

$$\cdot \begin{vmatrix} \tau^{l_1}, & \cdots, & \tau^{l_k}, & P_{m_1}(r\tau), & \cdots, & P_{m_{n-k}}(r\tau) \\ (\tau^{l_1})', & \cdots, & (\tau^{l_k})', & P'_{m_1}(r\tau), & \cdots, & P'_{m_{n-k}}(r\tau) \\ \cdots & \cdots & \cdots & \cdots & \cdots & \cdots \\ (\tau^{l_1})^{(n-1)}, & \cdots, & (\tau^{l_k})^{(n-1)}, & P_{m_1}^{(n-1)}(r\tau), & \cdots, & P_{m_{n-k}}^{(n-1)}(r\tau) \end{vmatrix}. \tag{1.10.9}$$

By Theorem 1.2.4 of Hua [1], i.e.

$$\frac{(-1)^{\frac{n(n-1)}{2}}}{1!2!\cdots(n-1)!} \begin{vmatrix} f_1(x), & \cdots, & f_n(x) \\ f'_1(x), & \cdots, & f'_n(x) \\ \cdots & \cdots & \cdots \\ f_1^{(n-1)}(x), & \cdots, & f_n^{(n-1)}(x) \end{vmatrix}_{x=r}$$

$$= \lim_{\substack{x_1 \to r \\ \cdots\cdots \\ x_n \to r}} \begin{vmatrix} f_1(x_1), & \cdots, & f_n(x_1) \\ \cdots & \cdots & \cdots \\ f_1(x_n), & \cdots, & f_n(x_n) \end{vmatrix} \Big/ \prod_{i<j} (x_i - x_j), \tag{1.10.10}$$

(1.10.9) can be expressed by

$$\rho^j(r)N(f) = r^{-(l_{k+1}+\cdots+l_n)}$$

$$\cdot \lim_{\substack{x_1 \to r \\ \cdots\cdots \\ x_n \to r}} \begin{vmatrix} x_1^{l_1}, & \cdots, & x_n^{l_1} \\ \cdots & \cdots & \cdots \\ x_1^{l_k}, & \cdots, & x_n^{l_k} \\ P_{m_1}(rx_1), & \cdots, & P_{m_1}(rx_n) \\ \cdots & \cdots & \cdots \\ P_{m_{n-k}}(rx_1), & \cdots, & P_{m_{n-k}}(rx_n) \end{vmatrix} \cdot \frac{1}{\prod_{i<j} (x_i - x_j)}. \tag{1.10.11}$$

By Lemma 1.10.3, we have $P_m(x) = \sum_{s=0}^{n-1} b_m^s x^s$. Thus

$$\begin{vmatrix} x_1^{l_1}, & \cdots, & x_n^{l_1} \\ \cdots & \cdots & \cdots \\ x_1^{l_k}, & \cdots, & x_n^{l_k} \\ P_{m_1}(rx_1), & \cdots, & P_{m_1}(rx_n) \\ \cdots & \cdots & \cdots \\ P_{m_{n-k}}(rx_1), & \cdots, & P_{m_{n-k}}(rx_n) \end{vmatrix}$$

$$= \sum_{0 \leqslant P_1 \neq P_2 \neq \cdots \neq P_{n-k} \leqslant n-1} b_{m_1}^{P_1} \cdots b_{m_{n-k}}^{P_{n-k}} r^{P_1+\cdots+P_{n-k}} \begin{vmatrix} x_1^{l_1}, & \cdots, & x_n^{l_1} \\ \cdots & \cdots & \cdots \\ x_1^{l_k}, & \cdots, & x_n^{l_k} \\ x_1^{P_1}, & \cdots, & x_n^{P_1} \\ \cdots & \cdots & \cdots \\ x_1^{P_{n-k}}, & \cdots, & x_n^{P_{n-k}} \end{vmatrix}$$

$$= \sum_{n-1 \geqslant q_1 > q_2 > \cdots > q_{n-k} \geqslant 0} \sum_{(P_1,\cdots,P_{n-k}) \in (q_1,\cdots,q_{n-k})} \delta_{(q_1,\cdots,q_{n-k})}^{(P_1,\cdots,P_{n-k})}$$

$$\cdot b_{m_1}^{P_1} \cdots b_{m_{n-k}}^{P_{n-k}} r^{q+\cdots+q_{n-k}} \cdot M_{(l_1,\cdots,l_k,q_1,\cdots,q_{n-k})}(x_1, \cdots, x_n),$$

where the notation $(p_1, \cdots, p_{n-k}) \in (q_1, \cdots, q_{n-k})$stands for that $(p_1, \cdots, p_{n-k})$ is a permutation of $(q_1, \cdots, q_{n-k})$ and $\delta_{(q)}^{(p)}$ is the signature of the permutation.

Besides, we have

$$\sum_{(p_1,\cdots,p_{n-k})\in(q_1,\cdots,q_{n-k})} \delta_{(q_1\cdots q_{n-k})}^{(p_1\cdots p_{n-k})} b_{m_1}^{p_1} \cdots b_{m_{n-k}}^{p_{n-k}}$$

$$= \begin{vmatrix} b_{m_1}^{q_1}, & \cdots, & b_{m_1}^{q_{n-k}} \\ \cdots & \cdots & \cdots \\ b_{m_{n-k}}^{q_1}, & \cdots, & b_{m_{n-k}}^{q_{n-k}} \end{vmatrix} = C_{m_1,\cdots,m_{n-k}}^{q_1,\cdots,q_{n-k}}, \tag{1.10.12}$$

and

$$\lim_{\substack{x_1\to r\\ \cdots\\ x_n\to r}} \frac{M_{(l_1,\cdots,l_k,q_1,\cdots,q_{n-k})}(x_1, \cdots, x_n)}{\prod_{i<j}(x_i - x_j)}$$

$$= N(l_1 - n + 1, l_2 - n + 2, \cdots, l_k - n + k, q_1 - n + k + 1,$$

$$q_2 - n + k + 2, \cdots, q_{n-k}) \cdot r^{l_1+\cdots+l_k+q_1+\cdots+q_{n-k}-\frac{n}{2}(n-1)}, \tag{1.10.13}$$

in which the corresponding coefficient is zero if some q_i is equal to some l_j. From (1.10.11), (1.10.12) and (1.10.13) it follows that

$$\rho^j(r)N(f) = r^{-(l_{k+1}+\cdots+l_n)} \sum_{n-1\geqslant q_1>q_2>\cdots\geqslant q_{n-k}\geqslant 0} C_{m_1,\cdots,m_{n-k}}^{q_1,\cdots,q_{n-k}}$$

$$\cdot N(l_1 - n + 1, \cdots, l_k - n + k, q_1 - n + k + 1, \cdots, q_{n-k})$$

$$\cdot r^{\sum_1^k l_i + \sum_1^{n-k} q_i - \frac{n(n-1)}{2}} = r^{\sum_{i=1}^n |l_i| - \frac{1}{2}n(n-1)} \cdot \sum_{n-1\geqslant q_1>q_2>\cdots\geqslant q_{n-k}\geqslant 0} C_{m_1,\cdots,m_{n-k}}^{q_1,\cdots,q_{n-k}}$$

$$\cdot N(l_1 - n + 1, l_2 - n + 2, \cdots, l_k - n + k, q_1 - n + k + 1, \cdots, q_{n-k}) \cdot r^{2(q_1+\cdots+q_{n-k})},$$

where $l_1 - n + 1 = f_1, \cdots, l_k - n + k = f_k$. Moreover, let $q_1 - n + k + 1 = \xi_{k+1}, \cdots, q_{n-k} = \xi_n$. Then $N(l_1 - n + 1, \cdots, l_k - n + k, q_1 - n + k + 1, \cdots, q_{n-k}) = N(f_1, \cdots, f_k, \xi_{k+1}, \cdots, \xi_n)$, $q_1 + \cdots + q_{n-k} = \xi_{k+1} + \cdots + \xi_n + \dfrac{(n-k)(n-k-1)}{2}$; and the condition

$$n - 1 \geqslant q_1 > \cdots > q_{n-k} \geqslant 0$$

becomes

$$k \geqslant \xi_{k+1} \geqslant \cdots \geqslant \xi_n \geqslant 0.$$

Therefore, we finally get

$$\rho^f(r)N(f) = r^{\sum_1^n |l_i| - \frac{1}{2}n(n-1)+(n-k)(n-k-1)} \cdot \sum_{k \geqslant \xi_{k+1} \geqslant \cdots \geqslant \xi_n \geqslant 0}$$

$$\cdot N(f_1, \cdots, f_k, \xi_{k+1}, \cdots, \xi_n) C_{m_1,\cdots,m_{n-k}}^{\xi_{k+1}+n-k-1,\cdots,\xi_n} r^{2(\xi_{k+1}+\cdots+\xi_n)}$$

$$= r^{f_1+\cdots+f_k-f_{k+1}-\cdots-f_n} \cdot \sum_{k \geqslant \xi_{k+1} \geqslant \cdots \geqslant \xi_n \geqslant 0} N(f_1, \cdots, f_k, \xi_{k+1}, \cdots, \xi_n)$$

$$\cdot C_{m_1,\cdots,m_{n-k}}^{\xi_{k+1}+n-k-1,\cdots,\xi_n} r^{2(\xi_{k+1}+\cdots+\xi_n)}. \tag{1.10.14}$$

To sum up, it yields

Theorem 1.10.3 *If* $l_1 > l_2 > \cdots > l_k \geqslant 0 > l_{k+1} > \cdots > l_n$, *then* $\rho^f(r)$ *can be expressed by* (1.10.14), *where*

$$l_i = f_i + n - i,$$

$$m_1 = |l_{k+1}|, \cdots, m_{n-k} = |l_n|,$$

and the coefficients are

$$C_{m_1,\cdots,m_{n-k}}^{q_1,\cdots,q_{n-k}} = \begin{vmatrix} b_{m_1}^{q_1}, & \cdots, & b_{m_1}^{q_{n-k}} \\ \cdots & \cdots & \cdots \\ b_{m_{n-k}}^{q_1}, & \cdots, & b_{m_{n-k}}^{q_{n-k}} \end{vmatrix},$$

with

$$b_m^q = (-1)^q \frac{(n+m-1)\cdots(n-q)}{q!(m-1)!(m+q)!}.$$

Chapter 2. Cesàro Summations of Fourier Series on Unitary Groups

§ 2.1 Cesàro Summations

Let $u(U)$ be an integrable function on a unitary group U_n of order n, whose Fourier series is given by

$$\sum_{f_1 \geqslant f_2 \geqslant \cdots \geqslant f_n} \operatorname{tr}(C_{f_1 \cdots f_n} \Phi'_{f_1 \cdots f_n}(U)), \tag{2.1.1}$$

where

$$C_{f_1 \cdots f_n} = \int_{U_n} u(V) \overline{\Phi_{f_1 \cdots f_n}(V)} \dot{V}.$$

Hua Luogeng [2] defined the Abel mean of (2.1.1) as

$$\sum_{f_1 \geqslant f_2 \geqslant \cdots \geqslant f_n} \rho^f(r) \operatorname{tr}(C_{f_1 \cdots f_n} \Phi'_{f_1 \cdots f_n}(U)), \tag{2.1.2}$$

and proved that: if $u(U)$ is continuous on U_n, then the limit of (2.1.2) exists and is equal to $u(U)$ as $r \to 1$. This is to say that the Fourier series (2.1.1) of $u(U)$ is Abel-summable to itself.

In the preceding chapter, we have given the explicit expression of $\rho^f(r)$ by using a complicated technique of the matrix integral. In the present chapter, we study Cesàro summations under the motivation of the Abel summation. In § 2.2, the definition of Cesàro summations and its kernel will be given. In § 2.3 and § 2.4, the following Riesz-type theorem will be proved: If $u(U)$ is continuous on U_n, then the Fourier series (2.1.1) of $u(U)$ is (c, α)-summable to itself for $\alpha > \frac{n-1}{n}$.

By the definition of Cesàro summations, not only the kernel can be expressed explicitly by matrices, but the coefficients of the summations and some related integral constants can also be calculated explicitly. In order to understand general Cesàro summations concretely, we should carefully stuby as an example the most typical and the most important case of Cesàro summations, i.e., Fejér summation. By demonstrating this example, we can determine coefficients and constants of general Cesàro summations in the same way. This topic will be treated in § 2.5—§ 2.7, and the method used there is a further application of the matrix integral method in the preceding chapter.

The still further development along this idea. e.g., approximation theory on general compact groups etc., will be discussed in Chapter 4.

The material of this chapter is taken from Sheng Kung (Gong Sheng) [2].

§ 2.2 Definition and Kernels of Cesàro Summations

Definition 2.2.1 The Cesàro (c, α)-sum of the Fourier series (2.1.1) of an integrable function $u(U)$ over U_n is defined by

$$\sum_{nN\geqslant f_1\geqslant f_2\geqslant\cdots\geqslant f_n\geqslant -nN} B^{\alpha}_{f_1\cdots f_n}\operatorname{tr}(C_{f_1\cdots f_n}\Phi'_{f_1\cdots f_n}(U)), \tag{2.2.1}$$

where

$$B^{\alpha}_{f_1\cdots f_n} = \frac{1}{CN(f)}\int_{U_n} \chi_{f_1\cdots f_n}(\bar{V})\mathrm{K}^{\alpha}_N(V)\dot{V}. \tag{2.2.2}$$

$\mathrm{K}^{\alpha}_N(V)$ in (2.2.2) is called *Cesàro kernel*, which is equal to

$$\frac{\det^n\left[\sum_{k=0}^{N} A^{\alpha-1}_{N-k}V^k(I-\bar{V}'^{2k+1})\right]}{B^{\alpha}_N(2A^{\alpha}_N)^{n^2}\det^n(I-\bar{V}')}, \tag{2.2.3}$$

where

$$B^{\alpha}_N = \frac{1}{C}\int_{U_n}\frac{\det^n\left[\sum_{k=0}^{N} A^{\alpha-1}_{N-k}V^k(I-\bar{V}'^{2k+1})\right]}{(2A^{\alpha}_N)^{n^2}\det(I-\bar{V}')}\dot{V}, \tag{2.2.4}$$

and

$$A^{\alpha}_N = C^{N+\alpha}_N = \frac{(\alpha+N)\cdots(\alpha+1)}{N!},$$

in which the condition $\alpha > -1$, naturally, is needed.

First of all, we prove

Theorem 2.2.1 *The Cesàro (c, α)-sum (2.2.1) of the Fourier series (2.1.1) of any integrable function $u(U)$ over U_n can be expressed as*

$$\frac{1}{C}\int_{U_n} u(VU)\mathrm{K}^{\alpha}_N(V)\dot{V}, \tag{2.2.5}$$

where $\mathrm{K}^{\alpha}_N(V)$ is defined by (2.2.3).

Proof. Assume that $e^{i\theta_1},\cdots,e^{i\theta_n}$ are characteristic roots of $\bar{V}$. Then

$$\sum_{nN\geq f_1\geq f_2\geq\cdots\geq f_n\geq -nN} B^{\alpha}_{f_1\cdots f_n}N(f)M_{f_1\cdots f_n}(\bar{V}) = \mathrm{K}^{\alpha}_N(V)D(e^{i\theta_1},\cdots,e^{i\theta_n}). \quad (2.2.6)$$

For $M_{f_1\cdots f_n}(\bar{V})$, readers are referred to Hua [1].

We now turn to prove (2.2.6). Regarding the two sides of (2.2.6) as functions of $e^{i\theta_1}, e^{i\theta_2}, \cdots, e^{i\theta_n}$ and comparing the corresponding coefficients, we immediately obtain

$$B^{\alpha}_{f_1\cdots f_n}N(f) = \frac{1}{(2\pi)^n}\int_0^{2\pi}\cdots\int_0^{2\pi} e^{i(l_1\theta_1+\cdots+l_n\theta_n)}\mathrm{K}^{\alpha}_N(V) \cdot D(e^{-i\theta_1},\cdots,e^{-i\theta_n})d\theta_1\cdots d\theta_n,$$

as both sides of (2.2.6) are polynomials in $e^{i\theta_1}, e^{i\theta_2},\cdots e^{i\theta_n}$.

By the technique of the matrix integral which was used repeatedly in Chapter 1, the above expression is just (2.2.2). Hence (2.2.6) holds.

Dividing the two sides of (2.2.6) by $D(e^{i\theta_1},\cdots,e^{i\theta_n})$, we get

$$\sum_{nN\geq f_1\geq\cdots\geq f_n\geq -nN} B^{\alpha}_{f_1\cdots f_n}N(f)\chi_{f_1\cdots f_n}(\bar{V}) = \mathrm{K}^{\alpha}_N(V). \quad (2.2.7)$$

From § 0.2 it can be found that (2.2.1) turns out to be

$$\frac{1}{C}\int_{U_n} u(VU) \sum_{nN\geq f_1\geq\cdots\geq f_n\geq -nN} B^{\alpha}_{f_1\cdots f_n}N(f)\chi_{f_1\cdots f_n}(\bar{V})\dot{V},$$

which, by (2.2.7), is just (2.2.5).

It can be seen immediately that the relation between the Cesàro (c,α)-sum defined above and the Abel sum defined by Hua Luogeng and the relation between (the (c,α)-kernel and the Poisson-Hua kernel are the same as in the case of the ordinary Fourier series. This is to say that the (C,α)-kernel becomes the Poisson-Hua kernel as α tends to infinity.

The Fourier series of (2.1.1) of $u(U)$ is said to be (c,α)-summable if the limit of (2.2.1) exists as N tends to infinity.

§2.3 Semi-Positivity of Cesàro Kernels

Now, we begin to prove the following Riesz-type Theorem.

Theorem 2.3.1 *Let $u(U)$ be continuous on U_n. Then its Fourier series (2.1.1) is (c,α)-summable to itself provided $\alpha>(n-1)/n$.*

To this end, we first show that the Cesaro kernel is semi-positive for $\alpha > (n-1)/n$. In other words, we shall show the following:

Theorem 2.3.2 *The (c, α)-kernel $K_N^\alpha(V)$ is semi-positive, i.e.,*

$$\frac{1}{C}\int_{U_n} |K_N^\alpha(V)|\dot{V} \leqslant M, \tag{2.3.1}$$

where $\alpha > (n-1)/n$, M depends only on n and α and is independent of N.

Proof. (2.3.1) is obvious for even n, as $K_N^\alpha(V)$ itself is nonnegative, thus it suffices to take $M = 1$.

For odd n we apply induction to n. when $n = 1$, (2.3.1) obviously holds. Suppose (2.3.1) holds when the order of a unitary group is less than n. Then from, we shall prove that the conclusion still holds for any unitary group of order n.

It can be readily seen that $|B_N^\alpha|$ does not tend to zero as $N \to \infty$. So we only need to prove that

$$I = \frac{1}{(2\pi)^n} \int\cdots\int_{\pi\geqslant\theta_1\geqslant\cdots\geqslant\theta_n\geqslant-\pi} |\sigma_N^\alpha(\theta_1)\cdots\sigma_N^\alpha(\theta_n)|^n \cdot |D(e^{i\theta_1}, \cdots, e^{i\theta_n})|^2 d\theta_1\cdots d\theta_n \tag{2.3.2}$$

is bounded, where

$$\sigma_N^\alpha(\theta) = \frac{1}{2A_N^\alpha \sin\frac{1}{2}\theta} \operatorname{Im}\left\{\sum_{k=0}^{N} A_{N-k}^{\alpha-1} e^{i(k+\frac{1}{2})\theta}\right\}.$$

Let us first put $\delta \geqslant \frac{1}{N}$ and divide the integral domain as follows

R_1: $\delta \geqslant \theta_1 \geqslant \theta_2 \geqslant \cdots \geqslant \theta_n \geqslant -\pi$,

R_2: $\pi \geqslant \theta_1 \geqslant \delta \geqslant \theta_2 \geqslant \cdots \geqslant \theta_n \geqslant -\pi$,

....

R_n: $\pi \geqslant \theta_1 \geqslant \theta_2 \geqslant \cdots \geqslant \delta \geqslant \theta_n \geqslant -\pi$,

R_{n+1}: $\pi \geqslant \theta_1 \geqslant \theta_2 \geqslant \cdots \geqslant \theta_n \geqslant \delta$.

By I_j we denote the part of (2.3.2) which corresponds to the sub-domain R_j. Thus we have

$$I = I_1 + I_2 + \cdots + I_{n+1}.$$

First we consider I_2. Again divide R_2 in to the following subdomains

$$R_{2n}:\ \pi \geqslant \theta_1 \geqslant \delta \geqslant \theta_2 \geqslant \cdots \geqslant \theta_n \geqslant -\delta,$$

$$R_{2\overline{n-1}}:\ \pi \geqslant \theta_1 \geqslant \delta \geqslant \theta_2 \geqslant \cdots \geqslant -\delta \geqslant \theta_n \geqslant -\pi,$$

$$\cdots\cdots\cdots\cdots\cdots\cdots\cdots\cdots\cdots\cdots\cdots$$

$$R_{22}:\ \pi \geqslant \theta_1 \geqslant \delta \geqslant \theta_2 \geqslant -\delta \geqslant \cdots \geqslant \theta_n \geqslant -\pi,$$

$$R_{21}:\ \pi \geqslant \theta_1 \geqslant \delta \geqslant -\delta \geqslant \theta_2 \geqslant \cdots \geqslant \theta_n \geqslant -\pi.$$

By I_{ij} we mean the corresponding part over R_{ij} of (2.3.2). We get

$$\begin{aligned} I_{2n} &= \frac{1}{(2\pi)^n} \int\cdots\int_{R_{2n}} |\sigma_N^\alpha(\theta_1)\cdots\sigma_N^\alpha(\theta_n)|^n \\ &\quad \cdot |D(e^{i\theta_1},\cdots,e^{i\theta_n})|^2 d\theta_1\cdots d\theta_n \\ &= \frac{1}{(2\pi)^n} \int\cdots\int_{\delta\geqslant\theta_2\geqslant\cdots\geqslant\theta_n\geqslant-\delta} |\sigma_N^\alpha(\theta_2)\cdots\sigma_N^\alpha(\theta_n)|^n \\ &\quad \cdot |D(e^{i\theta_2},\cdots,e^{i\theta_n})|^2 d\theta_2\cdots d\theta_n \\ &\quad \cdot \int_\delta^\pi |\sigma_N^\alpha(\theta_1)|^n \prod_{j=2}^n |e^{i\theta_1}-e^{i\theta_j}|^2 d\theta_1. \end{aligned}$$

However

$$\begin{aligned} \prod_{j=2}^n |e^{i\theta_1}-e^{i\theta_j}|^2 &= \prod_{j=2}^n |(1-e^{i\theta_1})-(1-e^{i\theta_j})|^2 \\ &= \sum_{m_1=0}^{n-1}\sum_{l_1=0}^{n-2}\sum_{m_2=0}^{1}\sum_{l_2=0}^{1}\cdots\sum_{m_n=0}^{1}\sum_{l_n=0}^{1} (-1)^{2(n-1)-m_1-l_1}(1-e^{i\theta_1})^{m_1} \\ &\quad \cdot (1-e^{-i\theta_1})(1-e^{i\theta_2})^{m_2}(1-e^{-i\theta_2})^{l_2}\cdots \\ &\quad \cdot (1-e^{i\theta_n})^{m_n}(1-e^{-i\theta_n})^{l_n}. \end{aligned}$$

Therefore

$$\begin{aligned} I_{2n} &\leqslant \sum_{\substack{s_1=0\\ s_1+s_2+\cdots+s_n=2(n-1)}}^{2(n-1)} \sum_{s_2=0}^{2}\cdots\sum_{s_n=0}^{2} a_{s_1\cdots s_n} \int\cdots\int_{\delta\geqslant\theta_2\geqslant\cdots\geqslant\theta_n\geqslant-\delta} |1-e^{i\theta_2}|^{s_2}\cdots \\ &\quad \cdots |1-e^{i\theta_n}|^{s_n} |\sigma_N^\alpha(\theta_2)|^n\cdots|\sigma_N^\alpha(\theta_n)|^n \\ &\quad \cdot |D(e^{i\theta_2},\cdots,e^{i\theta_n})|^2 d\theta_1\cdots d\theta_n \int_\delta^\pi |1-e^{i\theta_1}|^{s_1} \\ &\quad \cdot |\sigma_N^\alpha(\theta_1)|^n d\theta_1, \end{aligned}$$

where $a_{s_1\cdots s_n}$ refer to absolute constants depending only on $s_1, s_2, \cdots, s_n$.

In order to estimate the foregoing integrals, we take any one of them,

$$I_{s_1\cdots s_n} = \int_\delta^\pi |1-e^{i\theta_1}|^{s_1} |\sigma_N^\alpha(\theta_1)|^n d\theta_1$$

$$\cdot \int \cdots \int_{\delta \geq \theta_2 \geq \cdots \geq \theta_n \geq -\delta} |1 - e^{i\theta_2}|^{s_2} \cdots |1 - e^{i\theta_n}|^{s_n} |\sigma_N^a(\theta_2)|^n \cdots$$
$$\cdot |\sigma_N^a(\theta_n)|^n |D(e^{i\theta_2}, \cdots, e^{i\theta_n})|^2 d\theta_2 \cdots d\theta_n \tag{2.3.3}$$

as an example. Since $\delta \geq \frac{1}{N}$, we have

$$|\sigma_N^a(\theta)| \leq aN^{-a}|\theta|^{-a-1}, \tag{2.3.4}$$

where a denotes an absolute constant. Henceforth a is frequently used to express various absolute constants.

Hence

$$I_{s_1 \cdots s_n} \leq a \int_\delta^\pi N^{-\alpha n} \theta^{-\alpha n - n + s_1} d\theta N^{n-1} \delta^{s_2 + \cdots + s_n},$$

where we employ the induction assumption that

$$\int \cdots \int_{\delta \geq \theta_2 \geq \cdots \geq \theta_n \geq -\delta} |\sigma_N^a(\theta_2) \cdots \sigma_N^a(\theta_n)|^{n-1}$$
$$|D(e^{i\theta_2}, \cdots, e^{i\theta_n})|^2 d\theta_2 \cdots d\theta_n$$

is bounded. As $\alpha > \frac{n-1}{n}$ and $0 \leq s_1 \leq 2(n-1)$, we get

$$-\alpha n - n + s_1 \leq -(n-1) - n + s_1 < -1.$$

As a result, we have

$$I_{s_1 \cdots s_n} \leq aN^{-\alpha n + n - 1} \delta^{-\alpha n - n + s_1 + 1 + s_2 + \cdots + s_n} = O((N\delta)^{-\alpha n + n - 1}),$$

because $s_1 + s_2 + \cdots + s_n = 2(n-1)$.

Thus

$$I_{2n} = O((N\delta)^{-\alpha n + n - 1}).$$

Now we turn to $I_{\overline{2n-1}}$. Here

$$I_{\overline{2n-1}} = \frac{1}{(2\pi)^n} \int \cdots \int_{\delta \geq \theta_2 \geq \cdots \geq \theta_n \geq -\delta} |\sigma_N^a(\theta_2) \cdots \sigma_N^a(\theta_n)|^n$$
$$\cdot |D(e^{i\theta_2}, \cdots, e^{i\theta_n})|^2 d\theta_2 \cdots d\theta_{n-1} \int_\delta^\pi |\sigma_N^a(\theta_n)|^n \prod_{j=2}^{n-1} |e^{i\theta_j} - e^{i\theta_n}|^2$$
$$\cdot d\theta_n \int_\delta^\pi |\sigma_N^a(\theta_1)|^n \prod_{j=2}^{n} |e^{i\theta_1} - e^{i\theta_j}|^2 d\theta_1.$$

As in the case of I_{2n} we also have

$$\prod_{j=2}^{n} |e^{i\theta_1} - e^{i\theta_j}|^2 \prod_{k=2}^{n-1} |e^{i\theta_k} - e^{i\theta_n}|^2$$

$$= \prod_{j=2}^{n} |(1-e^{i\theta_1})-(1-e^{i\theta_j})|^2 \cdot \prod_{k=2}^{n-1} |(1-e^{i\theta_k})-(1-e^{i\theta_n})|^2$$

$$\leqslant \sum_{m_1=0}^{2n-2} \sum_{m_2=0}^{4} \cdots \sum_{m_{n-1}=0}^{4} \sum_{m_n=0}^{2n-2} \cdot b_{m_1\cdots m_n} \cdot |1-e^{i\theta_1}|^{m_1} |1-e^{i\theta_2}|^{m_2} \cdots |1-e^{i\theta_n}|^{m_n},$$

where $b_{m_1\cdots m_n}$ are absolute constants depending only on $m_1, \cdots, m_n$. However

$$\int_{\delta}^{\pi} |1-e^{i\theta_1}|^{m_1} |\sigma_N^{\alpha}(\theta_1)|^n d\theta_1 \int_{-\pi}^{-\delta} |1-e^{i\theta_n}|^{m_n} |\sigma_N^{\alpha}(\theta_n)|^n d\theta_n$$

$$\cdot \int_{\delta > \theta_2 > \cdots > \theta_{n-1} > -\delta} \cdots \int |1-e^{i\theta_2}|^{m_2} \cdots |1-e^{i\theta_{n-1}}|^{m_{n-1}} |\sigma_N^{\alpha}(\theta_2)|^n$$

$$\cdots |\sigma_N^{\alpha}(\theta_{n-1})|^n \cdot |D(e^{i\theta_2}, \cdots, e^{i\theta_{n-1}})|^2 \cdot d\theta_2 \cdots d\theta_{n-1}$$

$$\leqslant \int_{\delta}^{\pi} N^{-\alpha n} \theta_1^{-\alpha n - n + m_1} d\theta_1 \cdot \int_{-\pi}^{-\delta} N^{-\alpha n} \theta_n^{-\alpha n - n + m_n} d\theta_n \cdot \delta^{4n-6-m_1-m_n} N^{2(n-2)}$$

$$= O(N^{-\alpha n} \delta^{-\alpha n - n + m_1 + 1} N^{-\alpha n} \delta^{-\alpha n - n + m_n + 1} \delta^{4n-6-m_1-m_n} N^{2(n-2)})$$

$$= O((N\delta)^{2(-\alpha n + n - 2)}).$$

By the induction assumption that

$$\int_{\delta > \theta_2 > \cdots > \theta_{n-1} > -\delta} \cdots \int |\sigma_N^{\alpha}(\theta_2)|^{n-2} \cdots |\sigma_N^{\alpha}(\theta_{n-1})|^{n-2} \cdot |D(e^{i\theta_2}, \cdots, e^{i\theta_{n-1}})|^2 d\theta_2 \cdots d\theta_{n-1}$$

is bounded, we obtain

$$I_{\overline{2n-1}} = O((N\delta)^{2(-\alpha n + n - 2)}).$$

Similarly

$$I_{\overline{2n-j}} = O((N\delta)^{(j+1)(-\alpha n + n - j - 1)}).$$

Finally

$$I_2 = O((N\delta)^{-\alpha n + n - 1} + O((N\delta)^{2(-\alpha n + n - 2)}) + \cdots + \cdots + O((N\delta)^{(j+1)(-\alpha n + n - j - 1)}) + \cdots + O((N\delta)^{-\alpha n^2}).$$

The following can be verified via the same method:

$$I_3 = O((N\delta)^{2(-\alpha n + n - 2)}) + O((N\delta)^{3(-\alpha n + n - 3)}) + \cdots + O((N\delta)^{-\alpha n^2}), \cdots\cdots$$

and

$$I_{n+1} = O((N\delta)^{-\alpha n^2}).$$

Now, we proceed to treat the integral I_1 over R_1. Again divide R_1 into sub-domains as follows

$$s_1:\ \delta \geqslant \theta_1 \geqslant \cdots\cdots \geqslant \theta_{n-1} \geqslant \theta_n \geqslant -\delta,$$

$$s_2:\ \delta \geqslant \theta_1 \geqslant \cdots\cdots \geqslant \theta_{n-1} \geqslant -\delta \geqslant \theta_n \geqslant -\pi,$$

$$\cdots\cdots\cdots\cdots\cdots\cdots\cdots\cdots\cdots\cdots\cdots\cdots$$

$$s_{n+1}:\ -\delta \geqslant \theta_1 \geqslant \cdots\cdots \geqslant \theta_n \geqslant -\pi.$$

By J_k we denote the part of (2.3.2) corresponding to the domain S_k. As in the case of I_{jk}, it can be shown that

$$J_2 = O((N\delta)^{-\alpha n+n-1}) + O((N\delta)^{2(-\alpha n+n-2)}) + \cdots + O((N\delta)^{-\alpha n^2}),$$

$$J_3 = O((N\delta)^{2(-\alpha n+n-2)}) + \cdots + O((N\delta)^{-\alpha n^2}),$$

$$\cdots\cdots\cdots\cdots\cdots\cdots\cdots\cdots\cdots\cdots\cdots\cdots$$

$$J_{n+1} = O((N\delta)^{-\alpha n^2}).$$

Finally we consider J_1. We have

$$\begin{aligned} J_1 = & \frac{1}{(2\pi)^n} \int\cdots\int_{\delta\geqslant\theta_1\geqslant\cdots\geqslant\theta_n\geqslant-\delta} |\sigma_N^\alpha(\theta_1)\cdots\sigma_N^\alpha(\theta_n)|^n \\ & \cdot |D(e^{i\theta_1},\cdots,e^{i\theta_n})|^2 d\theta_1\cdots d\theta_n \\ \leqslant & \int\cdots\int_{\delta\geqslant\theta_2\geqslant\cdots\geqslant\theta_n\geqslant-\delta} |\sigma_N^\alpha(\theta_2)\cdots\sigma_N^\alpha(\theta_n)|^n |D(e^{i\theta_2},\cdots,e^{i\theta_n})|^2 d\theta_2\cdots d\theta_n \\ & \cdot \int_{\theta_2}^{\delta} |\sigma_N^\alpha(\theta_1)|^n \prod_{j=2}^{n} |e^{i\theta_1} - e^{i\theta_j}|^2 d\theta_1 \\ \leqslant & \, a(N\delta)^{2(n-1)} \int\cdots\int_{\delta\geqslant\theta_2\geqslant\cdots\geqslant\theta_n\geqslant-\delta} |\sigma_N^\alpha(\theta_2)\cdots\sigma_N^\alpha(\theta_n)|^{n-1} \\ & \cdot |D(e^{i\theta_2},\cdots,e^{i\theta_n})|^2 d\theta_2\cdots d\theta_n. \end{aligned}$$

By the induction assumption, the preceding expression is just

$$O((N\delta)^{2(n-1)}).$$

To sum up, we get

$$\begin{aligned} I = & \, O((N\delta)^{2(n-1)}) + O((N\delta)^{-\alpha n+n-1}) \\ & + O((N\delta)^{2(-\alpha n+n-2)}) + \cdots + O((N\delta)^{-\alpha n^2}). \end{aligned}$$

Taking $\delta = \dfrac{1}{N}$, we obtain

$$I = O(1),$$

which proves (2.3.1).

§ 2.4 Proof of Riesz-Type Theorem

Now we turn to prove Theorem 2.3.1.

From § 2.2, it is known that

$$\frac{1}{C}\int_{U_n} \mathbf{K}_N^{\alpha}(N)\dot{V} = 1. \tag{2.4.1}$$

In § 2.2, it has already been proved that the Cesaro (C,α)-sum (which is simply denoted by $\sum_N^{\alpha}(U)$) can be expressed by (2.2.5). Then, by (2.4.1), we obtain

$$\sum_N^{\alpha}(U) - u(U) = \frac{1}{C}\int_{U_n}(u(VU) - u(U))\mathbf{K}_N^{\alpha}(V)\dot{V}. \tag{2.4.2}$$

In order to prove that

$$\sum_N^{\alpha}(U) \to u(U), \quad \text{as} \to N,$$

it is sufficient to show that

$$\int_{U_n}(u(VU) - u(U))\mathbf{K}_N^{\alpha}(V)\dot{V} \tag{2.4.3}$$

tends to zero as $N\to\infty$. In fact, (2.4.3) is just

$$I = \frac{1}{(2\pi)^n B_N^{\alpha}}\int\cdots\int_{\pi\geqslant\theta_1\geqslant\cdots\geqslant\theta_n\geqslant-\pi}\varphi(e^{i\theta_1},\cdots,e^{i\theta_n})$$
$$\cdot(\sigma_N^{\alpha}(\theta_1)\cdots\sigma_N^{\alpha}(\theta_n))^n|D(e^{i\theta_1},\cdots,e^{i\theta_n})|^2 d\theta_1\cdots d\theta_n, \tag{2.4.4}$$

where

$$\varphi(e^{i\theta_1},\cdots,e^{i\theta_n}) = \frac{1}{w_n'}\int_{[U_n]}(u(VU) - u(U))[\dot{V}]. \tag{2.4.5}$$

Imitating the proof of Theorem 2. 3. 2, we divide the integral domain

$$\pi\geqslant\theta_1\geqslant\cdots\geqslant\theta_n\geqslant-\pi$$

into sub-domains $R_1, R_2, \cdots, R_{n+1}$. I_j' denotes the corresponding part over R_j of the integral in (2.4.4). As $u(U)$ is continuous on U_n, we have

$$|\varphi(e^{i\theta_1},\cdots,e^{i\theta_n})| \leqslant L$$

on U_n, where L is an absolute constant. Next, we turn to treat $I_2', \cdots, I_{n+1}'$ for $\delta\geqslant\frac{1}{N}$ as we did in § 2.3.

First consider I_2'. We divide R_2 into $R_{2n}, R_{\overline{2n-1}}, \cdots, R_{22}, R_{21}$ via a similar treatment with I_2 in the preceding section. By I_{ij}' we denote the part of (2.4.4) corresponding to the sub-domain R_{ij}. Then

$$I'_{2n} = \frac{1}{B_N^\alpha (2\pi)^n} \int \cdots \int_{R2n} \varphi(e^{i\theta_1}, \cdots, e^{i\theta_n})(\sigma_N^\alpha(\theta_1) \cdots \sigma_N^\alpha(\theta_n))^n$$

$$\cdot |D(e^{i\theta_1}, \cdots, e^{i\theta_n})|^2 d\theta_1 \cdots d\theta_n. \qquad (2.4.6)$$

From (2.4.6), it follows that

$$|I'_{2n}| \leqslant \frac{L}{|B_N^\alpha| (2\pi)^n} \int \cdots \int_{\delta \geqslant \theta_2 \geqslant \cdots \geqslant \theta_n \geqslant -\delta} |\sigma_N^\alpha(\theta_2)|^n \cdots |\sigma_N^\alpha(\theta_n)|^n$$

$$\cdot |D(e^{i\theta_2}, \cdots, e^{i\theta_n})|^2 d\theta_2 \cdots d\theta_n \int_\delta^\pi |\sigma_N^\alpha(\theta_1)|^n$$

$$\cdot \prod_{j=2}^n |e^{i\theta_1} - e^{i\theta_j}|^2 d\theta_1.$$

As above, we turn to study $I_{s_1 \cdots s_n}$ defined by (2.3.3). In virtue of (2.3.4), it follows that

$$I_{s_1 \cdots s_n} \leqslant a \int_\delta^\pi N^{-\alpha n} \theta_1^{-\alpha n - n + s_1} d\theta_1 \cdot N^{n-1} \delta^{s_2 + \cdots + s_n}$$

$$\cdot \int \cdots \int_{\delta \geqslant \theta_2 \geqslant \cdots \geqslant \theta_n \geqslant -\delta} |\sigma_N^\alpha(\theta_2) \cdots \sigma_N^\alpha(\theta_n)|^{n-1}$$

$$\cdot |D(e^{i\theta_2}, \cdots, e^{i\theta_n})|^2 d\theta_2 \cdots d\theta_n.$$

By Theorem 2. 3. 2, we have

$$I_{s_1 \cdots s_n} = O((N\delta)^{-\alpha n + n - 1}).$$

Thus

$$|I'_{2n}| = O((N\delta)^{-\alpha n + n - 1}) + O((N\delta)^{2(-\alpha n + n - 2)})$$
$$+ \cdots + O((N\delta)^{-\alpha n^2}).$$

Similarly

$$|I'_{\overline{2n-1}}| = O((N\delta)^{2(-\alpha n + n - 2)}) + \cdots + O((N\delta)^{-\alpha n^2}).$$

Therefore

$$|I'_2| = O((N\delta)^{-\alpha n + n - 1}) + \cdots + O((N\delta)^{-\alpha n^2}).$$

In the same way, it can be shown that

$$|I'_3| = O((N\delta)^{2(-\alpha n + n - 2)}) + \cdots + O((N\delta)^{-\alpha n^2}),$$

....

$$|I'_{n+1}| = O((N\delta)^{-\alpha n^2}).$$

Consequently

$$|I'_2| + \cdots + |I'_{n+1}| = O((N\delta)^{-\alpha n + n - 1}) + \cdots$$
$$+ O((N\delta)^{j(-\alpha n + n - j)}) + \cdots + O((N\delta)^{-\alpha n^2}).$$

For estimating I_1', as in the case of I_n in § 2.3, we partition R_1 into $n+1$ parts $s_1, \cdots, s_{n+1}$ and denote the corresponding part over s_k in (2.4.4) by J_k'. The similar method gives us

$$|J_2'| + \cdots + |J_{n+1}'| = O((N\delta)^{-\alpha n+n-1}) + O((N\delta)^{2(-\alpha n+n-2)}) + \cdots + O((N\delta)^{-\alpha n^2}).$$

Finally let us study J_1', where

$$J_1' = \frac{1}{(2\pi)^n B_N^\alpha} \int_{\delta \geqslant \theta_1 \geqslant \cdots \geqslant \theta_n \geqslant -\delta} \cdots \int \varphi(e^{i\theta_1}, \cdots, e^{i\theta_n}) \cdot (\sigma_N^\alpha(\theta_1) \cdots \sigma_N^\alpha(\theta_n))^n |D(e^{i\theta_1}, \cdots, e^{i\theta_n})|^2 d\theta_1 \cdots d\theta_n.$$

Since $u(U)$ is continuous on U_n, for any given $\eta > 0$, there must exist a δ, sufficiently small such that

$$|\varphi(e^{i\theta_1}, \cdots, e^{i\theta_n})| < \eta$$

for $\delta \geqslant \theta_1 \geqslant \theta_2 \geqslant \cdots \geqslant \theta_n \geqslant -\delta$. Thus, by Theorem 2.3.2, we have

$$\frac{1}{C} \int_{U_n} |\mathrm{K}_N^\alpha(V)| \dot{V} < M$$

for $\alpha > \frac{n-1}{n}$. So, for any given $\varepsilon > 0$, we can choose a δ sufficiently small such that

$$|J_1'| < \varepsilon/2.$$

Having chosen $\delta = \delta(\varepsilon)$ for any $\varepsilon > 0$, we can always choose an N sufficiently large such that

$$|I_2'| + \cdots + |I_{n+1}'| + |J_2'| + \cdots + |J_{n+1}'| < \varepsilon/2,$$

because $\alpha n - n + 1 > 0$. At last we confirm

$$|I| < \varepsilon$$

for any given $\varepsilon > 0$, which proves Theorem 2.3.1.

§ 2.5 Fejér Summation

We now carry out some detailed studies of the simplest and the most important case of (c, α) summation, the Fejér summation, so that we could have a clear understanding of (c, α) summations. It can be shown that both the coefficients and the integral constant of (c, α) summations can be determined with a reference to the method used in searching those of the Fejér summation.

By the definition in § 2.2 it can be readily seen that the Fejér kernel is

$$\frac{1}{B_N(N+1)^{n^2}}\left|\frac{\det(I-V^{N+1})}{\det(I-V)}\right|^{2n}, \tag{2.5.1}$$

where

$$B_N = B_N^1 = \frac{1}{C(N+1)^{n^2}} \cdot \int_{U_n}\left|\frac{\det(I-V^{N+1})}{\det(I-V)}\right|^{2n} \dot{V}. \tag{2.5.2}$$

It is very interesting that the Fejér kernel is also positive definite and has a very simple form. From this it is derived that for $\alpha \geqslant 1$ every (c, α) kernel is positive definite.

By definition, the coefficient of the Fejér summation is given as

$$B_{f_1\cdots f_n} = B^1_{f_1\cdots f_n} = \frac{1}{B_N N(f)(N+1)^{n^2}C} \cdot \int_{U_n}\chi_{f_1\cdots f_n}(\bar{V})\left|\frac{\det(I-V^{N+1})}{\det(I-V)}\right|^{2n} \dot{V}. \tag{2.5.3}$$

The Fejér sum of the Fourier series (2.1.1) of $u(U)$ becomes into

$$\sum_{nN\geqslant f_1\geqslant\cdots\geqslant f_n\geqslant -nN} B_{f_1\cdots f_n}\operatorname{tr}(C_{f_1\cdots f_n}\Phi'_{f_1\cdots f_n}(U)) = \frac{1}{B_N(N+1)^{n^2}C}\int_{U_n}u(VU)\left|\frac{\det(I-V^{N+1})}{\det(I-V)}\right|^{2n} \dot{V}. \tag{2.5.4}$$

Hence the Riesz-type Theorem 2.3.1 reduces to the following Fejér-type Theorem: If $u(U)$ is continuous on U_n, then the Fourier series (2.1.1) of $u(U)$ is Fejér-summable to itself.

Since the Fejér kernel is positive definite, the proof of the Fejér-type Theorem mentioned above can be verified much more easily than that of Theorem 2.3.1. There is no need to prove Theorem 2.3.2 in advance. The details are omitted (cf. the proof of Theorem 1.3.1).

§ 2.6 Explicit Expression of Coefficients

From this section on, we will use the complicated and skilful technique of matrix integral to give a concrete expression of the coefficients and integral constants of the Fejer summation. This procedure would tell us how to deduce the explicit expression of the general $B^{\alpha}_{f_1\cdots f_n}$ and the value of B^{α}_N.

Applying the technique used frequently in Chapter 1, from (2.5.2) and

$$D(e^{-i\theta_1},\cdots,e^{-i\theta_n})=(-1)^{\frac{1}{2}n(n-1)}D(1-e^{-i\theta_1},\cdots,1-e^{-i\theta_n}),$$

we obtain (cf. §§ 1.6—1.8)

$$B_{f_1\cdots f_n}N(f)=(N+1)^{-n^2}(-1)^{\frac{1}{2}n(n-1)}B_N^{-1}$$

$$\cdot\begin{vmatrix} \frac{1}{2\pi}\int_0^{2\pi}\frac{e^{il_1\theta}e^{inN\theta}(1-e^{-i(N+1)\theta})^{2n}d\theta}{(1-e^{-i\theta})^{2n}}, \cdots, \frac{1}{2\pi}\int_0^{2\pi}\frac{e^{il_n\theta}e^{inN\theta}(1-e^{-i(N+1)\theta})^{2n}d\theta}{(1-e^{-i\theta})^{2n}} \\ \frac{1}{2\pi}\int_0^{2\pi}\frac{e^{il_1\theta}e^{inN\theta}(1-e^{-i(N+1)\theta})^{2n}d\theta}{(1-e^{-i\theta})^{2n-1}}, \cdots, \frac{1}{2\pi}\int_0^{2\pi}\frac{e^{il_n\theta}e^{inN\theta}(1-e^{-i(N+1)\theta})^{2n}d\theta}{(1-e^{-i\theta})^{2n-1}} \\ \cdots\cdots\cdots\cdots \\ \frac{1}{2\pi}\int_0^{2\pi}\frac{e^{il_1\theta}e^{inN\theta}(1-e^{-i(N+1)\theta})^{2n}d\theta}{(1-e^{-i\theta})^{n+1}}, \cdots, \frac{1}{2\pi}\int_0^{2\pi}\frac{e^{il_n\theta}e^{inN\theta}(1-e^{-i(N+1)\theta})^{2n}d\theta}{(1-e^{-i\theta})^{n+1}} \end{vmatrix} \quad (2.6.1)$$

Write

$$a_p^q=\frac{1}{2\pi}\int_0^{2\pi}\frac{e^{iq\theta}(1-e^{-i(N+1)\theta})^{2n}d\theta}{(1-e^{-i\theta})^p}. \quad (2.6.2)$$

Then (2.6.1) becomes into

$$B_{f_1\cdots f_n}N(f)=(N+1)^{-n^2}(-1)^{\frac{1}{2}n(n-1)}B_N^{-1}$$

$$\cdot\begin{vmatrix} a_{2n}^{l_1+nN}, & a_{2n}^{l_2+nN}, & \cdots, & a_{2n}^{l_n+nN} \\ a_{2n-1}^{l_1+nN}, & a_{2n-1}^{l_2+nN}, & \cdots, & a_{2n-1}^{l_n+nN} \\ \cdots & \cdots & \cdots & \cdots \\ a_{n+1}^{l_1+nN}, & a_{n+1}^{l_2+nN}, & \cdots, & a_{n+1}^{l_n+nN} \end{vmatrix}. \quad (2.6.3)$$

Lemma 2.6.1 *When $p\leqslant 2n$, we have*

$$a_p^q=\frac{(2n)!}{(p-1)!}\sum_{\substack{k=0\\ k+(N+1)s=q}}^{\infty}\sum_{s=0}^{2n}\frac{(-1)^s(p+k-1)!}{k!s!(2n-s)!}. \quad (2.6.4)$$

Proof. Taking $z=re^{-i\theta}$, $0<r<1$, we get

$$a_p^q=\frac{1}{2\pi}\int_0^{2\pi}\frac{e^{iq\theta}(1-e^{-i(N+1)\theta})^{2n}d\theta}{(1-e^{-i\theta})^p}.$$

By $p\leqslant 2n$ and the Lebesgue subordinate theorem, the above expression is equal to

$$\lim_{r\to 1}\frac{1}{2\pi}\int_0^{2\pi}\frac{e^{iq\theta}(1-z^{N+1})^{2n}d\theta}{(1-z)^p}. \quad (2.6.5)$$

However, we have

$$(1-z)^{-p}=\sum_{k=0}^{\infty}\frac{(p+k-1)!}{k!(p-1)!}z^k,$$

and

$$(1 - z^{N+1})^{2n} = \sum_{k=0}^{2n} \frac{(-1)^s(2n)!}{s!(2n-s)!} z^{(N+1)s},$$

hence

$$\frac{(1 - z^{N+1})^{2n}}{(1-z)p} = \sum_{k=0}^{\infty} \sum_{s=0}^{2n} \frac{(-1)^s(p+k-1)!(2n)!}{k!(p-1)!s!(2n-s)!} z^{k+(N+1)s}.$$

From (2.6.2) and (2.6.5), (2.6.4) follows immediately.

a_p^q can also be written as

$$\frac{(2n)!}{(p-1)!} \sum_{\substack{s=0 \\ q-(N+1)s\geqslant 0}}^{2n} \frac{(-1)^s(p+q-(N+1)s-1)!}{(q-(N+1)s)!s!(2n-s)!}.$$

By Lemma 2.6.1, (2.6.3) is just

$$B_{l_1\cdots l_n}N(f) = \frac{((2n)!)^n(-1)^{\frac{1}{2}n(n-1)}}{(2n-1)!(2n-2)!\cdots n!(N+1)^{n^2}B_N}$$

$$\cdot \begin{vmatrix} \sum\limits_{\substack{s_1=0 \\ k_1\geqslant 0}}^{2n} \frac{(-1)^{s_1}(2n+k_1-1)!}{k_1!s_1!(2n-s_1)!}, \cdots, \sum\limits_{\substack{s_n=0 \\ k_n\geqslant 0}}^{2n} \frac{(-1)^{s_n}(2n+k_n-1)!}{k_n!s_n!(2n-s_n)!} \\ \sum\limits_{\substack{s_1=0 \\ k_1\geqslant 0}}^{2n} \frac{(-1)^{s_1}(2n-1+k_1-1)!}{k_1!s_1!(2n-s_1)!}, \cdots, \sum\limits_{\substack{s_n=0 \\ k_n\geqslant 0}}^{2n} \frac{(-1)^{s_n}(2n-1+k_n-1)!}{k_n!s_n!(2n-s_n)!} \\ \cdots\cdots\cdots\cdots\cdots\cdots \\ \sum\limits_{\substack{s_1=0 \\ k_1\geqslant 0}}^{2n} \frac{(-1)^{s_1}(n+1+k_1-1)!}{k_1!s_1!(2n-s_1)!}, \cdots, \sum\limits_{\substack{s_n=0 \\ k_n\geqslant 0}}^{2n} \frac{(-1)^{s_n}(n+1+k_n-1)!}{k_n!s_n!(2n-s_n)!} \end{vmatrix}, \tag{2.6.6}$$

where

$$k_1 = l_1 + nN - (N+1)s_1,$$

$$\cdots\cdots$$

$$k_n = l_n + nN - (N+1)s_n.$$

Simplifying (2.6.6), we get that the right side is equal to

$$\frac{((2n)!)^n(-1)^{\frac{1}{2}n(n-1)}}{(2n-1)!(2n-2)!\cdots n!(N+1)^{n^2}B_N}$$

$$\cdot \sum_{\substack{s_1=0 \\ k_1\geqslant 0}}^{2n} \cdots \sum_{\substack{s_n=0 \\ k_n\geqslant 0}}^{2n} \frac{(-1)^{s_1+\cdots+s_n}}{k_1!s_1!(2n-s_1)!\cdots k_n!s_n!(2n-s_n)!}$$

$$\cdot \begin{vmatrix} (2n-1+k_1)!, \cdots, (2n-1+k_n)! \\ (2n-2+k_1)!, \cdots, (2n-2+k_n)! \\ \cdots\cdots\cdots\cdots \\ (n+k_1)!, \cdots, (n+k_n)! \end{vmatrix}$$

$$= \frac{(n!)^n(-1)^{\frac{1}{2}n(n-1)}}{(2n-1)!\cdots n!(N+1)^{n^2}B_N}$$

$$\cdot \sum_{\substack{s_1=0\\k_1\geqslant 0}}^{2n}\cdots\sum_{\substack{s_n=0\\k_n\geqslant 0}}^{2n} C_{s_1}^{2n}C_n^{n+k_1}\cdots C_{s_n}^{2n}C_n^{n+k_n}(-1)^{s_1+\cdots+s_n}$$

$$\cdot \begin{vmatrix} (2n-1+k_1)\cdots(n+1+k_1),\cdots,(2n-1+k_n)\cdots(n+1+k_n) \\ \cdots\cdots\cdots\cdots \\ (n+2+k_1)(n+1+k_1),\cdots,(n+2+k_n)(n+1+k_n) \\ n+1+k_1,\cdots,n+1+k_n \\ 1,\cdots,1 \end{vmatrix}.$$

The determinant on the right side is equal to

$$\lim_{x\to 1}(-1)^{\frac{1}{2}n(n-1)} \begin{vmatrix} (x^{-(n+1+k_1)})^{(n-1)},\cdots,(x^{-(n+1+k_n)})^{(n-1)} \\ \cdots\cdots\cdots\cdots \\ (x^{-(n+1+k_1)})', \quad \cdots, \quad (x^{-(n+1+k_n)})' \\ x^{-(n+1+k_1)}, \quad \cdots, \quad x^{-(n+1+k_n)} \end{vmatrix},$$

which, by a theorem given in Hua [1], is just

$$\lim_{\substack{x_1\to 1\\ \cdots \\ x_n\to 1}} \frac{(-1)^{\frac{1}{2}n(n-1)}(n-1)!\cdots 2!\,1!}{D(x_1,\cdots,x_n)}$$

$$\cdot \begin{vmatrix} x_1^{-(n+1+k_1)},\cdots,x_n^{-(n+1+k_1)} \\ x_1^{-(n+1+k_2)},\cdots,x_n^{-(n+1+k_2)} \\ \cdots\cdots\cdots\cdots \\ x_1^{-(n+1+k_n)},\cdots,x_n^{-(n+1+k_n)} \end{vmatrix}$$

$$= \lim_{\substack{x_1\to 1\\ \cdots \\ x_n\to 1}} \frac{(-1)^{\frac{1}{2}n(n-1)}(n-1)!\cdots 2!\,1!}{D(x_1,\cdots,x_n)}$$

$$\cdot \begin{vmatrix} x_1^{(N+1)s_1-l_1},\cdots,x_n^{(N+1)s_1-l_1} \\ x_1^{(N+1)s_2-l_2},\cdots,x_n^{(N+1)s_2-l_2} \\ \cdots\cdots\cdots\cdots \\ x_1^{(N+1)s_n-l_n},\cdots,x_n^{(N+1)s_n-l_n} \end{vmatrix}.$$

On the basis of the definition of $N(g_1,\cdots,g_n)$, the right side amounts to

$$\begin{aligned} &(n-1)!\cdots 2!\,1!\,N((N+1)s_1-l_1,\ (N+1)s_2 \\ &\quad -l_2-1,\cdots,(N+1)s_n-l_n-n+1) \\ &= (n-1)!\cdots 2!\,1!\,N((N+1)s_1-f_1-n+1, \\ &\qquad (N+1)s_2-f_2-n+1,\cdots,(N+1)s_n-f_n-n+1) \end{aligned}$$

$$= (n-1)!\cdots 2!1!N((N+1)s_1 - f_1, (N+1)s_2 - f_2, \cdots, (N+1)s_n - f_n).$$

Finally, we have

$$B_{f_1\cdots f_n} = \frac{(-1)^{\frac{1}{2}n(n-1)}(n!)^n(n-1)!\cdots 2!1!}{(N+1)^{n^2}(2n-1)!\cdots n!N(f)} \cdot \sum_{\substack{s_1=0\\k_1\geq 0}}^{2n}\cdots\sum_{\substack{s_n=0\\k_n\geq 0}}^{2n}(-1)^{s_1+\cdots+s_n}C_{s_1}^{2n}C_n^{n+k_1}\cdots C_{s_n}^{2n}C_n^{n+k_n} \cdot N((N+1)s_1 - f_1, \cdots, (N+1)s_n - f_n),$$

where

$$k_j = l_j + nN - (N+1)s_j, \quad j = 1, 2, \cdots, n.$$

§ 2.7 Calculation of Integral Constants

According to the definition of $B_{f_1\cdots f_n}$, we have

$$B_{0\cdots 0} = 1.$$

Thus

$$B_N = \frac{(-1)^{\frac{1}{2}n(n-1)}(n!)^n(n-1)!\cdots 2!1!}{(N+1)^{n^2}(2n-1)!\cdots n!} \cdot \sum_{\substack{s_1=0\\k_1\geq 0}}^{2n}\cdots\sum_{\substack{s_n=0\\k_n\geq 0}}^{2n}(-1)^{s_1+\cdots+s_n}C_{s_1}^{2n}C_n^{n+k_1}\cdots C_{s_n}^{2n}C_n^{n+k_n} \cdot N((N+1)s_1, \cdots, (N+1)s_n), \tag{2.7.1}$$

where

$$k_j = n - j + nN - (N+1)s_j = (N+1)(n-s_j) - j, \qquad j = 1, 2, \cdots, n.$$

On account of

$$N((N+1)s_1, \cdots, (N+1)s_n) = \frac{D((N+1)s_1 + n - 1, \cdots, (N+1)s_{n-1} + 1, (N+1)s_n)}{D(n-1, \cdots, 1, 0)} = \frac{D((N+1)s_1, \cdots, (N+1)s_n)}{D(n-1, \cdots, 1, 0)}\left(1 + o\left(\frac{1}{N}\right)\right),$$

we have

$$N((N+1)s_1, \cdots, (N+1)s_n) = (N+1)^{\frac{n(n-1)}{2}} \cdot \frac{D(s_1, \cdots, s_n)}{D(n-1, \cdots, 1, 0)}\left(1 + o\left(\frac{1}{N}\right)\right).$$

By the definition of k_j, a necessary condition for $k_j \geqslant 0$ when $N \geqslant n-1$ is that $s_j = 0, 1, \cdots, n-1$. By the habitual method, (2.7.1) becomes

$$B_N = \frac{(n!)^n (n-1)! \cdots 2! 1! (-1)^{\frac{n(n-1)}{2}}}{(N+1)^{\frac{1}{2}n(n+1)}(2n-1)! \cdots n!} \left(1 + O\left(\frac{1}{N}\right)\right)$$

$$\cdot \sum_{n-1 \geqslant s_1 \geqslant \cdots \geqslant s_n \geqslant 0} \begin{vmatrix} A_{s_1}^1, A_{s_1}^2, \cdots, A_{s_1}^n \\ A_{s_2}^1, A_{s_2}^2, \cdots, A_s^n \\ \cdots \cdots \cdots \cdots \cdots \\ A_{s_n}^1, A_{s_n}^2, \cdots, A_{s_n}^n \end{vmatrix} \frac{D(s_1, \cdots, s_n)}{D(n-1, \cdots, 1, 0)}, \quad (2.7.2)$$

where

$$A_{s_j}^k = (-1)^{s_j} C_{s_j}^{An} C_n^{n+(N+1)(n-s_j)-k}.$$

If s_l and $s_{l'}$ are equal in the determinant of (2.7.2), then the determinant vanishes. However, $n-1 \geqslant s_1 \geqslant s_2 \geqslant \cdots \geqslant s_n \geqslant 0$, thus the determinant vanishes with the exception of $s_1 = n-1$, $s_2 = n-2$, $\cdots$, $s_n = 0$ only. So, (2.7.2) becomes into

$$B_N = \frac{(n!)^n (n-1)! \cdots 2! 1! (-1)^{\frac{1}{2}n(n-1)}}{(N+1)^{\frac{1}{2}n(n+1)}(2n-1)! \cdots n!} \left(1 + O\left(\frac{1}{N}\right)\right)$$

$$\cdot \begin{vmatrix} A_{n-1}^1, A_{n-1}^2, \cdots, A_{n-1}^n \\ A_{n-2}^1, A_{n-2}^2, \cdots, A_{n-2}^n \\ \cdots \cdots \cdots \cdots \cdots \\ A_0^1, A_0^2, \cdots, A_0^n \end{vmatrix}$$

Therefore

$$B_N = \frac{(n!)^n (n-1)! \cdots 2! 1!}{(N+1)^{\frac{1}{2}n(n+1)}(2n-1)! \cdots n!}$$

$$\cdot \frac{((2n)!)^n}{(2n)! \cdots (n+1)!(n-1)! \cdots 1!} \left(1 + O\left(\frac{1}{N}\right)\right)$$

$$\cdot \begin{vmatrix} C_n^{n+(N+1)-1}, & C_n^{n+(N+1)-2}, \cdots, & C_n^{n+(N+1)-n} \\ \cdots \cdots \cdots \cdots \cdots \cdots \\ C_n^{n+(N+1)(n-1)-1}, C_n^{n+(N+1)(n-1)-2}, \cdots, C_n^{n+(N+1)(n-1)-n} \\ C_n^{n+(N+1)n-1}, & C_n^{n+(N+1)n-2}, \cdots, & C_n^{n+(N+1)n-n} \end{vmatrix}. \quad (2.7.3)$$

By Δ we denote the determinant on the right hand of (2.7.3). Then (2.7.3) is nothing but

$$B_N = \frac{(n!)^{n-1}((2n)!)^{n-1}\Delta}{(N+1)^{\frac{1}{2}n(n+1)}((2n-1)! \cdots (n+1)!)^2} \left(1 + O\left(\frac{1}{N}\right)\right). \quad (2.7.4)$$

Since

$$\sum_{s=0}^{p} C_t^{t+s} = C_{t+1}^{t+p+1},$$

Δ in (2.7.4) becomes

$$\begin{vmatrix} \sum_{k=0}^{(N+1)-1} C_{n-1}^{n-1+k}, & \sum_{k=0}^{(N+1)-2} C_{n-1}^{n-1+k}, \cdots, & \sum_{k=0}^{(N+1)-n} C_{n-1}^{n-1+k} \\ \cdots\cdots & & \\ \sum_{k=0}^{n(N+1)-1} C_{n-1}^{n-1+k}, & \sum_{k=0}^{n(N+1)-2} C_{n-1}^{n-1+k}, \cdots, & \sum_{k=0}^{n(N+1)-n} C_{n-1}^{n-1+k} \end{vmatrix}.$$

Subtracting each column from the adjacent column, we obtain

$$\begin{vmatrix} C_{n-1}^{n-1+(N+1)-1}, & C_{n-1}^{n-1+(N+1)-2}, \cdots, & C_{n-1}^{n-1+(N+1)-n+1}, & C_n^{N+1} \\ \cdots\cdots & & & \\ C_{n-1}^{n-1+n(N+1)-1}, & C_{n-1}^{n-1+n(N+1)-2}, \cdots, & C_{n-1}^{n-1+n(N+1)-n+1}, & C_n^{(N+1)n} \end{vmatrix}.$$

Repeating this procedure, finally we get

$$\Delta = \begin{vmatrix} C_1^{N+1}, & \cdots, & C_{n-1}^{N+1}, & C_n^{N+1} \\ \cdots\cdots & & & \\ C_1^{(n-1)(N+1)}, & \cdots, & C_{n-1}^{(n-1)(N+1)}, & C_n^{(n-1)(N+1)} \\ C_1^{n(N+1)}, & \cdots, & C_{n-1}^{n(N+1)}, & C_n^{n(N+1)} \end{vmatrix}.$$

By the conventional technique, Δ is equal to

$$\frac{Nn!(+1)^n}{n!(n-1)!\cdots 1!} \lim_{x\to 1} \begin{vmatrix} x^{(N+1)-1}, \cdots, (x^{(N+1)-1})^{(n-1)} \\ \cdots\cdots \\ x^{(n-1)(N+1)-1}, \cdots, (x^{(n-1)(N+1)-1})^{(n-1)} \\ x^{n(N+1)-1}, \cdots, (x^{n(N+1)-1})^{(n-1)} \end{vmatrix}$$

$$= \frac{n!(N+1)^n(n-1)!\cdots 1!}{n!(n-1)!\cdots 1!}$$

$$\cdot \lim_{\substack{x_1\to 1\\ \cdots \\ x_n\to 1}} \begin{vmatrix} x_1^{(N+1)-1}, \cdots, (x_1^{(N+1)-1})^{(n-1)} \\ \cdots\cdots \\ x_{n-1}^{(n-1)(N+1)-1}, \cdots, (x_{n-1}^{(n-1)(N+1)-1})^{(n-1)} \\ x_n^{n(N+1)-1}, \cdots, (x_n^{n(N+1)-1})^{(n-1)} \end{vmatrix} \frac{1}{D(x_1, \cdots, x_n)}$$

$$= (N+1)^n N(nN, (n-1)N, \cdots, N)$$

$$= (N+1)^n (N+1)^{\frac{1}{2}n(n-1)}$$

$$= (N+1)^{\frac{1}{2}n(n+1)}. \qquad (2.7.5)$$

From (2.7.4), we are led to

Theorem 2.7.1 *If $N \geqslant n-1$, then*

$$B_N = \frac{(n!)^{n-1}((2n)!)^{n-1}}{((2n-1)!\cdots(n+1)!)^2}\left(1+O\left(\frac{1}{N}\right)\right). \tag{2.7.6}$$

From this theorem and (2.6.7) we get

Theorem 2.7.2 *If $N \geqslant n-1$, then*

$$B_{f_1\cdots f_n} = \frac{(2n)!\cdots 2!1!(-1)^{\frac{1}{2}n(n-1)}\left(1+O\left(\frac{1}{N}\right)\right)}{N(f)((2n)!)^n n!(N+1)^{n^2}}$$

$$\cdot \sum_{\substack{s_1=0\\k_1\geqslant 0}}^{2n}\cdots\sum_{\substack{s_n=0\\k_n\geqslant 0}}^{2n}(-1)^{s_1+\cdots+s_n}C_{s_1}^{2n}C_n^{n+k_1}\cdots C_{s_n}^{2n}C_n^{n+k_n}$$

$$\cdot N((N+1)s_1-f_1,\cdots,(N+1)s_n-f_n). \tag{2.7.7}$$

By the definition of $N(f)$ and Theorem 2.7.2, it can be shown that

$$B_{f_1\cdots f_n} = 1+O\left(\frac{1}{N}\right).$$

§ 2.8 Some Remarks

(1) The condition "$u(U)$ is continuous on U_n" in the Riesz-type Theorem (i.e. Theorem 2.3.1) can be weakened and replaced by "$u(U)$ is integrarable and bounded on U_n". Under the latter weaker condition, $\sum_N^\alpha(U)$ tends to $u(U)$ as $N\to\infty$ at those points where $u(U)$ is continuous. This shows that the Riesz summability is a local property.

(2) In [2], Hua Luogeng defined the Abel summation (cf. Chapter 1) and proved a convergence theorem (i.e. Theorem 1.3.1). When r sufficiently approaches 1, the Abel sum

$$\sum_{N\geqslant f_1\geqslant\cdots\geqslant f_n\geqslant -N} \rho^f(r)\operatorname{tr}(C_{f_1\cdots f_n}\Phi'_{f_1\cdots f_n}(U))$$

can be taken as an approximate value of $u(U)$, which, however, contains a parameter r. From the Riesz-type theorem in this book, a convergence theorem follows again but r does not appear in the (c,α)-sum (2.2.1) which is taken as an approximate value of $u(U)$.

(3) Whether the condition $\alpha > \frac{n-1}{n}$ of the Riesz-type theorem, Theorem 2.3.1, can be further improved is an open problem. As is well

known, the result is the best possible for $n=1$. For $n>1$ the problem is worth studying.

(4) From the proof of the Riesz-type Theorem 2.3.1, we could deduce a more general convergence theorem. This will be discussed in Chapter 4 and the theorem will be applied to approximation theory on compact groups.

Chapter 3. Partial Sum of Fourier Series on Unitary Groups

§ 3.1 Dirichlet Kernel

Now Consider the partial sum

$$S_N^{(n)}(U) = \sum_{N \geq l_1 > l_2 > \cdots > l_n \geq -N} \operatorname{tr}(C_{f_1 \cdots f_n} \Phi'_{f_1 \cdots f_n}(U)) \tag{3.1.1}$$

of the Fourier series

$$\sum_{f_1 \geq f_2 \geq \cdots \geq f_n} \operatorname{tr}(C_{f_1 \cdots f_n} \Phi'_{f_1 \cdots f_n}(U)) \tag{3.1.2}$$

of $u(U)$, where

$$l_1 = f_1 + n - 1, \; l_2 = f_2 + n - 2, \cdots, \; l_n = f_n.$$

We begin with the following theorem (Gong Sheng [1]).

Theorem 3.1.1 *The partial sum* (3.1.1) *of the Fourier series* (3.1.2) *of an integrable function* $u(U)$ *can be expressed by*

$$\frac{1}{C} \int_{U_n} u(VU) \mathscr{D}_N(V) \dot{V},$$

where $\mathscr{D}(V)$ *is the Dirichlet kernel equal to*

$$\frac{1}{(n-1)! \cdots 2! D(\bar{\lambda}_1, \cdots, \bar{\lambda}_n)} \cdot D\left(\frac{\partial}{\partial \lambda_1}, \cdots, \frac{\partial}{\partial \lambda_n}\right) \cdot \frac{\det(\bar{V}'^N - V^{N+1})}{\det(I - V)}, \tag{3.1.3}$$

where $\lambda_1, \lambda_2, \cdots, \lambda_n$ *are characteristic roots of* V, D *signifies the Vandermonde determinant, and* $D\left(\frac{\partial}{\partial \lambda_1}, \cdots, \frac{\partial}{\partial \lambda_n}\right)$ *denotes the operator*

$$(-1)^{\frac{n(n-1)}{2}}\begin{vmatrix} 1, & 1, & \cdots, & 1 \\ \frac{\partial}{\partial\lambda_1}, & \frac{\partial}{\partial\lambda_2}, & \cdots, & \frac{\partial}{\partial\lambda_n} \\ \cdots & \cdots & \cdots & \cdots \\ \left(\frac{\partial}{\partial\lambda_1}\right)^{n-1}, & \left(\frac{\partial}{\partial\lambda_2}\right)^{n-1}, & \cdots, & \left(\frac{\partial}{\partial\lambda_n}\right)^{n-1} \end{vmatrix}. \tag{3.1.4}$$

Proof. By § 0.2, (3.1.2) means

$$\frac{1}{C}\int_{U_n} u(VU) \sum_{f_1\geqslant\cdots\geqslant f_n} N(f)\chi_{f_1\cdots f_n}(\bar{V})\dot{V}. \tag{3.1.5}$$

Thus (3.1.2) is just

$$\frac{1}{C}\int_{U_n} u(VU) \sum_{N\geqslant l_1>l_2>\cdots>l_n\geqslant -N} N(f)\chi_{f_1\cdots f_n}(\bar{V})\dot{V}.$$

Let us calculate

$$\sum_{N\geqslant l_1>l_2>\cdots>l_n\geqslant -N} N(f)\chi_{f_1\cdots f_n}(\bar{V}).$$

By (1.3.2) and (3.1.5), the Abel summability of the series (3.1.1) means

$$\frac{1}{C}\int_{U_n} u(VU) \sum_{f_1\geqslant f_2\geqslant\cdots\geqslant f_n} \rho^f(r)N(f)\chi_{f_1\cdots f_n}(\bar{V})\dot{V}. \tag{3.1.6}$$

From [2] and (3.1.6), it follows that

$$\sum_{f_1\geqslant f_2\geqslant\cdots\geqslant f_n} \rho^f(r)N(f)\chi_{f_1\cdots f_n}(\bar{V}) = \frac{(1-r^2)^{n^2}}{|\det(I-r\bar{V}')|^2}.$$

Thus, by the definition of $\chi_f(\bar{V})$, we find that if $e^{i\theta_1},\cdots,e^{i\theta_n}$ ($\theta_1\geqslant\theta_2\geqslant\cdots\geqslant\theta_n$) are characteristic roots of $\bar{V}$, then

$$\frac{1}{D(e^{i\theta_1},\cdots,e^{i\theta_n})}\sum_{f_1\geqslant f_2\geqslant\cdots\geqslant f_n} \rho^f(r)N(f)M_{f_1\cdots f_n}(e^{i\theta_1},\cdots,e^{i\theta_n})$$
$$= \frac{(1-r^2)^{n^2}}{|1-re^{i\theta_1}|^{2n}\cdots|1-re^{i\theta_n}|^{2n}},$$

where

$$M_{f_1\cdots f_n}(\chi_1,\cdots,\chi_n) = \begin{vmatrix} \chi_1^{l_1}, & \cdots, & \chi_n^{l_1} \\ \chi_1^{l_2}, & \cdots, & \chi_n^{l_2} \\ \cdots & \cdots & \cdots \\ \chi_1^{l_n}, & \cdots, & \chi_n^{l_n} \end{vmatrix}.$$

Consider the function

$$g(\theta_1,\cdots,\theta_n) = \frac{D(e^{i\theta_1},\cdots,e^{i\theta_n})(1-r^2)^{n^2}}{|1-re^{i\theta_1}|^{2n}\cdots|1-re^{i\theta_n}|^{2n}}, \tag{3.1.7}$$

in $\theta_1, \theta_2, \cdots, \theta_n$, and expand it into the multiple Fourier series

$$\sum_{\infty > v_1, \cdots, v_n > -\infty} a_{v_1 v_2 \cdots v_n} e^{i(v_1\theta_1 + \cdots + v_n\theta_n)}. \tag{3.1.8}$$

Examine its partial sum

$$\sum_{v_1=-N}^{N} \cdots \sum_{v_n=-N}^{N} a_{v_1 v_2 \cdots v_n} e^{i(v_1\theta_1 + \cdots + v_n\theta_n)}$$

$$= \frac{1}{(2\pi)^n} \int_0^{2\pi} \cdots \int_0^{2\pi} g(\psi_1, \cdots, \psi_n) \prod_{j=1}^{n} d_N(\psi_j - \theta_j) d\psi_1 \cdots d\psi_n, \tag{3.1.9}$$

where

$$d_N(\theta) = \frac{\sin(N + 1/2)\theta}{\sin(1/2)\theta}$$

is the Dirichlet Kernel of one variable.

Since

$$\frac{D(e^{i\theta_1}, \cdots, e^{i\theta_n})(1 - r^2)^{n^2}}{|1 - re^{i\theta_1}|^{2n} \cdots |1 - re^{i\theta_n}|^{2n}}$$

$$= \sum_{f_1 \geq \cdots \geq f_n} \rho^f(r) N(f) \cdot \left[\sum_{\mu} \delta^{1,2,\cdots,n}_{\mu_1,\mu_2 \cdots \mu_n} e^{i(l_{\mu_1}\theta_1 + l_{\mu_2}\theta_2 + \cdots + l_{\mu_n}\theta_n)}\right]$$

$$= \sum_{\infty > v_1, \cdots, v_n > -\infty} a_{v_1 \cdots v_n} e^{i(v_1\theta_1 + \cdots + v_n\theta_n)},$$

it can be seen that the left side of (3.1.9) is equal to

$$\sum_{N \geq l_1 > l_2 > \cdots > l_n \geq -N} \rho^f(r) N(f) M_{f_1 \cdots f_n}(e^{i\theta_1}, \cdots, e^{i\theta_n}). \tag{3.1.10}$$

Hence

$$\sum_{N \geq l_1 > l_2 > \cdots > l_n \geq -N} \rho^f(r) N(f) \chi_{f_1 \cdots f_n}([e^{i\theta_1}, \cdots, e^{i\theta_n}])$$

$$= \frac{1}{(2\pi)^n} \int_0^{2\pi} \cdots \int_0^{2\pi} \frac{D(e^{i\psi_1}, \cdots, e^{i\psi_n})(1 - r^2)^{n^2}}{D(e^{i\theta_1}, \cdots, e^{i\theta_n}) \prod_{v=1}^{n} |1 - re^{i\psi_v}|^{2n}}$$

$$\cdot \prod_{j=1}^{n} d_N(\psi_j - \theta_j) d\psi_1 \cdots d\psi_n. \tag{3.1.11}$$

Applying the same integral technique as before, we can show that the right side of (3.1.11) is exactly

$$\frac{(1 - r^2)^{n^2}}{D(e^{i\theta_1}, \cdots, e^{i\theta_n})(2\pi)^n} \cdot \int \cdots \int_{2\pi \geq \psi_1 \geq \psi_2 \geq \cdots \geq \psi_n \geq 0} \frac{D(e^{i\psi_1}, \cdots, e^{i\psi_n})}{\prod_{j=1}^{n} |1 - re^{i\psi_j}|^{2n}}$$

$$\cdot P(\psi_1, \cdots, \psi_n, \theta_1, \cdots, \theta_n) d\psi_1 \cdots d\psi_n, \tag{3.1.12}$$

where $P(\psi_1, \cdots, \psi_n, \theta_1, \cdots, \theta_n)$ is

$$\begin{vmatrix} d_N(\psi_1 - \theta_1), \cdots, d_N(\psi_1 - \theta_n) \\ d_N(\psi_2 - \theta_1), \cdots, d_N(\psi_2 - \theta_n) \\ \cdots\cdots\cdots\cdots\cdots\cdots \\ d_N(\psi_n - \theta_1), \cdots, d_N(\psi_n - \theta_n) \end{vmatrix}.$$

Integrating the integral of the right side of (3.1.12) over the coset of the unitary group, we obtain

$$\frac{1}{D(e^{i\theta_1}, \cdots, e^{i\theta_n})C} \int_{U_n} \frac{(1-r^2)^{n^2} P(\psi_1, \cdots, \psi_n, \theta_1, \cdots, \theta_n)}{|\det(I - r\overline{W}')|^{2n} D(e^{-i\psi_1}, \cdots, e^{-i\psi_n})} \dot{W}. \quad (3.1.13)$$

Set $r \to 1$. Then $\rho^f(r) \to 1$. Thus (3.1.10) is just $s_N^{(n)}(U)$ and by the theory of harmonic functions, (3.1.13) becomes

$$\frac{1}{D(e^{i\theta_1}, \cdots, e^{i\theta_n})} \lim_{\substack{\psi_1 \to 0 \\ \cdots \\ \psi_n \to 0}} \frac{P(\psi_1, \cdots, \psi_n, \theta, \cdots, \theta_n)}{D(\psi_1, \cdots, \psi_n)}, \quad (3.1.14)$$

as $r \to 1$, which is

$$\frac{1}{D(e^{i\theta_1}, \cdots, e^{i\theta_n})} \lim_{\substack{\psi_1 \to 0 \\ \cdots \\ \psi_n \to 0}} \frac{P(\psi_1, \cdots, \psi_n, \theta_1, \cdots, \theta_n)}{D(\psi_1, \cdots, \psi_n)}$$

$$\cdot \frac{D(\psi_1, \cdots, \psi_n)}{D(e^{-i\psi_1}, \cdots, e^{-i\psi_n})}. \quad (3.1.15)$$

By theorem 1.2.4 of [1], $\mathscr{D}_N(V)$ is equal to

$$\frac{i^{\frac{1}{2}n(n-1)}}{(n-1)!\cdots 2!1!D(e^{i\theta_1}, \cdots, e^{i\theta_n})} \begin{vmatrix} d_N(\theta_1), \cdots, d_N(\theta_N) \\ d_N'(\theta_1), \cdots, d_N'(\theta_N) \\ \cdots\cdots\cdots\cdots\cdots\cdots \\ d_N^{(n-1)}(\theta_1), \cdots, d_N^{(n-1)}(\theta_n) \end{vmatrix}. \quad (3.1.16)$$

From (3.1.16) and the definition (3.1.4) of $D\left(\frac{\partial}{\partial\lambda_1}, \cdots, \frac{\partial}{\partial\lambda_n}\right)$, (3.1.3) immediately follows. Otherwise, proceeding from (3.1.14) directly, (3.1.3) can also be deduced by using Theorem 1.2.4 of [1].

§ 3.2 Algebraic Proof of Dirichlet Kernel

In this section we shall state Hua's result. The author proved Theorem 3.1.1, and then Hua Luogeng gave the following simple algebraic proof, independent of the Poisson-Hua kernels.

By the definition, we admit

$$\sum_{N \geq l_1 > \cdots > l_n \geq -N} N(f) M_{l_1 \cdots l_n}(e^{i\theta_1}, \cdots, e^{i\theta_n})$$

$$= (-1)^{\frac{n(n-1)}{2}} \sum_{N \geq l_1 > l_2 > \cdots > l_n \geq -N} \begin{vmatrix} 1, & 1, & \cdots, & 1 \\ l_1, & l_2, & \cdots, & l_n \\ \cdots & \cdots & \cdots & \cdots \\ l_1^{n-1}, & l_2^{n-1}, & \cdots, & l_n^{n-1} \end{vmatrix}$$

$$\cdot \begin{vmatrix} \lambda_1^{l_1}, & \cdots, & \lambda_n^{l_n} \\ \lambda_1^{l_2}, & \cdots, & \lambda_n^{l_2} \\ \cdots & \cdots & \cdots \\ \lambda_1^{l_n}, & \cdots, & \lambda_n^{l_n} \end{vmatrix} \frac{1}{D(n-1, \cdots, 1, 0)}, \tag{3.2.1}$$

where $\lambda_1 = e^{i\theta_1}$, $\lambda_2 = e^{i\theta_2}, \cdots$, $\lambda_n = e^{i\theta_n}$.

Owing to the symmetry of the expansion of a determinant, the value of the right side of (3.2.1) does not change if $(l_1, l_2, \cdots, l_n)$ is carried out by another permutation. Hence

$$\sum_{N \geq l_1 > l_2 > \cdots > l_n \geq -N} N(f) M_{l_1 \cdots l_n}(e^{i\theta_1}, \cdots, e^{i\theta_n})$$

$$= (-1)^{\frac{n(n-1)}{2}} \frac{1}{n!} \sum_{l_1=-N}^{N} \cdots \sum_{l_n=-N}^{N} \begin{vmatrix} 1, & 1, & \cdots, & 1 \\ l_1, & l_2, & \cdots, & l_n \\ \cdots & \cdots & \cdots & \cdots \\ l_1^{n-1}, & l_2^{n-1}, & \cdots, & l_n^{n-1} \end{vmatrix}$$

$$\cdot \begin{vmatrix} \lambda_1^{l_1}, & \cdots, & \lambda_n^{l_1} \\ \lambda_1^{l_2}, & \cdots, & \lambda_n^{l_2} \\ \cdots & \cdots & \cdots \\ \lambda_1^{l_n}, & \cdots, & \lambda_n^{l_n} \end{vmatrix} \frac{1}{D(n-1, \cdots, 1, 0)}$$

$$= (-1)^{\frac{n(n-1)}{2}} \frac{1}{n!} \sum_{l_1=-N}^{N} \cdots \sum_{l_n=-N}^{N} \frac{1}{D(n-1, \cdots, 1, 0)}$$

$$\cdot \begin{vmatrix} a_{11}(\lambda), & a_{12}(\lambda), & \cdots, & a_{1n}(\lambda) \\ a_{21}(\lambda), & a_{22}(\lambda), & \cdots, & a_{2n}(\lambda) \\ \cdots & \cdots & \cdots & \cdots \\ a_{n1}(\lambda), & a_{n2}(\lambda), & \cdots, & a_{nn}(\lambda) \end{vmatrix}, \tag{3.2.2}$$

where

$$a_{ij}(\lambda) = l_1^{i-1}\lambda_j^{l_1} + l_2^{i-1}\lambda_j^{l_2} + \cdots + l_n^{i-1}\lambda_j^{l_n}.$$

Separate the numerator on the right-hand side of (3.2.2) into n^n determinants which have the form

$$\sum_{l_1=-N}^{N} \cdots \sum_{l_n=-N}^{N} \begin{vmatrix} \lambda_1^{l_{\nu_1}}, & \lambda_2^{l_{\nu_1}}, & \cdots, & \lambda_n^{l_{\nu_1}} \\ l_{\nu_2}\lambda_1^{l_{\nu_2}}, & l_{\nu_2}\lambda_2^{l_{\nu_2}}, & \cdots, & l_{\nu_2}\lambda_n^{l_{\nu_n}} \\ \cdots & \cdots & \cdots & \cdots \\ l_{\nu_n}^{n-1}\lambda_1^{l_{\nu_n}}, & l_{\nu_n}^{n-1}\lambda_2^{l_{\nu_n}}, & \cdots & l_{\nu_n}^{n-1}\lambda_n^{l_{\nu_n}} \end{vmatrix}, \tag{3.2.3}$$

where $1 \leqslant \nu_j \leqslant n$, $j = 1, 2, \cdots, n$. Obviously, all determinants vanish unless $(\nu_1, \nu_2, \cdots, \nu_n)$ is a permutation of $(1, 2, \cdots, n)$. Therefore

$$\sum_{N \geqslant l_1 > \cdots > l_n \geqslant -N} N(f) M_{f_1 \cdots f_n}(e^{i\theta_1}, \cdots, e^{i\theta_n}) = (-1)^{\frac{n(n-1)}{2}}$$

$$\cdot \sum_{l=-N}^{N} \cdots \sum_{l_n=-N}^{N} \begin{vmatrix} \lambda_1^{l_1}, & \lambda_2^{l_1}, & \cdots, & \lambda_n^{l_n} \\ l_2\lambda_1^{l_2}, & l_2\lambda_2^{l_2}, & \cdots, & l_n\lambda_n^{l_n} \\ \cdots & \cdots & \cdots & \cdots \\ l_n^{n-1}\lambda_1^{l_n}, & l_n^{n-1}\lambda_2^{l_n}, & \cdots, & l_n^{n-1}\lambda_n^{l_n} \end{vmatrix} \frac{1}{D(n-1, \cdots, 1, 0)}$$

$$= (-1)^{\frac{n(n-1)}{2}} \frac{1}{D(n-1, \cdots, 1, 0)}$$

$$\cdot \begin{vmatrix} \sum_{-N}^{N} \lambda_1^l, & \sum_{-N}^{N} \lambda_2^l, & \cdots, & \sum_{-N}^{N} \lambda_n^l \\ \sum_{-N}^{N} l\lambda_1^l, & \sum_{-N}^{N} l\lambda_2^l, & \cdots, & \sum_{-N}^{N} l\lambda_n^l \\ \cdots & \cdots & \cdots & \cdots \\ \sum_{-N}^{N} l^{n-1}\lambda_1^l, & \sum_{-N}^{N} l^{n-1}\lambda_2^l, & \cdots, & \sum_{-N}^{N} l^{n-1}\lambda_n^l \end{vmatrix}$$

$$= \frac{i^{\frac{1}{2}n(n-1)}}{D(n-1, \cdots, 1, 0)} \begin{vmatrix} d_N(\theta_1), & d_N(\theta_2), & \cdots, & d_N(\theta_n) \\ d'_N(\theta_1), & d'_N(\theta_2), & \cdots, & d'_N(\theta_n) \\ \cdots & \cdots & \cdots & \cdots \\ d_N^{(n-1)}(\theta_1), & d_N^{(n-1)}(\theta_2), & \cdots, & d_N^{(n-1)}(\theta_n) \end{vmatrix},$$

which proves Theorem 3.1.1. Evidently, this proof is simpler and more direct than that given in § 3.1. In Chapter 8, we shall use this method to calculate Dirichlet kernels of Fourier series on $SO(n)$. The method used in § 3.1 belongs to an effective way of dealing with the matrix integral.

§ 3.3 Partial Sum of Fourier Series

As we have known if $u(U)$ is an integrable function on unitary group U_n, then the partial sum

$$S_N^{(u)}(U) = \sum_{N \geqslant l_1 > l_2 > \cdots > l_n \geqslant -N} \operatorname{tr}(C_{f_1 \cdots f_n} \Phi'_{f_1 \cdots f_n}(U)) \tag{3.3.1}$$

of its Fourier series

$$\sum_{f_1 \geqslant f_2 \geqslant \cdots \geqslant f_n} \operatorname{tr}(C_{f_1 \cdots f_n} \Phi'_{f_1 \cdots f_n}(U))$$

can be expressed by

$$\frac{1}{C} \int_{U_n} u(VU) \mathscr{D}_N(V) \dot{V}, \tag{3.3.2}$$

where $\mathscr{D}_N(V)$ is the Dirichlet kernel expressed by (3.1.3) or by (3.1.16) for convenience. Let $\bar{V}$ be denoted by

$$\bar{V} = W \begin{pmatrix} e^{i\theta_1} & & \\ & \ddots & \\ & & e^{i\theta_n} \end{pmatrix} \bar{W}',$$

where $W \in U_n$. Take $\Gamma \in U_n$ such that

$$\Gamma \begin{pmatrix} e^{i\theta_1} & & \\ & \ddots & \\ & & e^{i\theta_n} \end{pmatrix} \bar{\Gamma}' = \begin{pmatrix} e^{i\theta_{\nu_1}} & & \\ & \ddots & \\ & & e^{i\theta_{\nu_n}} \end{pmatrix},$$

where $(\nu_1, \cdots, \nu_n)$ is a permutation of $(1, 2, \cdots, n)$. Write

$$\bar{V}^* = W\Gamma \begin{pmatrix} e^{i\theta_1} & & \\ & \ddots & \\ & & e^{i\theta_n} \end{pmatrix} \bar{\Gamma}'\bar{W}'.$$

Then, by the definition of integral over group,

$$\frac{1}{C}\int_{U_n} u(V^*U)\,\mathscr{D}_N(V)\dot{V} = \frac{1}{C}\int_{U_n} u(VU)\mathscr{D}_N(V)\dot{V}.$$

Work out all possible V^*. Add all $u(V^*U)$ together and divide the obtained sum by $n!$. Denote the result by $u^*(VU)$. Then

$$s_N^{(u)}(U) = \frac{1}{C}\int_{U_n} u^*(VU)\mathscr{D}_N(V)\dot{V}. \tag{3.3.3}$$

Letting

$$g(\theta_1, \cdots, \theta_n) = \frac{1}{\omega_n'}\int_{[V]} u^*(VU)[\dot{V}], \tag{3.3.4}$$

we get

$$S_N^{(u)}(U) = \frac{1}{(2\pi)^n} \int\cdots\int_{\pi \geqslant \theta_1 \geqslant \cdots \geqslant \theta_n \geqslant -\pi} g(\theta_1, \cdots, \theta_n)\mathscr{D}_N(V) \cdot |D(e^{i\theta_1}, \cdots, e^{i\theta_n})|^2 d\theta_1 \cdots d\theta_n.$$

By the definition of $u^*(U)$, $g(\theta_1, \cdots, \theta_n)$ is a symmetric function in $\theta_1, \cdots, \theta_n$. Therefore, by the above well-known technique,

$$S_N^{(u)}(V) = \frac{i^{\frac{n(n-1)}{2}}}{(2\pi)^n (n-1)! \cdots 2!1!} \int_0^{2\pi} \cdots \int_0^{2\pi} g(\theta_1, \cdots, \theta_n) \cdot D(e^{-i\theta_1}, \cdots, e^{-i\theta_n}) d_N(\theta_1) \cdot d_N'(\theta_2) \cdots d_N^{(n-1)}(\theta_n) d\theta_1 \cdots d\theta_n. \tag{3.3.5}$$

It can be shown that

$$\frac{1}{C}\int_{U_n} \mathscr{D}_N(V)V = 1. \tag{3.3.6}$$

In fact, we have

$$\frac{1}{C}\int_{U_n}\mathscr{D}_N(V)\dot{V}$$

$$=\frac{i^{\frac{n(n-1)}{2}}}{(2\pi)^n(n-1)!\cdots 2!1!}\int_0^{2\pi}\cdots\int_0^{2\pi}D(e^{-i\theta_1},\cdots,e^{-i\theta_n})$$

$$\cdot d_N(\theta_1)d_N'(\theta_2)\cdots d_N^{(n-1)}(\theta_n)d\theta_1\cdots d\theta_n. \tag{3.3.7}$$

Integrating the above (3.3.7) $\frac{n(n-1)}{2}$ times by parts, we obtain, in virtue of periodicity

$$\frac{(-i)^{\frac{n(n-1)}{2}}}{(2\pi)^n(n-1)!\cdots 2!1!}\int_0^{2\pi}\cdots\int_0^{2\pi}d_N(\theta_1)\cdots d_N(\theta_n)$$

$$\cdot\frac{\partial^0}{\partial\theta_1^0}\frac{\partial^1}{\partial\theta_2^1}\cdots\frac{\partial^{n-1}}{\partial\theta_n^{n-1}}D(e^{-i\theta_1},\cdots,e^{-i\theta_n})d\theta_1\cdots d\theta_n,$$

which indicates the value of the Nth partial sum of the multiple Fourier series of the function

$$\frac{(-i)^{\frac{n(n-1)}{2}}}{(n-1)!\cdots 2!1!}\frac{\partial^0}{\partial\theta_1^0}\frac{\partial^1}{\partial\theta_2^1}\cdots\frac{\partial^{n-1}}{\partial\theta_n^{n-1}}D(e^{-i\theta_1},\cdots,e^{-i\theta_n})$$

at $\theta_1=\cdots=\theta_n=0$ and this, obviously, is 1 (by the definition of D).

From (3.3.2) and (3.3.1), it follows that

$$S_N^{(n)}(U)-u(U)=\frac{1}{C}\int_{U_n}(u^*(VU)-u(U))\mathscr{D}(V)\dot{V},$$

which, by (3.3.5) and (3.3.7), is equal to

$$\frac{i^{\frac{n(n-1)}{2}}}{(2\pi)^n(n-1)!\cdots 2!1!}\int_0^{2\pi}\cdots\int_0^{2\pi}(g(\theta_1,\cdots,\theta_n)-u(U))$$

$$\cdot D(e^{-i\theta_1},\cdots,e^{-i\theta_n})d_N(\theta_1)d_N'(\theta_2)\cdots d_N^{(n-1)}(\theta_n)d\theta_1\cdots d\theta_n. \tag{3.3.8}$$

§ 3.4 Convergence Theorem of Fourier Series

Let $0<p<1$. If $u(U)$ is continuous on U_n and

$$|u(U)-u(V)|\leqslant A(\operatorname{tr}(U-V)(\overline{U-V})')^p,$$

where A is an absolute constant, then $u(U)$ is said to be satisfied with the Lipschitz condition. The whole class of functions is denoted by $C^p\cdot(U_n)$ and the class of functions having continuous derivatives of order $\frac{n(n+1)}{2}$ is denoted by $C^{\frac{n(n+1)}{2}}(U_n)$.

Let $u(U)\in C^{\frac{n(n-1)}{2}}(U_n)$. Integrating it $\frac{n(n-1)}{2}$ times by parts and using periodicity, we obtain from (3.3.8) that $s_N^{(n)}(U)-u(U)$ is equal to

$$\frac{(-i)^{\frac{n(n-1)}{2}}}{(2\pi)^n(n-1)!\cdots 2!1!}\int_0^{2\pi}\cdots\int_0^{2\pi}\frac{\partial^0}{\partial\theta_1^0}\frac{\partial^1}{\partial\theta_2^1}\cdots\frac{\partial^{n-1}}{\partial\theta_n^{n-1}}$$
$$\cdot[(g(\theta_1,\cdots,\theta_n)-u(U))D(e^{-i\theta_1},\cdots,e^{-i\theta_n})]$$
$$\cdot d_N(\theta_1)d_N(\theta_2)\cdots d_N(\theta_n)d\theta_1\cdots d\theta_n. \tag{3.4.1}$$

By $I_{\mu_1\cdots\mu_n}$, we mean

$$\frac{(-i)^{\frac{n(n-1)}{2}}}{(2\pi)^n(n-1)!\cdots 2!1!}\int_0^{2\pi}\cdots\int_0^{2\pi}\frac{\partial^{\mu_1}}{\partial\theta_1^{\mu_1}}\frac{\partial^{\mu_2}}{\partial\theta_2^{\mu_2}}\cdots\frac{\partial^{\mu_n}}{\partial\theta_n^{\mu_n}}$$
$$\cdot(g(\theta_1,\cdots,\theta_n)-u(U))\frac{\partial^{\nu_1}}{\partial\theta_1^{\nu_1}}\frac{\partial^{\nu_2}}{\partial\theta_2^{\nu_2}}\cdots\frac{\partial^{\nu_n}}{\partial\theta_n^{\nu_n}}$$
$$\cdot D(e^{-i\theta_1},\cdots,e^{-i\theta_n})d_N(\theta_1)\cdots d_N(\theta_n)d\theta_1\cdots d\theta_n, \tag{3.4.2}$$

where μ_k and ν_k are nonnegative integers such that

$$\mu_k+\nu_k=k-1,$$
$$\mu_k\geqslant 0,\ \nu_k\geqslant 0,\quad k=1,2,\cdots,n.$$

Thus

$$s_N^{(n)}(U)-u(U)=\sum_{\mu_1,\cdots,\mu_n}I_{\mu_1\cdots\mu_n}.$$

First consider $I_{0,\cdots,0}$, which, by the definition, is equal to

$$\frac{(-i)^{\frac{n(n-1)}{2}}}{(2\pi)^n(n-1)!\cdots 2!1!}\int_0^{2\pi}\cdots\int_0^{2\pi}(g(\theta_1,\cdots,\theta_n)-u(U))$$
$$\cdot\frac{\partial^0}{\partial\theta_1^0}\frac{\partial^1}{\partial\theta_2^1}\cdots\frac{\partial^{n-1}}{\partial\theta_n^{n-1}}D(e^{-i\theta_1},\cdots,e^{-i\theta_n})\cdot d_N(\theta_1)\cdots d_N(\theta_n)d\theta_1\cdots d\theta_n. \tag{3.4.3}$$

As in § 1.3 we partition the integral domain

$$\pi\geqslant\theta_1\geqslant\theta_2\geqslant\cdots\geqslant\theta_n\geqslant-\pi,$$

into $R_1, R_2,\cdots,R_{n+1}$. Denote the integral over R_j by I_j. First consider I_2. Since

$$\int_\delta^\pi f(t)d_N(t)dt=O\left(\frac{1}{\delta}\,\omega\left(\frac{1}{N}\right)\right)$$

and

$$\int_{-\pi}^{\pi}|d_N(\theta)|\,d\theta=O(\ln N),$$

where $\delta>0$ and ω is the modulus of continuity of f (cf. Nantanson [1] and K. K. Chen [1]), we have

$$I_2 = O\left(\frac{1}{\delta}\,\omega\left(\frac{1}{N}\right)\ln^{n-1}N\right).$$

As $u(U) \in C^{\frac{n(n-1)}{2}}$, we have for $n > 1$ that

$$I_2 = O\left(\frac{1}{\delta N}\ln^{n-1}N\right).$$

Similarly, we can verify the following

$$I_p = O\left(\frac{1}{\delta^{p-1}N}\ln^{n-p+1}N\right), \quad p = 2, 3, \cdots, n+1.$$

As in § 1.5, again partition R_1 into $s_1, \cdots, s_{n+1}$ and denote the integral (3.4.3) over s_j by J_j. As before, we can show that

$$J_p = O\left(\frac{1}{\delta^{p-1}N}\ln^{n-p+1}N\right), \quad p = 2, 3, \cdots, n+1.$$

Finally we consider J_1. Owing to $\delta \geqslant \theta_1 \geqslant \theta_2 \geqslant \cdots \geqslant \theta_n \geqslant -\delta$, we have

$$|g(\theta_1, \cdots, \theta_n) - u(U)| = \omega(u, \delta),$$

where ω is the modulus of continuity of $u(U)$. Hence

$$J_1 = O(\omega(u, \delta)\ln^n N).$$

Therefore the corresponding part over the domain

$$\pi \geqslant \theta_1 \geqslant \theta_2 \geqslant \cdots \geqslant \theta_n \geqslant -\pi$$

of the integral (3.4.3) is equal to

$$O(\delta \ln^n N) + O\left(\frac{1}{\delta N}\ln^{n-1}N\right) + \cdots +$$

$$\cdot\, O\left(\frac{1}{\delta^{p-1}N}\ln^{n-p+1}N\right) + \cdots + O\left(\frac{1}{\delta^n N}\right).$$

Treat the other sub-domains similarly so that we have

$$I_{0,\cdots,0} = O(\delta \ln^n N) + O\left(\frac{1}{\delta N}\ln^{n-1}N\right) + \cdots + O\left(\frac{1}{\delta^n N}\right).$$

Taking $\delta = (N\ln^n N)^{\frac{-1}{n+1}}$, we have

$$I_{0,\cdots 0} = O(N^{\frac{-1}{n+1}}\ln^{\frac{n}{n+1}}N).$$

Let us examine $I_{\mu_1,\cdots,\mu_n}$. When all $\mu_m (m = 1, 2, \cdots, n)$ are not zero, there are at least two $\nu_m (m = 1, 2, \cdots, n)$ equal to each other, because all $\mu_m (1 \leqslant m \leqslant n)$ are not zero and $0 \leqslant \mu_k \leqslant k-1$, say $\nu_j = \nu_k$. By the definition of $D(e^{-i\theta_1}, \cdots, e^{-i\theta_n})$, it is known that

$$\xi = \frac{\partial^{\nu_1}}{\partial \theta_1^{\nu_1}} \cdots \frac{\partial^{\nu_n}}{\partial \theta_n^{\nu_n}} D(e^{-i\theta_1}, \cdots, e^{-i\theta_n})$$

takes $(e^{-i\theta_j} - e^{-i\theta_k})$ as its factor. Thus

$$\xi = (e^{-i\theta_j} - e^{-i\theta_k}) F(e^{-i\theta_1}, \cdots, e^{-i\theta_n}).$$

Since

$$e^{-i\theta_j} - e^{-i\theta_k} = (1 - e^{-i\theta_k}) - (1 - e^{-i\theta_j}),$$

we have

$$\begin{aligned} I_{\mu_1,\cdots,\mu_n} = & \frac{(-i)^{\frac{n(n-1)}{2}}}{(2\pi)^n (n-1)!\cdots 2!1!} \int_0^{2\pi} \cdots \int_0^{2\pi} \frac{\partial^{\mu_1}}{\partial \theta_1^{\mu_1}} \cdots \frac{\partial^{\mu_n}}{\partial \theta_n^{\mu_n}} \\ & \cdot (g(\theta_1, \cdots, \theta_n) - u(U)) \\ & \cdot [(1 - e^{-i\theta_k}) - (1 - e^{-i\theta_j})] \\ & \cdot F(e^{-i\theta_1}, \cdots, e^{-i\theta_n}) d_N(\theta_1) \cdots d_N(\theta_n) d\theta_1 \cdots d\theta_n. \end{aligned}$$

If $u(U) \in C^{\frac{n(n-1)}{2}+p}(U_n)(0 < p < 1)$, then

$$I_{\mu_1,\cdots,\mu_n} = O\left(\frac{1}{N^1} \ln^{n-1} N\right),$$

because $F(e^{-i\theta_1}, \cdots, e^{-i\theta_n})$ obviously is differentiable. Therefore $s_N^{(u)}(U)$ is convergent and converges to $u(U)$. Thus we have

Theorem 3.4.1 (Sheng Kung (Gong Sheng) [3]) *If* $u(U) \in C^{\frac{n(n-1)}{2}+p}(U_n)(0 < p < 1)$, *then the partial sum* (3.3.3) *of the Fourier series of* $u(U)$ *is convergent and converges to* $u(U)$. *In addition, we have*

$$|s_N^{(u)}(U) - u(U)| \leqslant A \max\left[\left(\frac{\ln^{n^2} N}{N}\right)^{\frac{1}{n+1}}, \frac{1}{N^p} \ln^{n-1} N\right],$$

where A *is an absolute constant.*

§ 3.5 Another Definition of Summation and Its Kernels

The method used above in searching for Dirichlet kernels immediately suggests that a general summation be defined for the Fourier series of any integrable function on U_n and the corresponding kernels be explicitly determined.

Let $u(\theta)$ be an integrable function in $0 \leqslant \theta \leqslant 2\pi$ and its Fourier series be

$$\sum_{-\infty}^{\infty} a_p e^{ip\theta}. \tag{3.5.1}$$

T-summation is defined by

$$\tau_m = \sum_{-\infty}^{\infty} u_{mp} a_p e^{ip\theta}$$

and its kernel is

$$k_m(\theta) = \sum_{-\infty}^{\infty} u_{mp} e^{ip\theta} \tag{3.5.2}$$

(of course we should suppose that the kernel exists, i.e. (3.5.2) is convergent). If $\tau_m \to s$ whenever m tends to a limit, then (3.5.1) is said to be T-summable with sum s.

Thus we are led to the following

Definition 3.5.1 Let $u(U)$ be an integrable function on U_n and its Fourier series be

$$\sum_{f_1 \geqslant \cdots \geqslant f_n} \operatorname{tr}(C_{f_1 \cdots f_n} \Phi'_{f_1 \cdots f_n}(U)). \tag{3.5.3}$$

Introduce sum

$$\tau_m = \sum_{f_1 \geqslant \cdots \geqslant f_m} u_{ml_1} u_{ml_2} \cdots u_{ml_n} \operatorname{tr}(C_{f_1 \cdots f_n} \Phi'_{f_1 \cdots f_n}(U)).$$

If $\tau_m \to s$ whenever m tends to a limit, then (3.5.3) is said *to be T-summable with sum s*.

Example 3.5.1 (The Cesàro summation of order α, $\alpha > -1$)

$$\tau_N = \sum_{N \geqslant l_1 > \cdots > l_n \geqslant -N} A^{\alpha}_{l_1} \cdots A^{\alpha}_{l_n} \operatorname{tr}(C_{f_1 \cdots f_n} \Phi'_{f_1 \cdots f_n}(U)),$$

where

$$A^{\alpha}_{ij} = \frac{\Gamma(\alpha + N - |l_i| + 1)\Gamma(N+1)}{\Gamma(\alpha + N + 1)\Gamma(N - |l_i| + 1)}.$$

Example 3.5.2 (The Abel-Poisson summation)

$$\tau_r = \sum_{f_1 \geqslant f_2 \geqslant \cdots \geqslant f_n} r^{|l_1| + \cdots + |l_n|} \operatorname{tr}(C_{f_1 \cdots f_n} \Phi'_{f_1 \cdots f_n}(U)).$$

Following the proof of Theorem 3.1.1, we can obtain

Theorem 3.5.1 *The kernel $K_m(V)$ of T-summation defined by Definition 3.5.1 is expressed as*

$$\frac{1}{(n-1)! \cdots 2!1!} \frac{D\left(\frac{\partial}{\partial \lambda_1}, \cdots, \frac{\partial}{\partial \lambda_n}\right)}{D(\bar{\lambda}_1, \cdots, \bar{\lambda}_n)} k_m(\theta_1) \cdots k_m(\theta_n) \tag{3.5.4}$$

where $\lambda_1 = e^{-i\theta_1}, \cdots, \lambda_n = e^{-i\theta_n}$ indicate characteristic roots of $\bar{V}$, D is Vandermonde determinant and $k_m(\theta) = \sum_{-\infty}^{\infty} u_{mp} e^{ip\theta}$ means the kernel of T-summation of Fourier series of one variable.

Proof. As we have known, the multiple Fourier series of (3.1.7) is (3.1.8). The sum

$$\sum_{\infty>v_1,\cdots,v_n>-\infty} a_{v_1\cdots v_n} u_{mv_1} u_{mv_2}\cdots u_{mv_n} e^{i(v_1\theta_1+\cdots+v_n\theta_n)} \tag{3.5.5}$$

of (3.1.8) exactly turns out

$$\frac{1}{(2\pi)^n}\int_0^{2\pi}\cdots\int_0^{2\pi} g(\psi_1,\cdots,\psi_n)\prod_{j=1}^{n} k_m(\psi_j-\theta_j)d\psi_1\cdots d\psi_n,$$

where $k_m(\theta)=\sum_{-\infty}^{\infty} u_{mp}e^{ip\theta}$ and g is defined by (3.1.7).

It is found out that (3.5.5) is just

$$\sum_{l_1>l_2>\cdots>l_n} u_{ml_1}\cdots u_{ml_n}\rho^f(r)N(f)M_{f_1\cdots f_n}(e^{i\theta_1},\cdots,e^{i\theta_n}).$$

Thus

$$\sum_{l_1>l_2>\cdots>l_m} u_{ml_1}\cdots u_{ml_n}\rho^f(r)N(f)\chi_{f_1\cdots f_n}([e^{i\theta_1},\cdots,e^{i\theta_n}])$$

$$=\frac{1}{(2\pi)^n}\int_0^{2\pi}\cdots\int_0^{2\pi}\frac{D(e^{i\psi_1},\cdots,e^{i\psi_n})(1-r^2)^{n^2}}{D(e^{i\theta_1},\cdots,e^{i\theta_n})\prod_{v=1}^{n}|1-re^{i\psi_v}|^{2n}}$$

$$\cdot\prod_{j=1}^{n} k_m(\psi_j-\theta_j)d\psi_1\cdots d\psi_n. \tag{3.5.6}$$

By the well-known technique, the right-hand side of (3.5.6) is reduced to

$$\frac{(1-r^2)^{n^2}}{D(e^{i\theta_1},\cdots,e^{i\theta_n})}\frac{1}{(2\pi)^n}\int\cdots\int_{\psi_1\geq\cdots\geq\psi_n}\frac{D(e^{i\psi_1},\cdots,e^{i\psi_n})}{\prod_{j=1}^{n}|1-re^{i\psi_j}|^{2n}}$$

$$\cdot\, Q(\psi_1,\cdots,\psi_n,\theta_1,\cdots,\theta_n)d\psi_1\cdots d\psi_n. \tag{3.5.7}$$

However $Q(\psi_1,\cdots,\psi_n,\theta_1,\cdots,\theta_n)$ is

$$\begin{vmatrix} k_m(\psi_1-\theta_1),\cdots,k_m(\psi_1-\theta_n)\\ k_m(\psi_2-\theta_1),\cdots,k_m(\psi_2-\theta_n)\\ \cdots\cdots\cdots\cdots\cdots\\ k_m(\psi_n-\theta_1),\cdots,k_m(\psi_n-\theta_n)\end{vmatrix}.$$

Integrating (3.5.7) over the coset of unitary group leads to

$$\frac{1}{D(e^{i\theta_1},\cdots,e^{i\theta_n})}\frac{1}{C}\int_{U_n}\frac{(1-r^2)^{n^2}}{|\det(I-r\overline{W}')|^{2n}}$$

$$\cdot\frac{Q(\psi_1,\cdots,\psi_n,\theta_1,\cdots,\theta_n)}{D(e^{-i\psi_1},\cdots,e^{-i\psi_n})}\dot{W}.$$

Let $r \to 1$. Owing to $\rho^f(r) \to 1$, we apply the same method as in § 3.1. The conclusion follows.

Indeed, another proof of Theorem 3.5.1 which is based on the method in § 3.2 can be deduced. The details are omitted.

Example 3.5.3 The Cesàro kernel $F_N^\alpha(V)$ of order $\alpha(\alpha > -1)$ comes to

$$\frac{1}{(n-1)!\cdots 2!1!D(\lambda_1,\cdots,\lambda_n)}$$

$$D\left(\frac{\partial}{\partial\lambda_1},\cdots,\frac{\partial}{\partial\lambda_n}\right)\cdot\frac{\det\left[\sum_{k=0}^{N} A_{N-k}^{\alpha-1}V^k(I-\bar{V}'^{2k+1})\right]}{(2A_N^\alpha)^n\det(I-V)}, \tag{3.5.8}$$

which is different from the one defined in § 2.2. In particular, the Fejér kernel is

$$\frac{1}{(N+1)^n(n-1)!\cdots 2!1!D(\lambda_1,\cdots,\lambda_n)}$$

$$\cdot D\left(\frac{\partial}{\partial\lambda_1},\cdots,\frac{\partial}{\partial\lambda_n}\right)\cdot\left|\frac{\det(I-V^{N+1})}{\det(I-V)}\right|^2, \tag{3.5.9}$$

which is also defferent from the one defined in § 2.2.

Example 3.5.4 The Poisson kernel is expressed by

$$\frac{1}{(n-1)!\cdots 2!1!}\cdot\frac{D\left(\frac{\partial}{\partial\lambda_1},\cdots,\frac{\partial}{\partial\lambda_n}\right)}{D(\lambda_1,\cdots,\lambda_n)}\cdot\frac{(1-r^2)^n}{|\det(I-rV)|^2}, \tag{3.5.10}$$

which is different from the Poisson-Hua kernel defined in § 1.3.

From this definition and the results on the multiple Fourier series (cf. Zygmund [1]), it immediately follows that if $u(U)\in C^{\frac{n(n-1)}{2}}(U_n)$, then, in the sense mentioned above, its Fourier series is both Abel-summable and Cesàro-summable of order α $(\alpha > 0)$ to itself.

§ 3.6 Absolute Convergence of Fourier Series

Let $u(U)$ be integrable over unitary group U_n of order n. Then (3.1.1) is its Fourier series, where

$$C_{f_1\cdots f_n} = \int_{U_n} u(V)\,\overline{\Phi_{f_1\cdots f_n}(V)}\,\dot{V},$$

$$\Phi_{f_1\cdots f_n}(U) = \sqrt{\frac{N(f)}{C}}\,A_{f_1\cdots f_n}(U),$$

and $A_{f_1\cdots f_n}(U)$ is the unitary representation of U_n with signature $(f_1,\cdots f_n)$. Thus

$$C_f\bar{C}_f' = \iint u(V)u(U)\,\overline{\Phi_f(V)}\,\Phi_f'(U)\dot{V}\dot{U}$$
$$= \frac{N(f)}{C}\iint u(V)u(U)\,\overline{A_f(V)}A_f'(U')\dot{V}\dot{U}.$$

Putting $U\bar{V}' = w$, we get

$$\mathrm{tr}(C_f\bar{C}_f') = \frac{N(f)}{C}\iint u(V)u(WV)\chi_f(W)\dot{W}\dot{V}.$$

Let

$$g(w) = \int u(V)u(WV)\dot{V},$$

Consequently

$$\mathrm{tr}(C_f\bar{C}_f') = \frac{N(f)}{C}\int g(W)\chi_f(W)\dot{W}.$$

Assuming that $e^{i\theta_1},\cdots, e^{i\theta_n}$ are characteristic roots of W and setting

$$h(\theta_1,\cdots,\theta_n) = \frac{1}{\omega_n'}\int_{[\dot{W}]} g(W)[\dot{W}],$$

we get

$$\mathrm{tr}(C_f\bar{C}_f') = \frac{N(f)}{(2\pi)^n}\int\cdots\int_{\theta_1\geqslant\cdots\geqslant\theta_n} h(\theta_1,\cdots,\theta_n)M_f(e^{i\theta_1},\cdots,e^{i\theta_n})$$
$$\cdot D(e^{-i\theta_1},\cdots,e^{-i\theta_n})d\theta_1\cdots d\theta_n.$$

As before, replacing h by h^* and using the conventional technique, we obtain that

$$\mathrm{tr}(C_f\bar{C}_f') = \frac{N(f)}{(2\pi)^n}\int_{-\pi}^{\pi}\cdots\int_{-\pi}^{\pi} h^*(\theta_1,\cdots,\theta_n)e^{i(l_1\theta_1+\cdots+l_n\theta_n)}$$
$$\cdot D(e^{-i\theta_1},\cdots,e^{-i\theta_n})d\theta_1\cdots d\theta_n.$$

Putting

$$H(\theta_1,\cdots,\theta_n) = h^*(\theta_1,\cdots,\theta_n)D(e^{-i\theta_1},\cdots,e^{-i\theta_n})$$
$$= D(e^{-i\theta_1},\cdots,e^{-i\theta_n})\frac{1}{\omega_n'}\int_{[W]}\int_{U_n} u(V)u^*(WV)\dot{V}[\dot{W}], \tag{3.6.1}$$

we have

$$\frac{\mathrm{tr}(C_f\bar{C}_f')}{N(f)} = \frac{1}{(2\pi)^n}\int_{-\pi}^{\pi}\cdots\int_{-\pi}^{\pi} H(\theta_1,\cdots,\theta_n)$$
$$\cdot e^{i(l_1\theta_1+\cdots+l_n\theta_n)}d\theta_1\cdots d\theta_n, \tag{3.6.2}$$

where $l_1 > l_2 > \cdots > l_n$. If $(i_1,\cdots,i_n)$ is a permutation of $(1, 2,\cdots, n)$, it is readily shown that

$$\frac{1}{(2\pi)^n}\int_{-\pi}^{\pi}\cdots\int_{-\pi}^{\pi} H(\theta_1,\cdots,\theta_n)e^{i(l_{i_1}\theta_1+\cdots+l_{i_n}\theta_n)}$$
$$=\delta^{12\cdots n}_{i_1 i_2\cdots i_n}\frac{\operatorname{tr}(C_f\bar{C}_f')}{N(f)}.$$

Thus the multiple Fourier series of any integrable function $H(\theta_1,\cdots,\theta_n)$ of several variables equals

$$\sum_{l_1,\cdots,l_n} d_{l_1\cdots l_n}e^{-i(l_1\theta_1+\cdots+l_n\theta_n)},\tag{3.6.3}$$

where

$$d_{l_1\cdots l_n}=\begin{cases}\dfrac{\operatorname{tr}(C_f\bar{C}_f')}{N(f)}, \text{ if } l_1>l_2>\cdots>l_n;\\ 0, \text{ if at least two among } l_1,\cdots,l_n \text{ are equal;}\\ \delta^{12\cdots n}_{i_1 i_2\cdots i_n}\dfrac{\operatorname{tr}(C_f\bar{C}_f')}{N(f)}, \text{ if } (i_1,\cdots,i_n) \text{ is a permutation of } (1,\\ \qquad 2,\cdots,n\end{cases}$$

Thus (3.6.3) can be written as

$$\sum_{l_1>\cdots>l_n}\frac{\operatorname{tr}(C_f\bar{C}_f')}{N(f)}M_{l_1\cdots l_n}(e^{-i\theta_1},\cdots,e^{-i\theta_n}).\tag{3.6.4}$$

In the same way it can be shown that the multiple Fourier series of the function $D\left(\frac{\partial}{\partial\theta_1},\cdots,\frac{\partial}{\partial\theta_n}\right)H(\theta_1,\cdots,\theta_n)$ is

$$D(n-1,\cdots,1,0)\sum_{l_1>\cdots>l_n}(-i)^{\frac{n(n-1)}{2}}$$
$$\cdot\operatorname{tr}(C_f\cdot\bar{C}_f')M_{l_1\cdots l_n}(e^{-i\theta_1},\cdots,e^{-i\theta_n}).\tag{3.6.5}$$

Let $f(\theta_1,\cdots,\theta_n)$ be a continuous function defined on $-\pi\leqslant\theta_1,\cdots,\theta_n\leqslant\pi$, and suppose

$$A^{(k)}(f;\theta_1,\cdots,\theta_n;h_1,\cdots,h_n)$$
$$=f(\theta_1,\cdots,\theta_{k-1},\theta_k+h_k,\theta_{k+1},\cdots,\theta_n)-f(\theta_1,\cdots,\theta_n),$$
$$A^{(j,k)}(f;\theta_1,\cdots,\theta_n;h_1,\cdots,h_n)$$
$$=A^{(j)}(A^{(k)}(f;\theta_1,\cdots,\theta_n;h_1,\cdots,h_n)).\ (j<k)$$

Besides, $A^{(j,k,l)}(f;\theta_1,\cdots,\theta_n;h_1,\cdots,h_n)$, for $j<k<l$ etc., can be similarly defined. Put

$$A(f;\theta_1,\cdots,\theta_n;h_1,\cdots,h_n)=A^{(1,2,\cdots,n)}(f;\theta_1,\cdots,\theta_n;h_1,\cdots,h_n).$$

We call

$$\omega(f;\theta_1,\cdots,\theta_n;h_1,\cdots,h_n)$$
$$=\sup_{\substack{|\delta_1|\leqslant h_1,\cdots,|\delta_n|\leqslant h_n\\ |\theta_1|\leqslant\pi,\cdots,|\theta_n|\leqslant\pi}}|A(f;\theta_1,\cdots,\theta_n;\delta_1,\cdots,\delta_n)|$$

and

$\omega_p(f; h_1, \cdots, h_n) =$

$$\left[\sup_{|\delta_1|\leqslant h_1,\cdots,|\delta_n|\leqslant h_n} \int_{-\pi}^{\pi} \cdots \int_{-\pi}^{\pi} |A(f; \theta_1, \cdots, \theta_n; \delta_1, \cdots, \delta_n)|^p d\theta_1 \cdots d\theta_n\right]^{\frac{1}{p}} \quad (p \geqslant 1)$$

the modulus of continuity and the modulus of integral of order p of a continuous function $f(\theta_1, \cdots, \theta_n)$ defined on $-\pi \leqslant \theta_1, \cdots, \theta_n \leqslant \pi$, respectively.

If, for any $h_1 > 0, \cdots, h_n > 0$, there exists α such that

$$\omega_p(f; h_1, \cdots, h_n) \leqslant K(h_1 \cdots h_n)^\alpha$$

holds valid (K being an absolute constant), then $f(\theta_1, \cdots, \theta_n)$ is said to be "satisfying the Lipschitz condition (p, α)". The set consisting of all functions satisfying such condition is denoted by Lip (p, α).

We now are in the position to prove the following three theorems.

Theorem 3.6.1 Let $u(U) \in C^{\frac{n(n-1)}{2}}$ and

$$D\left(\frac{\partial}{\partial\theta_1}, \cdots, \frac{\partial}{\partial\theta_n}\right) H(\theta_1, \cdots, \theta_n) \in \mathrm{Lip}(2, \alpha),$$

where $H(\theta_1, \cdots, \theta_n)$ *is defined by* (3.6.1). *Then, when* $\alpha > \frac{1}{r} - \frac{1}{2}$, *the series*

$$\sum_{f_1 \geqslant \cdots \geqslant f_n} |tr(C_{f_1\cdots f_n} \overline{C}'_{f_1\cdots f_n})|^r \tag{3.6.6}$$

is convergent for $1 > r > \frac{2}{3}$.

Theorem 3.6.2 *Let* $u(U) \in C^{\frac{3n(n-1)}{2}}$ *and*

$$\left(D\left(\frac{\partial}{\partial\theta_1}, \cdots, \frac{\partial}{\partial\theta_n}\right)\right)^3 H(\theta_1, \cdots, \theta_n) \in \mathrm{Lip}(2, \alpha),$$

where $H(\theta_1, \cdots, \theta_n)$ *is defined by* (3.6.1). *Then, when* $\alpha > \frac{1}{r} - \frac{1}{2}$, *the series*

$$\sum_{f_1 \geqslant \cdots \geqslant f_n} \left(\sum_{j,k=1}^{N(f)} |C_f^{(j,k)}||\varphi_f^{(jk)}|\right)^{2r} \tag{3.6.7}$$

is convergent for $1 > r > \frac{2}{3}$.

Theorem 3.6.3 *Let* $u(U) \in C^{2n(n-1)}$ *and*

$$\left(D\left(\frac{\partial}{\partial\theta_1}, \cdots, \frac{\partial}{\partial\theta_n}\right)\right)^5 H(\theta_1, \cdots, \theta_n) \in \mathrm{Lip}(2, \alpha).$$

Then the Fourier series of $u(U)$ *is absolutely convergent for* $\alpha > \frac{1}{2}$.

We now proceed to prove these results.

Since the multiple Fourier series of the function

$$D\left(\frac{\partial}{\partial\theta_1}, \cdots, \frac{\partial}{\partial\theta_n}\right) H(\theta_1, \cdots, \theta_n)$$

takes the form of (3.6.5) and

$$D\left(\frac{\partial}{\partial\theta_1}, \cdots, \frac{\partial}{\partial\theta_n}\right) H(\theta_1, \cdots, \theta_n) \in \mathrm{Lip}(2, \alpha),$$

the series (3.6.6), by the theorem on absolute convergence of multiple Fourier series (cf., for example, theorem 2 in Musielak [2]), is convergent provided $\alpha > \frac{1}{r} - \frac{1}{2}$. The proof of Theorem 3.6.1 is completed.

From the Schwarz inequality, it follows that

$$|\mathrm{tr}(C_{f_1\cdots f_n} \Phi'_{f_1\cdots f_n}(U))|^2 \leqslant \frac{(N(f))^2}{C} \mathrm{tr}(C_{f_1\cdots f_n} \bar{C}'_{f_1\cdots f_n}). \tag{3.6.8}$$

As before it can be shown that the multiple Fourier series of the function

$$\left(D\left(\frac{\partial}{\partial\theta_1}, \cdots, \frac{\partial}{\partial\theta_n}\right)\right)^3 H(\theta_1, \cdots, \theta_n)$$

is

$$(D(n-1, \cdots, 1, 0))^3 \sum_{f_1 \geqslant \cdots \geqslant f_n} (-i)^{\frac{3n(n-1)}{2}} \cdot \mathrm{tr}(C_f \bar{C}'_f) M_f(e^{-i\theta_1}, \cdots, e^{-i\theta_n}).$$

Since

$$\left(D\left(\frac{\partial}{\partial\theta_1}, \cdots, \frac{\partial}{\partial\theta_n}\right)\right)^3 H(\theta_1, \cdots, \theta_n) \in \mathrm{Lip}(2, \alpha),$$

by the theorem on absolute convergence of multiple Fourier series, it immediately follows that, if $\alpha > \frac{1}{r} - \frac{1}{2}$, the series

$$\sum_{f_1 \geqslant \cdots \geqslant f_n} (N(f))^2 \mathrm{tr}(C_{f_1\cdots f_n} \bar{C}'_{f_1\cdots f_n})$$

is convergent. This implies the convergence of the series (3.6.7).

On account of the Schwarz inequality, we have

$$\left|\sum_{j,k=1}^{N(f)} \left| c_f^{(jk)} \| \varphi_f^{(jk)} \|^2 \leqslant \frac{(N(f))^2}{C} \operatorname{tr}(C_{f_1\cdots f_n} \bar{C}'_{f_1\cdots f_n}). \right.\right. \tag{3.6.9}$$

This proves Theorem 3.6.2.

Similarly, by the Schwarz inequality, we obtain that

$$\left| \sum_{f_1 \geqslant \cdots \geqslant f_n} N(f) (\operatorname{tr}(C_f \bar{C}'_f))^{\frac{1}{2}} \right| \leqslant \left(\sum_{f_1 \geqslant \cdots \geqslant f_n} N^4(f) \operatorname{tr}(C_f \bar{C}'_f) \right)^{\frac{1}{2}} \cdot \left(\sum_{f_1 \geqslant \cdots \geqslant f_n} \frac{1}{(N(f))^2} \right)^{\frac{1}{2}}, \tag{3.6.10}$$

and that the multiple Fourier series of the function

$$\left(D\left(\frac{\partial}{\partial \theta_1}, \cdots, \frac{\partial}{\partial \theta_n} \right) \right)^5 H(\theta_1, \cdots, \theta_n)$$

is

$$(D(n-1, \cdots, 1, 0))^4 \cdot \sum_{l_1 > l_2 > \cdots > l_n} (N(f))^4 \operatorname{tr}(C_f \bar{C}'_f) M_{f_1\cdots f_n}(e^{-i\theta_1}, \cdots, e^{-i\theta_n}).$$

In the same way, applying the theorem on absolute convergence of the multiple Fourier series leads to the following conclusion: if

$$\left(D\left(\frac{\partial}{\partial \theta_1}, \cdots, \frac{\partial}{\partial \theta_n} \right) \right)^5 H(\theta_1, \cdots, \theta_n) \in \operatorname{Lip}(2, \alpha),$$

then the series

$$\sum_{l_1 > \cdots > l_n} (N(f))^4 \operatorname{tr}(C_f \bar{C}'_f)$$

is convergent for $\alpha > \frac{1}{2}$. From (3.6.9) and (3.6.10), Theorem 3.6.3 follows.

From the above proof, it can be seen that by using some other theorems on absolute convergence of the multiple Fourier series such as the extended forms of the Zygmund theorem and the Szasz theorem, we can establish various sufficient conditions for the absolute convergence of the Fourier series on unitary groups.

Finally it should be pointed out that the conditions used in all the theorems of this Chapter, in particular the one on the order of differentiability, may be weakend.

Both the convergence and the absolute convergence discussed here are performed on the basis of "cubical summation". The spherical summation, which will be treated in Chapter 5, is different from the cubical summation. By using the idea of the present chapter and referring to the method of Chapter 5 and the results of Chandrasekharan-Minakshisundaran [1] in Chapter 9. we can deduce the results on the absolute convergence of "spherical summation".

Chapter 4. On Peter-Weyl Theorem

§ 4.1 Peter-Weyl Theorem

Let G be a compact Lie group, f be a continuous function on G. Then, for any given $\varepsilon > 0$, there must exist a function g defined on the representative ring of G such that

$$|f(\sigma) - g(\sigma)| < \varepsilon$$

holds for any $\sigma \in G$. This is a famous theorem due to F. Peter and H. Weyl (cf. Chevally [1]or Peter-Weyl [1]).

The conclusion is also valid for general compact topological groups (cf., for example, Peter-Weyl [1] and L. S. Pontriagin [1]) and this is an extremely important existence theorem in the approxi mation theory on compact topological groups.

In[2], Hua Luogeng proved a convergence theorem related to the continuous functions on unitary groups and indicated its significance for general compact groups.

Here, based on the preceding chapters, a preliminary inquiry into the approximation theory on compact topological groups will be made. We will not only find out the function $g(\sigma)$ defined on the representative ring of G explicitly, but also indicate the error bound between f and g.

In §4.2, the modulus of continuity of any continuous function on a compact topological group G is defined. In§4.3, the approximant and its degree of any continuous function pertaining to Lip p on a compact topological group G via Cesàro means are studied. For example, by means of the results in Chapter2, it is verified that, *if* $f(\sigma) \in$ Lipp (for its definition see §4.2), then there is a finite linear expression $g_N(\sigma)$ defined on the representative ring of G such that

$$|f(\sigma) - g_N(\sigma)| < AN^{-p}, \ A \text{ is anabsolute constant,}$$

holds and $g_N(\sigma)$ can be given explicitly (cf. (4.3.8)).

In fact we will show how to construct other approximants and how to determine the degree of the approximation error. This is included in

§ 4.4. As space is limited, the heavy and complicated calculation will be omitted there.

In § 4.5, we give an example of dealing with the problem on interpolation on unitary groups. We will end this chapter with an application of some results to the approximation theory of holomorphic functions in the hyperbolic space for matrices of several complex variables.

§ 4.2 Continuous Functions on Compact Topological Groups

Let G be a compact topological group and $\sigma \in G$. If $f(\sigma)$ is a continuous function on G,then $f(\sigma)$ must be uniformly continuous on G. In other words, for every $\varepsilon > 0$, there exists a neighborhood V of unit element e of G such that

$$|f(x) - f(y)| < \varepsilon \tag{4.2.1}$$

holds whenever $xy^{-1} \in V$, $x \in G$ and $y \in G$. Equivalently speaking, for every $\varepsilon > 0$, there exists a neighborhood V' of the unit element e of G such that (4.2.1) holds valid provided $x^{-1}y \in V'$, $x \in G$ and $y \in G$(cf. Pontriagin L. S. [1]).

For any compact topological group G, there is at least one faithful representation. This representation can certainly be unitary one and is isomorphic to G. Thus, in a sense of isomophism, any compact topological group can be imbedded in a unitary group as a subgroup. Namely, if $\sigma \in G$, then the unitary represe-ntation $U(\sigma)$ of σ makes G be isomorphic to $\{U(\sigma) | \sigma \in G\}$.

Let $W = \{U \in U_n | U = U(\sigma), \sigma \in G\}$,and then $W \subset U_n$. Define continuous function $u(U) = f(\sigma)$ on W.$u(U)$ not only can be defined on W, but also can be continuously extended on the whole U_n(cf. Chevalley[1] and Pontriagin[1]). However, it should be noticed that there may be more than one faithful representation, and so is W.

Assume that $U$and V are two points in W, $U = (u_{ij})$ and $V = (v_{ij})$, $1 \leqslant i, j \leqslant n$. Then the square of the Euclidean distance $d(U, V)$ between U and V is

$$\sum_{i,j=1}^{n} |u_{ij} - v_{ij}|^2.$$

Namely

$$\operatorname{tr}((U - V)\overline{(U - V)}') = \operatorname{tr}(2I - U\bar{V}' - V\bar{U}').$$

If $e^{i\theta_1}, \cdots, e^{i\theta_n}$ refer to the characteristic roots of $U\bar{V}'$, then the square of $d(U, V)$ is

$$2\sum_{j=1}^{n} (1 - \cos\theta_j). \tag{4.2.2}$$

Let $u(U)$ be a continuous function on U_n. Then

$$\omega(u, \delta) = \max_{d(U,V)\leqslant\delta} |u(U) - u(V)|$$

is called the modulus of continuity of $u(U)$.

If $f(\sigma)$ is continuous on a compact topological group G and $\sigma \in G$, then G has at least one faithful unitary represen-tation, and, in a sense of isomorphism, G can be imbedded in some unitary group (say a unitary group of order m) as its subgroup. As before, denote

$$W = \{U \in U_m | U = U(\sigma), \sigma \in G\}.$$

Then, for a faithful unitary representation of order m,we have

$$\omega'_m(f, \sigma) = \max_{x\in G, y\in G} |f(x) - f(y)|.$$

$$d(U(x), U(y)) \leqslant \delta$$

Furthermore, suppose that $f(\sigma)$ is continuous on G and

$$|f(x) - f(y)| \leqslant A(d(U(x), U(y)))p',$$

where A is an absolute constant. Among the possible faithful unitary representations, choose one of them. The corresponding exponent $p'$is maximum and is denoted by p. Then we refer to $f(\sigma)$ as "satisfying the Lipschitz condition" and denote this by $f(\sigma) \in \text{Lip } p$.

§ 4.3 Approximation by Cesàro Means

In Chapter 2, we carefully studied the Cesàro (c, α) mean of the Fourier series on unitary group U_n and proved: if $u(U)$ is continuous on U_n, then the Fourier series of $u(U)$ is (c, α)-summable to itself for $\alpha > \frac{n-1}{n}$.

In the process of proving this theorem, actually, we showed that the absolute value of the difference between $u(U)$ and the Nth term $\sum_N^\alpha(U)$ of the Cesàro (c, α) mean of its Fourier series is no greater than

$$O(\omega(u, \delta)) + O\left(\frac{1}{(N\delta)^{\alpha n-n+1}}\right)$$
$$+ O\left(\frac{1}{(N\delta)^{2(\alpha n-n+2)}}\right) + \cdots + O\left(\frac{1}{(N\delta)^{\alpha n^2}}\right).$$

Thus, if $u(U) \in \text{Lip}\alpha$, we obtain by taking $\delta = N^{\frac{n-\alpha n-1}{p+\alpha n-n+1}}$ that

$$|u(U) - \sum_N^\alpha(U)| \leqslant AN^{\frac{-p(\alpha n-n+1)}{p+\alpha n-n}},$$

where A is an absolute constant and $\alpha > \frac{n-1}{n}$.

This result is rather rough. Recalling the proof of Theorem 2.3.1 leads to sharper results.

Theorem 4.3.1 *Assume that $u(U)$ is a continuous function on U_n and belongs to* Lip p, $0 < p \leqslant 1$. *Then the Nth term $\Sigma_N^\alpha(U)$ of the (C, α) mean of its Fourier series satisfies*

$$(1)\quad |u(U) - \Sigma_N^\alpha(U)| \leqslant A_1 N^{-p}, \tag{4.3.1}$$

if $\alpha n - n + 1 > p$;

$$(2)\quad |u(U) - \Sigma_N^\alpha(U)| \leqslant A_2 N^{-p} \ln N, \tag{4.3.2}$$

if $\alpha n - n + 1 = p$;

$$(3)\quad |u(U) - \Sigma_N^\alpha(U)| \leqslant A_3 N^{-\alpha n + n - 1}, \tag{4.3.3}$$

if $\alpha n - n + p < p$. *A_1, A_2 and A_3 are absolute constants.*

Proof. In virtue of (2.3.3), by $I'_{s_1,\cdots,s_n}$ we denote

$$\int_\delta^\pi |\varphi(e^{i\theta_1}, \cdots, e^{i\theta_n})| \, |1 - e^{i\theta_1}|^{s_1} |\sigma_N^\alpha(\theta_1)|^n d\theta_1$$

$$\cdot \int \cdots \int_{\delta \geqslant \theta_1 \geqslant \cdots \geqslant \theta_n \geqslant -\delta} |1 - e^{i\theta_2}|^{s_2} \cdots |1 - e^{i\theta_n}|^{s_n} |\sigma_N^\alpha(\theta_2)|^n \cdots$$

$$\cdot |\sigma_N^\alpha(\theta_n)|^n |D(e^{i\theta_1}, \cdots, e^{i\theta_n})|^2 d\theta_2 \cdots d\theta_n, \tag{4.3.4}$$

where φ, σ_N^α and D were defined in Chapter 2.

Owing to $u(U) \in$ Lip p and, for $\delta \geqslant \dfrac{1}{N}$, we have

$$|\sigma_N^\alpha(\theta)| \leqslant a_1 N^{-\alpha} \theta^{-\alpha-1}.$$

Hence, we conclude that $I'_{s_1,\cdots,s_n}$ is no greater than

$$a_2 \int_\delta^\pi N^{-\alpha n} \theta_1^{-\alpha n - n + s_1 + p} d\theta_1 N^{n-1} \delta^{s_2 + \cdots + s_n}$$

$$\cdot \int \cdots \int_{\delta \geqslant \theta_2 \geqslant \cdots \geqslant \theta_n \geqslant -\delta} |\sigma_N^\alpha(\theta_2) \cdots \sigma_N^\alpha(\theta_n)|^{n-1} |D(e^{i\theta_2}, \cdots, e^{i\theta_n})|^2$$

$$\cdot d\theta_2 \cdots d\theta_n + a_3 \delta^p \int_\delta^\pi N^{-\alpha n} \theta_1^{-\alpha n - n + s_1} d\theta_1 N^{n-1} \delta^{s_2 + \cdots + s_n}$$

$$\cdot \int \cdots \int_{\delta \geqslant \theta_2 \geqslant \cdots \geqslant \theta_n \geqslant -\delta} |\sigma_N^\alpha(\theta_2) \cdots \sigma_N^\alpha(\theta_n)|^{n-1} |D(e^{i\theta_1}, \cdots, e^{i\theta_n}|^2$$

$$\cdot d\theta_2 \cdots d\theta_n, \tag{4.3.5}$$

where a_1, a_2 and a_3 are distinct absolute constants.

(1) When $\alpha n - n + 1 > p$, we have by $s_1 \leqslant 2n - 2$ that

$$-\alpha n - n + s_1 + p < -\alpha n - n + s_1 + \alpha n - n + 1$$
$$= -2n + 1 + s_1 \leqslant -1.$$

Thus, from (4.3.5), it yields

$$I'_{s_1, \cdots, s_n} = O((\delta N)^{-\alpha n + n - 1} \delta^p).$$

Taking $\delta = \frac{1}{N}$, we get(4.3.1).

(2) When $\alpha n - n + 1 = p$, we have that

$$-\alpha n - n + s_1 + p = -\alpha n - n + s_1 + \alpha n - n + 1$$
$$= -2n + 1 + s_1.$$

So, by(4.3.5), we note

$$I'_{2n-2, s_2, \cdots, s_n} = O(N^{-\alpha n + n - 1} \ln \delta) + O((\delta N)^{-\alpha n + n - 1} \delta^p).$$

If $s_1 \neq 2n - 2$, there exists

$$I'_{s_1, \cdots, s_n} = O((\delta N)^{-\alpha n + n - 1} \delta^p).$$

Namely

$$|u(U) - \Sigma_N^{\alpha}(U)| \leqslant a_4 N^{-p} \ln \delta + a_5 N^{-p},$$

where a_4 and a_5 are absolute constants. Setting $\delta = \frac{1}{N} \ln N$, we obtain (4.3.2).

(3) When $\alpha n - n + 1 < p$, we have

$$-\alpha n - n + s_1 + p > -\alpha n - n + s_1 + \alpha n - n + 1$$
$$= -2n + 1 + s_1,$$

Thus

$$I'_{2n-2, s_2, \cdots, s_n} = O(N^{-\alpha n + n - 1}) + O((N\delta)^{-\alpha n + n - 1} \delta^p).$$

Therefore

$$|u(U) - \Sigma_N^{\alpha}(U)| \leqslant a_6 N^{-\alpha n + n - 1} + a_7 (N\delta)^{-\alpha n + n - 1}$$
$$\cdot \delta^p \ln \delta + a_8 (N\delta)^{-\alpha n + n - 1} \cdot \delta^p.$$

As $p > \alpha n - n + 1$ and $\delta \geqslant \frac{1}{N}$, (4.3.3) follows.

Recall that the Nth term of the Cesàro (c, α) mean of the Fourier series for an integrable function $u(U)$ over U_n is

$$\sum_{nN \geqslant f_1 \geqslant \cdots \geqslant f_n \geqslant -nN} B^{\alpha}_{f_1 \cdots f_n} \operatorname{tr}(C_{f_1 \cdots f_n} \Phi'_{f_1 \cdots f_n}(U)), \tag{4.3.6}$$

where

$$B_{f_1\cdots f_n}=\frac{1}{CN(f)}\int_{U_n}\chi_{f_1\cdots f_n}(\bar{V})K_N^{\alpha}(V)\dot{V}, \tag{4.3.7}$$

and $K_N^{\alpha}(V)$ is the Cesàro (C,α) kernel defined by(2.2.3).

If G is a compact topological group, $f(\sigma)$ is a continuous function on G and $f(\sigma)\in \mathrm{Lip}\ p$, then we can construct a function

$$g_N(\sigma)=\sum_{nN\geqslant f_1\geqslant\cdots\geqslant f_n\geqslant -nN} B^{\alpha}_{f_1\cdots f_n}\mathrm{tr}(C_{f_1\cdots f_n}\Phi'_{f_1\cdots f_n}(U(\sigma))), \tag{4.3.8}$$

defined on the representative ring of G,where $B_{f_1\cdots f_n}$ was defined by(4.3.7). Thus Theorem 4.3.1 can be put to the following form.

Theorem 4.3.2 *Let G be a compact topological group, $f(\sigma)$ be a continuous function on G and $f(\sigma)\in \mathrm{Lip}\ p\ (\sigma\in G)$. Define the function $g_N(\sigma)$ on the representative ring of G by (4.3.8). Then it follows that*

(1) $|f(\sigma)-g_N(\sigma)|\leqslant A_4N^{-q}$, *if* $p\neq \alpha n-n+1$.

Here $q=\min(p,\alpha n-n+1)$ *and A_4 is an absolute constant.*

(2) $|f(\sigma)-g_N(\sigma)|\leqslant A_2N^{-p}\ln N$, *if* $p=\alpha n-n+1$.

As a corollary of the above theorem, for given p, taking sufficiently large α such that $\alpha n-n+1>p$, we are led to

Theorem 4.3.3 *Let G be compact topological group, $f(\sigma)$ be a continuous function on G and $f(\sigma)\in \mathrm{Lip}\ p(\sigma\in G)$. Then the function $g_N(\sigma)$ can be constructed on the representative ring of G,such that*

$$|f(\sigma)-g_N(\sigma)|\leqslant A_5N^{-p},$$

where A_5 is an absolute constant.

As another corollary of Theorem 4.3.2, in particular, taking $\alpha=1$ in the theorem (i.e. taking the Fejér means), we arrive at

Theorem 4.3.4 *Let G be a compact topological group, $f(\sigma)$ be a continuous function on G and $f(\sigma)\in \mathrm{Lip}\ p(\sigma\in G)$. Set*

$$g_N(\sigma)=\sum_{nN\geqslant f_1\geqslant\cdots f_n\geqslant -nN} B_{f_1\cdots f_n}\mathrm{tr}(C_{f_1\cdots f_n}\Phi'_{f_1\cdots f_n}(U(\sigma)), \tag{4.3.9}$$

on the representative ring of G,where, $B_{f_1\cdots f_n}$ is defined by(2.5.2). When $N\geqslant n-1$, $B_{f_1\cdots f_n}$ can be expressed explicitly by(2.7.7). Then

(1) $|f(\sigma)-g_N(\sigma)|<A_6N^{-p}$ (4.3.10)

for $p<1$;

(2) $|f(\sigma)-g_N(\sigma)|<A_7\ln N/N$ (4.3.11)

for $p=1$. *Here A_6 and A_7 are absolute constants.*

By the way, S. N. Bernstein's theorem of the Fourier series of the function of one variable is a special case of Theorem 4.3.4.

§ 4.4 Some General Corollaries

By means of the Cesàro summation of the Fourier series of a function $u(U)$ defined on unitary groups, the approximation theorem on compact topological groups can be deduced. Thus both the function defined on the representative ring of G and the bound of the approximation error can be explicitly given. The function thus obtained is not necessarily the best approximation. Besides, the approximation foumulae based on other summabilities and the estimate of their errors can be worked out in the same way. To this end, in the present section we will give some general results as corollaries of the method used above. For example, the following general convergence theorem can be proved.

Let $K_N(\theta)$ be the kernel of T-summation of Fourier series of one variable, $V \in U_n$, $e^{i\theta_1}, \cdots, e^{i\theta_n}$ be characteristic roots of $\bar{V}$ and

$$D = \frac{1}{C}\int_{U_n} (K_N(\theta_1)\cdots K_N(\theta_n))^n \dot{V}. \tag{4.4.1}$$

T-means of the Fourier series of $u(U)$ are defined by

$$\sum_{nN \geqslant f_1 \geqslant \cdots \geqslant f_n \geqslant -nN} D_{f_1\cdots f_n} \operatorname{tr}(C_{f_1\cdots f_n} \Phi'_{f_1\cdots f_n}(U)), \tag{4.4.2}$$

where

$$D_{f_1\cdots f_n} = \frac{1}{DN(f)C}\int_{U_n} \chi_{f_1\cdots f_n}(\bar{V})(K_N(\theta_1)\cdots K_N(\theta_n))^n \dot{V}.$$

(4.4.2) can be written as

$$\int_{U_n} u(VU) T_N(\bar{V}) \dot{V},$$

where

$$T_N(\bar{V}) = \frac{1}{DC}(K_N(\theta_1)\cdots K_N(\theta_n))^n. \tag{4.4.3}$$

If, in addition, we assume that

(i) When $\pi \geqslant \theta \geqslant \delta$, for any fixed $\delta > 0$ we have

$$|K_N(\theta)| = 0(N^{-\eta}) \quad (\eta > 0), \tag{4.4.4}$$

(ii) For any θ we have

$$|K_N(\theta)| = 0(N^{\xi}) \quad (\xi > 0), \tag{4.4.5}$$

then we have the following

Theorem 4.4.1 *Let $u(U)$ be continuous on U_n. Assume that $T_N(\bar{V})$ defined by* (4.4.3) *satisfies*

$$\int_{U_m} |T_N(\bar{V})|\dot{V} \leqslant H_m \quad (m = 1, 2, \cdots, n), \tag{4.4.6}$$

where H_m are absolute constants depending only on m, and $K_N(\theta)$ satisfies (4.4.4) *and* (4.4.5) *with $\eta \geqslant \xi$. Then T-means* (4.4.2) *of its Fourier series converges to itself.*

The proof is omitted here and readers are referred to Theorem 2.3.1.

As an example of the theorem mentioned above, we define the Jackson means of the Fourier series of a function $u(U)$ defined on U_n by

$$\begin{aligned} J_N(U) &= \sum_{2nN \geqslant f_1 \geqslant \cdots \geqslant f_n \geqslant -2nN} J_{f_1 \cdots f_n} \operatorname{tr}(C_{f_1 \cdots f_n} \Phi'_{f_1 \cdots f_n}(U)) \\ &= \frac{J}{C} \int_{U_n} u(VU) \left| \frac{\det(I - V^{N+1})}{\det(I - V)} \right|^{4n} \dot{V}, \end{aligned} \tag{4.4.7}$$

where

$$J = \left(\frac{3}{N(2N^2 + 1)} \right)^{n^2}.$$

The value of $J_{f_1 \cdots f_n}$, certainly, can be expressed by integral. Here (4.4.6) is satisfied, as $H_m = 1$, $m = 1, 2, \cdots, n$. Furthermore, we have

$$K_N(\theta) = 0(N^{-3}\delta^{-4}), \quad \text{if } \pi \geqslant \theta \geqslant \delta,$$

and

$$K_N(\theta) = 0(N), \quad \text{for any } \theta.$$

Thus all conditions of Theorem 4.4.1 are satisfied. So (4.4.7) is convergent and converges to $u(U)$.

Indeed the magnitude of the error between T-means and $u(U)$ can be estimated. For example, it is liable to deduce roughly that

$$|J_N(U) - u(U)| \leqslant AN^{-\frac{p(n+2)}{4n+p-1}},$$

if $u(U) \in \text{Lip } p$.

The estimate of the error between T-summation satisfying the conditions of Theorem 4.4.1 and $u(U)$ is omitted here. These results, of course, can be extended to compact topological groups.

§ 4.5 An Example of Interpolation on Unitary Groups

How to pose general interpolation problem on unitary groups is worthy to be studied. Here we cite only the simplest example.

Given a function $u(U)$ defined on U_n, what we need to do is to find out a finite linear expression which has the same values as $u(U)$ at the following $n\,(2N+1)^n$ points

$$U_{p_1\cdots p_n} = V_1 \begin{pmatrix} e^{\frac{2p_1\pi i}{2N+1}} & & \\ & \ddots & \\ & & e^{\frac{2p_n\pi i}{2N+1}} \end{pmatrix} \bar{V}'_0,$$

where p_j are positive integers with $1 \leqslant p_j \leqslant 2N+1$, $j = 1, 2, \cdots, n$, and V_0 is a fixed unitary matrix. In other words, the problem under consideration is a simple interpolation one with equidistant.

It is easy to answer this problem, as all the $n(2N+1)^n$ points, no matter how large the number N is taken, are located on the same manifold of dimension n and U_n itself is of dimension n^2.

Consider

$$\frac{\det^n(I - U^{2N+1})}{A^0 C \det^{nN} U \det^n(I-U)}$$

$$= \frac{1}{A^0 C} \prod_{j=1}^{n} \frac{\sin^n\left(N+\frac{1}{2}\right)\theta_j}{\sin^n \theta_j/2},$$

where

$$A^0 = \frac{1}{C}\int_{U_n} \det^{nN} V \frac{\det^n(I - \bar{V}'^{2N+1})}{\det^n(I-\bar{V}')} \dot{V},$$

and $e^{i\theta_1}, \cdots, e^{i\theta_n}$ are characteristic roots of U.

It can be shown that

$$\sum_{nN \geqslant f_1 \geqslant \cdots \geqslant f_n \geqslant -nN} A^0_{f_1\cdots f_n} N(f) \chi_{f_1\cdots f_n}(U) = \frac{\det^n(I - U^{2N+1})}{A^0 \det^{nN} U \det^n(I-U)},$$

where

$$A^0_{f_1\cdots f_n} = \frac{1}{A^0 N(f) C}\int_{U_n} \chi_{f_1\cdots f_n}(\bar{V}) \frac{\det^{nN} V \det^n(I - \bar{V}'^{2N+1})}{\det^n(I-\bar{V}')} \dot{V}.$$

Set

$$\varphi_{p_1\cdots p_n}(U) = \frac{\det{}^n(I - U^{2N+1} U^{2N+1}_{p_1\cdots p_n})}{(2N+1)^{n^2} \det^{nN} U \cdot U_{p_1\cdots p_n} \det^n(I - U U_{p_1\cdots p_n})}.$$

By this definition, it can be seen that

(i)

$$\varphi_{p_1\cdots p_n}(U) = A^0 C (2N+1)^{n^2}$$

$$\cdot \sum_{nN \geqslant f_1 \geqslant \cdots \geqslant f_n \geqslant -nN} A^0_{f_1\cdots f_n} \operatorname{tr}(\Phi^1_{f_1\cdots f_n}(U_{p_1\cdots p_n}) \Phi^1_{f_1\cdots f_n}(U)).$$

Namely, $\varphi_{P_1\cdots P_n}(U)$ is a finite linear expression and can also be written as a finite linear expression of those unitary representations of which the indices are in $nN \geqslant f_1 \geqslant \cdots \geqslant f_n \geqslant -nN$.

(ii)

$$\varphi_{P_1\cdots P_n}(u_{q_1\cdots q_n}) = \begin{cases} 1, \text{ if } (P_1,\cdots,P_n) = (q_1,\cdots,q_n) \\ 0, \text{ if}(P_1,\cdots,P_n) \neq (q_1,\cdots,q_n). \end{cases}$$

Thus the function

$$P_N(U) = \sum_{P_1,P_2,\cdots,P_n} u(U_{P_1\cdots P_n})\varphi_{P_1\cdots P_n}(U)$$

is just the desired function. More clearly, this is

$$P_N(U) = \frac{A^0 C}{(2N+1)^{n^2}}$$

$$\cdot \sum_{nN\geqslant f_1\geqslant\cdots\geqslant f_n\geqslant -nN} A^0_{f_1\cdots f_n}\operatorname{tr}[E_{f_1\cdots f_n}\Phi^1_{f_1\cdots f_n}(U)],$$

where

$$E_{f_1\cdots f_n} = \sum_{P_1,P_2,\cdots,P_n} u(U_{P_1,\cdots P_n})\Phi^1_{f_1\cdots f_n}(U_{P_1\cdots P_n}).$$

Certainly, there may be other methods of constructing $P_N(U)$. For example, applying the first method of S. N. Bernstein, we can find out another $P_N(U)$ by using the Fejér kernel of Chapter 2 (cf. Chap. IV vol. 3 of Natanson[1]).

§ 4.6 Approximation on Hyperbolic Space of Matrices of Several Complex Variables

The approximation on the hyperbolic space of matrices of several complex variables can be immediately deduced from the preceding results. The classical domain R_I of Class I of several complex variables is defined by

$$I - Z\bar{Z}' > 0, \quad Z = Z^{(m,n)}.$$

When $m = n$, in particular, unitary group U_n of order n is its characteristic manifold.

For example, from the results of Chapter 3, the following result is true:

Theorem 4.6.1 *Let $u(Z)$ be holomorphic both on domain*

$$I - Z\bar{Z}' > 0, \quad Z = Z^{(n)}$$

and on its characteristic manifold. Then the first Nth Taylor expansion $S_N^{(n)}(Z)$ *of* $u(Z)$ *satisfies*

$$|u(Z) - s_N^{(n)}(Z)| < A_8\left(\frac{\ln^{n^2} N}{N}\right)^{\frac{1}{n+1}}$$

Where A_8 *is an absolute constant. The first Nth Taylor expansion means the representation* $\Phi_{f_1 \cdots f_n}(Z)$ *of* Z *with* $N \geqslant l_1 > \cdots > l_n \geqslant 0$.

Again for example, from Theorem 4.3.4 we obtain immediately the following

Theorem 4.6.2 *Let* $u(Z)$ *be holomorphic in* R_I *and belong to* Lip p *on* U_n. *Then the Fejér mean* $\Sigma_N^{(r)}(Z)$ *of the Taylor expansion of* $u(Z)$ *satisfies*

$$|u(Z) - \Sigma_N^{(r)}(Z)| < A_9 N^{-\rho}, \quad \text{if } 0 < \rho < 1,$$

$$|u(Z) - \Sigma_N^{(r)}(Z)| < A_{10} \ln N / N, \quad \text{if } P = 1,$$

Where A_9 *and* A_{10} *are absolute constants,*

$$\Sigma_N^{(r)}(Z) = \sum_{nN \geqslant f_1 \geqslant \cdots \geqslant f_n \geqslant 0} B_{f_1 \cdots f_n} \operatorname{tr}(C_{f_1 \cdots f_n} \Phi'_{f_1 \cdots f_n}(Z))$$

and $B_{f_1 \cdots f_n}$ *is defined by* (2.5.2).

The other results, certainly, can be established on R_I. We will not repeat them here.

Chapter 5. Spherical Summation of Fourier Series on Unitary Groups

§ 5.1 Introduction

Assume that $u(U)$ is integrable over the unitary group U_n of order n. Then it has the Fourier series

$$\sum_{f_1 \geqslant \cdots \geqslant f_n} \operatorname{tr}(C_{f_1 \cdots f_n} \Phi'_{f_1 \cdots f_n}(U)). \tag{5.1.1}$$

In the preceding chapters, what was studied was limited to the "cubical" summation, i. e. problems drawn from

$$\sum_{N \geqslant l_1 > \cdots > l_n \geqslant -N} \operatorname{tr}(C_{f_1 \cdots f_n} \Phi'_{f_1 \cdots f_n}(U)). \tag{5.1.2}$$

A series of results obtained in the preceding chapters possesses a succinct kernel feature. In this chapter, we consider the "spherical" summation. The so-called spherical summation is to consider the Fourier series as

$$\sum_{m=0}^{\infty} \sum_{\substack{f_1 \geqslant \cdots \geqslant f_n \\ l_1^2 + \cdots + l_n^2 = m}} \operatorname{tr}(C_{f_1 \cdots f_n} \Phi'_{f_1 \cdots f_n}(U)). \tag{5.1.3}$$

Starting from this point of view, we carry on investigation. If the region considered is a topological product of n unit circles rather than a unitary group, the latter corresponds to the study of multiple Fourier series represented by S. Bochner [1]. And the former corresponds to the study of multiple Fourier series represented by A. Zygmund [1].

In this chapter, we first give the integral expression for general spherical means of the Fourier series of any integrable function $u(U)$ over the unitary group U_n of order n. Naturally, the definition of spherical means itself needs some conditions. Starting from this point, we verified that if $u(U)$ is continuous on U_n, then its Fourier series is Riesz-summable of order δ to itself for $\delta \geqslant \frac{n^2-1}{2}$.

In the meantime, we proved that

(1) If $u(U)$ is continuous on U_n, then its Fourier series is Gauss-Sommerfeld summable to itself.

(2) If $u(U)$ is continuous on U_n, then its Fourier series is Abel-summable to itself.

We should notice that all summations mentioned here are "spherical", i.e. we start from (5.1.3). These are totally different from those defined before. Hereafter, the summation always means "spherical" summation unless otherwise specified. Finally, a convergence theorem concerning the general summation is given.

When the "cubical" summation is considered, the corresponding kernel is very simple but the coefficients of the summation are rather complicated. This can be avoided in the case of "spherical" summation.

This chapter is based on Sheng Kung (Gong Sheng) [5].

§ 5.2 Spherical Summation of Fourier Series

Let $\varphi(t)$ be a fixed function defined on $0 \leqslant t < \infty$ with $\varphi(0) = 1$. The so-called "spherical" summation means the limiting case of a certain mean

$$\sum_{m=0}^{\infty} \phi\left(\frac{\sqrt{m}}{R}\right) \sum_{\substack{f_1 \geqslant \cdots \geqslant f_n \\ l_1^2 + \cdots + l_n^2 = m}} \operatorname{tr}\left(C_{f_1 \cdots f_n} \Phi'_{f_1 \cdots f_n}(U)\right), \tag{5.2.1}$$

as $R \to \infty$, where

$$\phi(t) = \varphi(t) / \varphi\left(\sqrt{\frac{(2n-1)n(n-1)}{6R^2}}\right).$$

The Fourier series of $u(U)$ is said to be so-and-so spherical summable to its limit if the limit of (5.2.1) exists, as $R \to \infty$.

The most interesting and concrete examples of $\varphi(t)$ are the following functions:

(1) $\varphi(t) = e^{-t}$——the Abel summation,

(2) $\varphi(t) = e^{-t^2}$——the Gauss-Sommerfeld Summation,

(3) $\varphi(t) = \begin{cases} (1-t^2)^\delta, & \text{if } 0 \leqslant t < 1 \\ 0, & \text{if } 1 \leqslant t \end{cases}$——the Riesz summation of order δ.

More clearly, the Abel summation of the Fourier series of $u(U)$ is the limiting case of the Abel means

$$\exp\left(\sqrt{\frac{(2n-1)n(n-1)}{6R^2}}\right) \sum_{\substack{f_1 \geqslant \cdots \geqslant f_n \\ m = l_1^2 + \cdots + l_n^2}} e^{-\frac{\sqrt{m}}{R}} \operatorname{tr}\left(C_{f_1 \cdots f_n} \Phi'_{f_1 \cdots f_n}(U)\right), \tag{5.2.2}$$

as R→∞; the Gauss-Sommerfeld summability is the one of the Gauss-Sommerfeld means

$$\exp\left(\frac{(2n-1)n(n-1)}{6R^2}\right) \sum_{\substack{f_1 \geq \cdots \geq f_n \\ m=l_1^2+\cdots+l_n^2}} e^{-\frac{m}{R^2}} \operatorname{tr}(C_{f_1\cdots f_n}\Phi'_{f_1\cdots f_n}(U)) \tag{5.2.3}$$

as R→∞; and the Riesz summability of order δ is the one of the Riesz means of order δ

$$\left(1-\frac{(2n-1)n(n-1)}{6R^2}\right)^{-\delta} \sum_{\substack{f_1 \geq \cdots \geq f_n \\ m=l_1^2+\cdots+l_n^2}} \left(1-\frac{m}{R^2}\right)^{\delta} \operatorname{tr}(C_{f_1\cdots f_n}\Phi'_{f_1\cdots f_n}(U)), \tag{5.2.4}$$

as R→∞.

We are going to prove the following three theorems:

Theorem 5.2.1 *If $u(U)$ is continuous on U_n, then its Fourier series is Abel-summable to itself, i.e. (5.2.2) converges to $u(U)$ itself as $R\to\infty$.*

Theorem 5.2.2 *If $u(U)$ is continuous on U_n, then its Fourier series is Gauss-Sommerfeld-summable to itself, i. e. (5.2.3) converges to $u(U)$ itself as R→∞.*

Theorem 5.2.3 *If $u(U)$ is continuous on U_n, then its Fourier series is order δ Riesz-summable to itself for $\delta > (n^2-1)/2$, i. e., for $\delta > (n^2-1)/2$, (5.2.4) converges to $u(U)$ itself as $R\to\infty$.*

From the process of the proofs of these theorems, it would be found that all these summabilities, namely, the Abel-summability, the Gauss-Sommerfeld summability and the Riesz-summability of order δ $(\delta > (n^2-1)/2)$ are "local" properties.

§ 5.3 Integral Expression

Now consider (5.2.1) and denote it by $S_R^\phi(U)$ which can be written as

$$\frac{1}{C}\int_{U_n} u(WU) \sum_{\substack{f_1 \geq \cdots \geq f_n \\ m=l_1^2+\cdots+l_n^2}} \phi\left(\frac{\sqrt{m}}{R}\right) N(f)\chi_{f_1\cdots f_n}(\overline{W})\dot{W}. \tag{5.3.1}$$

By $e^{i\theta_1}, \cdots, e^{i\theta_n}$, we denote characteristic roots of $\overline{W}$. Thus from Chapter 1, it follows that

$$\sum_{f_1 \geq \cdots \geq f_n} r^{|l_1|+\cdots+|l_n|} N(f)\chi_{f_1\cdots f_n}(\overline{W})$$

$$= \frac{i^{\frac{n(n-1)}{2}}}{(n-1)!\cdots 2!1!D(e^{i\theta_1},\cdots e^{i\theta_n})} \cdot \begin{vmatrix} p_r(\theta_1),\cdots, p_r(\theta_n) \\ p_r'(\theta_1),\cdots, p_r'(\theta_n) \\ \cdots\cdots\cdots \\ p_r^{(n-1)}(\theta_1),\cdots, p_r^{(n-1)}(\theta_n) \end{vmatrix},$$

where $p_r(\theta) = \dfrac{1-r^2}{1-2r\cos\theta + r^2}$.

By $g(\theta_1,\cdots\theta_n)$, we denote

$$\frac{i^{\frac{n(n-1)}{2}}}{(n-1)!\cdots 2!1!} \begin{vmatrix} p_r(\theta_1),\cdots, p_r(\theta_n) \\ p_r'(\theta_1),\cdots, p_r'(\theta_n) \\ \cdots\cdots\cdots \\ p_r^{(n-1)}(\theta_1),\cdots, p_r^{(n-1)}(\theta_n) \end{vmatrix}$$

$$= \sum_{f_1\geqslant\cdots\geqslant f_n} r^{|l_1|+\cdots+|l_n|} N(f) M_{f_1\cdots f_n}([e^{i\theta_1},\cdots e^{i\theta_n}]). \tag{5.3.2}$$

Let the Fourier series of $g(\theta_1,\cdots,\theta_n)$ be

$$\sum_{\infty>v_1,\cdots v_n>-\infty} a_{v_1\cdots v_n} e^{i(v_1\theta_1+\cdots+v_n\theta_n)}.$$

By Bochner [1], we admit

$$\sum_{m=0}^{\infty} \sum_{m=l_1^2+\cdots+l_n^2} \phi\left(\frac{\sqrt{m}}{R}\right) a_{l_1\cdots l_n} e^{i(l_1\theta_1+\cdots+l_n\theta_n)}$$

$$= R\int_0^\infty g_\theta(t) H_\varphi(tR)\,dt, \tag{5.3.3}$$

if $\varphi(t)$ is absolutely continuous on any finite interval and

$$\int_0^\infty |\varphi(t)|\, t^{\frac{n-1}{2}}\,dt < \infty, \tag{5.3.4}$$

where

$$g_\theta(t) = \frac{\Gamma\left(\frac{n}{2}\right)}{2(\pi)^{\frac{n}{2}}} \int_\sigma g(\theta_1 + t\eta_1,\cdots\theta_n + t\eta_n)\,d\sigma_\eta, \tag{5.3.5}$$

$$H_\phi(tR) = \frac{(tR)^{\frac{n}{2}}}{2^{\frac{n}{2}-1}\,\Gamma\left(\frac{n}{2}\right)} \int_0^\infty \varphi(u) u^{\frac{n}{2}} J_{\frac{n-1}{2}}(utR)\,du. \tag{5.3.6}$$

$J_\mu(s)$ belongs to the Bessel function of the first kind of order μ. σ denotes the sphere $\eta_1^2 + \cdots + \eta_n^2 = 1$ and $d\sigma_n$ represents the volume element on the sphere. On the other hand, the left side of (5.3.3) is

$$\sum_{m=0}^{\infty} \sum_{\substack{f_1 \geq \cdots \geq f_n \\ m = l_1^2 + \cdots + l_n^2}} \phi\left(\frac{\sqrt{m}}{R}\right) r^{|l_1| + \cdots + |l_n|} N(f) M_{f_1 \cdots f_n}([e^{i\theta_1}, \cdots e^{i\theta_n}]).$$

Thus we obtain

$$\sum_{m=0}^{\infty} \sum_{m = l_1^2 + \cdots + l_n^2} \phi\left(\frac{\sqrt{m}}{R}\right) r^{|a_1| + \cdots + |a_n|} N(f) \chi_{f_1 \cdots f_n}(\overline{W})$$

$$= \frac{R}{D(e^{i\theta_1}, \cdots e^{i\theta_n})} \int_0^\infty g_\theta(t) H_\phi(tR)\, dt. \tag{5.3.7}$$

Let $F(r, R, \overline{W})$ denote the above value. Then it can be put into

$$\frac{R\Gamma\left(\frac{n}{2}\right)}{2(\pi)^{\frac{n}{2}} D(e^{i\theta_1}, \cdots e^{i\theta_n})} \int_0^\infty \int_\sigma g(\theta + t\eta) H_\phi(tR)\, d\sigma_\eta\, dt \tag{5.3.8}$$

(cf. Chapter IV of Chandrasekharan & Minakshisundaran [1]), where $\theta = (\theta_1, \cdots, \theta_n)$ and $\eta = (\eta_1, \cdots, \eta_n)$. (5.3.8) can be rewritten as

$$\frac{R\Gamma\left(\frac{n}{2}\right)}{2\pi^{\frac{n}{2}} D(e^{i\theta_1}, \cdots e^{i\theta_n})} \int_{-\infty}^{\infty} \cdots \int g(\theta + \xi)$$

$$\cdot \frac{H_\varphi(|\xi| R)}{|\xi|^{n-1}}\, d\xi_1 \cdots d\xi_n, \tag{5.3.9}$$

where $\xi = (\xi_1, \xi_2, \cdots, \xi_n)$ and $|\xi| = (\xi_1^2 + \cdots + \xi_n^2)^{1/2}$.

(5.3.9) can also be put into

$$\frac{\Gamma\left(\frac{n}{2}\right) R \cdot i^{\frac{n(n-1)}{2}}}{2\pi^{\frac{n}{2}} D(e^{i\theta_1}, \cdots e^{i\theta_n})(n-1)! \cdots 2!1!}$$

$$\cdot \sum_{i_0, i_1, \cdots, i_n} \int_{-\infty}^{\infty} \cdots \int \delta_{i_0, i_1, \cdots, i_{n-1}}^{0, 1, \cdots, n-1} p_r^{(i_0)}(\theta_1 + \xi_1) \cdots$$

$$\cdot P_r^{(i_n)}(\theta_n + \xi_n) \cdot \frac{H_\varphi(|\xi| R)}{|\xi|^{n-1}}\, d\xi_1 \cdots d\xi_n. \tag{5.3.10}$$

In the integral mentioned above, we differenciate p_r with respect to θ and then replace θ by $\theta + \xi$. In fact, if we differentiate p_r with respect to ξ and replace ξ by $\theta + \xi$, we would obtain the same result.

Apply integration by parts to cach of the preceding integrals and let H_φ satisfy the following property

$$\left.\frac{\partial^{j_1}}{\partial\xi_1^{j_1}}\cdots\frac{\partial^{j_n}}{\partial\xi_n^{j_n}}\frac{H_\varphi(|\xi|R)}{|\xi|^{n-1}}\right|_{|\xi|=\infty}=0, \tag{5.3.11}$$

where $0\leqslant j_1,\cdots,j_n\leqslant n-1$. Then (5.3.10) becomes

$$\frac{\Gamma\left(\frac{n}{2}\right)i^{\frac{n(n-1)}{2}}R}{2\pi^{\frac{n}{2}}D(e^{i\theta_1},\cdots e^{i\theta_n})(n-1)!\cdots 2!1!}\cdot\int_{-\infty}^{\infty}\cdots\int p_r(\theta_1+\xi_1)\cdots p_r(\theta_n+\xi_n)$$

$$\cdot D\left(\frac{\partial}{\partial\xi_1},\cdots\frac{\partial}{\partial\xi_n}\right)\frac{H_\varphi(|\xi|R)}{|\xi|^{n-1}}d\xi_1\cdots d\xi_n, \tag{5.3.12}$$

where

$$D\left(\frac{\partial}{\partial\xi_1},\cdots\frac{\partial}{\partial\xi_n}\right)=(-1)^{\frac{n(n-1)}{2}}\begin{vmatrix}1, & 1, & \cdots & 1,\\ \frac{\partial}{\partial\xi_1}, & \frac{\partial}{\partial\xi_2}, & \cdots & \frac{\partial}{\partial\xi_n}\\ \cdots & \cdots & \cdots & \cdots\\ \frac{\partial^{n-1}}{\partial\xi_1^{n-1}}, & \frac{\partial^{n-1}}{\partial\xi_2^{n-1}}, & \cdots & \frac{\partial^{n-1}}{\partial\xi_n^{n-1}}\end{vmatrix}.$$

Write

$$s_R^\phi(r,U)=\sum_{\substack{f_1\geqslant\cdots\geqslant f_n\\ m=l_1^2+\cdots+l_n^2<R^2}}\phi\left(\frac{\sqrt{m}}{R}\right)r^{|l_1|+\cdots+|l_n|}\mathrm{tr}(C_{f_1\cdots f_n}\Phi'_{f_1\cdots f_n}(U)).$$

Then $s_R^\phi(r,U)$ is equal to

$$\frac{1}{C}\int_{U_n}u(WU)F(r,R,\overline{W})\dot{W}, \tag{5.3.13}$$

where $F(r,R,\overline{W})$ is defined by (5.3.8).

As $F(r,R,\overline{W})$ is a symmetric function of the characteristic roots of $\overline{W}$, we can symmetrize u with respect to characteristic roots of $\overline{W}$ and, as before, we obtain another function $u^*(WU)$. It can be shown that (5.3.13) is also equal to

$$\frac{1}{C}\int_{U_n}u^*(WU)F(r,R,\overline{W})\dot{W}.$$

Let

$$\phi_U(e^{-i\theta_1},\cdots e^{-i\theta_n})=\frac{1}{w_n'}\int_{[\dot{W}]}u^*(WU)[\dot{W}].$$

Obviously, ϕ_U is a symmetric function of $\theta_1,\cdots,\theta_n$, thus we obtain

$$s_R^\phi(r,U)=\frac{1}{(2\pi)^n}\int\cdots\int_{\pi\geqslant\theta_1\geqslant\cdots\geqslant\theta_n\geqslant-\pi}\phi_U(e^{-i\theta_1},\cdots e^{-i\theta_n})$$

$$\cdot F(r,R,\overline{W})|D(e^{i\theta_1},\cdots e^{i\theta_n})|^2d\theta_1\cdots d\theta_n. \tag{5.3.14}$$

Inserting (5.3.12) into (5.3.14) and applying the Fubini Theorem, we get

$$s_R^{\Phi}(r,U)=\frac{i^{\frac{n(n-1)}{2}}R\Gamma\left(\frac{n}{2}\right)}{2\pi^{\frac{n}{2}}(2\pi)^n(n-1)!\cdots 2!1!}$$

$$\cdot\int_{-\infty}^{\infty}\cdots\int_{-\infty}^{\infty}\int\cdots\int_{\pi\geqslant\theta_1\geqslant\cdots\geqslant\theta_n\geqslant-\pi}\phi_U(e^{-i\theta_1},\cdots e^{-i\theta_n})$$

$$\cdot D(e^{-i\theta_1},\cdots,e^{-i\theta_n})p_r(\theta_1+\xi_1)\cdots P_r(\theta_n+\xi_n)$$

$$\cdot D\left(\frac{\partial}{\partial\xi_1},\cdots\frac{\partial}{\partial\xi_n}\right)\frac{H_\Phi(|\xi|R)}{|\xi|^{n-1}}d\theta_1\cdots d\theta_n d\xi_1\cdots d\xi_n. \tag{5.3.15}$$

In the preceding integral, varied permutations of the order of θ being made and then ξ being endowed with the corresponding permutation, it follows that

$$s_R^{\Phi}(r,U)=\frac{i^{\frac{n(n-1)}{2}}R\Gamma\left(\frac{n}{2}\right)}{2\pi^{\frac{n}{2}}n!(n-1)!\cdots 2!1!(2\pi)^n}$$

$$\cdot\int_{-\infty}^{\infty}\cdots\int_{-\infty}^{\infty}\int_{-\pi}^{\pi}\cdots\int\phi_U(e^{-i\theta_1},\cdots e^{-i\theta_n})$$

$$\cdot D(e^{-i\theta_1},\cdots e^{-i\theta_n})p_r(\theta_1+\xi_1)\cdots p_r(\theta_n+\xi_n)$$

$$\cdot D\left(\frac{\partial}{\partial\xi_1},\cdots\frac{\partial}{\partial\xi_n}\right)\frac{H_\Phi(|\xi|R)}{|\xi|^{n-1}}d\theta_1\cdots d\theta_n d\xi_1\cdots d\xi_n. \tag{5.3.16}$$

From Lebesgue's subordinate theorem and Zygmund's Abel summability theorem for multiple Fourier series (cf. Chandrasekharan-Minakshisundaran [1]), it follows that (5.3.16) tends to

$$\frac{\Gamma\left(\frac{n}{2}\right)i^{\frac{n(n-1)}{2}}R}{n!(n-1)!\cdots 2!2\pi^{\frac{n}{2}}}$$

$$\cdot\int_{-\infty}^{\infty}\cdots\int_{-\infty}^{\infty}D(e^{i\xi_1},\cdots e^{i\xi_n})\phi_U(e^{i\xi_1},\cdots e^{i\xi_n})$$

$$\cdot D\left(\frac{\partial}{\partial\xi_1},\cdots\frac{\partial}{\partial\xi_n}\right)\frac{H_\Phi(|\xi|R)}{|\xi|^{n-1}}d\xi_1\cdots d\xi_n, \tag{5.3.17}$$

as $r\to 1$. Thus (5.3.17) also can be expressed by

$$\frac{\Gamma\left(\frac{n}{2}\right)i^{\frac{n(n-1)}{2}}R}{2\pi^{\frac{n}{2}}(n-1)!\cdots 2!}\int\cdots\int_{\infty>\xi_1\geqslant\xi_2\geqslant\cdots\geqslant-\infty}D\left(\frac{\partial}{\partial\xi_1},\cdots\frac{\partial}{\partial\xi_n}\right)\frac{H_\Phi(|\xi|R)}{|\xi|^{n-1}}$$

$$\cdot D(e^{i\xi_1},\cdots e^{i\xi_n})\phi_U(e^{i\xi_1},\cdots e^{i\xi_n})d\xi_1\cdots d\xi_n. \tag{5.3.18}$$

Finally we get

Theorem 5.3.1 *Let $u(U)$ be integrable on U_n. Then a certain kind of spherical means* (5.2.1) *of its Fourier series can be expressed by*

$$\frac{\Gamma\left(\frac{n}{2}\right)\cdot R}{2\pi^{\frac{n}{2}}(n-1)!\cdots 2!}\int\cdots\int_{\infty>\xi_1\geqslant\cdots\geqslant\xi_n>-\infty} D(\lambda_1,\cdots\lambda_n)\phi_U(\lambda_1\cdots\lambda_n)$$
$$\cdot D\left(\frac{\partial}{\partial\xi_1},\cdots\frac{\partial}{d\xi_n}\right)\frac{H_\Phi(|\xi|R)}{|\xi|^{n-1}}\,d\xi_1\cdots d\xi_n, \tag{5.3.19}$$

where

$$\phi_U(\lambda_1,\cdots\lambda_n)=\frac{1}{w_n'}\int_{[W]}u(WU)[\dot{W}],$$

$\lambda_1,\cdots,\lambda_n$ *with* $\lambda_j=e^{i\xi_j}$ *are characteristic roots of* W, $\Phi(t)$ *satisfies* (5.3.4) *and* (5.3.11), *and is absolutely continuous on every finite interval.*

(5.3.19) can also be expressed by

$$\frac{i^{\frac{n(n-1)}{2}}R}{n!(n-1)!\cdots 2!1!}\int_0^\infty h_U(t)t^{n-1}dt. \tag{5.3.20}$$

where

$$h_U(t)=\frac{\Gamma\left(\frac{n}{2}\right)}{2\pi^{\frac{n}{2}}}\int_\sigma D(e^{it\eta_1},\cdots e^{it\eta_n})\phi_U(e^{it\eta_1},\cdots e^{it\eta_n})$$
$$\cdot D\left(\frac{\partial}{\partial\xi_1},\cdots\frac{\partial}{\partial\xi_n}\right)\frac{H_\Phi(|\xi|R)}{|\xi|^{n-1}}\bigg|_{\xi=t\eta}d\sigma_\eta. \tag{5.3.21}$$

σ denotes the sphere $\eta_1^2+\cdots+\eta_n^2=1$ (cf. Chapter IV in Chandrasekharan & Minakshisundaran [1]).

§ 5.4 Expressions for Riesz Means

From the results in the preceding section, it immediately follows that(Bochner [1], Chandrasekharan & Minakshisundaran [1])

(1) $$H_\Phi=H_R^\Phi(u)=\frac{(n-1)!}{2^{n-2}\left(\Gamma\left(\frac{n}{2}\right)\right)^2}\frac{v^{n-1}}{(1+v^2)^{\frac{n+1}{2}}}$$
$$\cdot\exp\left(\sqrt{\frac{(2n-1)\cdot n\cdot(n-1)}{6R^2}}\right) \tag{5.4.1}$$

for $\varphi(t)=e^{-t}$, i. e. Abel's summation;

(2) $$H_\Phi=H_R^\Phi(u)=\frac{u^{n-1}R^{-\frac{v^2}{4}}}{2^{n-1}\Gamma(n/2)}\exp\left(\frac{(2n-1)n(n-1)}{6R^2}\right) \tag{5.4.2}$$

for $\varphi(t)=e^{-t^2}$, i.e. Gauss-Sommerfeld's summation;

(3)
$$H_t = H_R^\delta(u) = \frac{2^{\delta-\frac{n}{2}+1}\Gamma(\delta+1)}{\Gamma\left(\frac{n}{2}\right)} u^{n-1}V_{\delta+\frac{n}{2}}(u) \cdot\left(1-\frac{(2n-1)n(n-1)}{6R^2}\right)^{-\delta} \tag{5.4.3}$$

for $\varphi(t)=\begin{cases}(1-t^2)^\delta, & \text{if } 0\leqslant t<1,\\ 0, & \text{if } 1\leqslant t,\end{cases}$ i. e. the Riesz summation of order δ. Here

$$V_S(v)=\frac{J_S(v)}{v^S}$$

and J_S is the Bessel function of the first kind of order s.

Obviously, all these three functions $\varphi(t)$ are absolutely continuous on any finite interval and all satisfy the condition (5.3.4), i. e.

$$\int_0^\infty |\varphi(t)|t^{\frac{n-1}{2}}dt<\infty$$

$\left(\text{when}\,\delta>\frac{n^2-1}{2},\ \varphi(t)\text{ in the Riesz summability satisfies (5.3.4)}\right)$.

The conditions (5.3.11) corresponding to these three cases are as follows,

(1) when $|\xi|=\infty$,

$$\frac{\partial^{j_1}}{\partial\xi_1^{j_1}}\cdots\frac{\partial^{j_n}}{\partial\xi_n^{j_n}}\cdot\frac{1}{(1+|\xi|^2R^2)^{\frac{n+1}{2}}}$$

vanish for $0\leqslant j_1,\cdots,j_n\leqslant n-1$.

(2) When $|\xi|=\infty$,

$$\frac{\partial^{j_1}}{\partial\xi_1^{j_1}}\cdots\frac{\partial^{j_n}}{\partial\xi_n^{j_n}}e^{-\frac{|\xi|^2R^2}{4}}$$

vanish for $0\leqslant j_1,\cdots,j_n\leqslant n-1$.

(3) When $|\xi|=\infty$,

$$\frac{\partial^{j_1}}{\partial\xi_1^{j_1}}\cdots\frac{\partial^{j_n}}{\partial\xi_n^{j_n}}V_{\delta+\frac{n}{2}}(|\xi|R)$$

vanish for $0\leqslant j_1,\cdots,j_n\leqslant n-1$.

Here (1) and (2) are obvious. As to (3), noticing that

$$\frac{d}{xdx}V_\mu(x)=-V_{\mu+1}(x),$$

and

$$J_\mu(t) = o(t^{-\frac{1}{2}}),$$

as $t \to \infty$, we confirm (5.3.11).

Therefore, the Abel-means (5.2.2) of the Fourier series of $u(U)$ can be expressed by

$$\frac{i^{\frac{n(n-1)}{2}} R^n \exp\left(\sqrt{\frac{(2n-1)n(n+1)}{6R^2}}\right)}{2\pi^{\frac{n}{2}}(n-2)!\cdots 2!1!2^{n-2}\Gamma\left(\frac{n}{2}\right)}$$

$$\cdot \int\cdots\int_{\infty>\xi_1\geqslant\cdots\geqslant\xi_n>-\infty} D\left(\frac{\partial}{\partial\xi_1},\cdots\frac{\partial}{\partial\xi_n}\right)\cdot\frac{1}{(1+|\xi|^2R^2)^{\frac{n+1}{2}}}$$

$$\cdot D(e^{i\xi_1},\cdots e^{i\xi_n})\phi_U(e^{i\xi_1},\cdots e^{i\xi_n})d\xi_1\cdots d\xi_n, \tag{5.4.4}$$

which also can be expressed by (5.3.20). Here $h_U(t)$ is equal to

$$\exp\left(\sqrt{\frac{(2n-1)n(n+1)}{6R^2}}\right)\frac{(n-1)!\,R^{n-1}}{2^{n-1}\,\pi^{\frac{n}{2}}\Gamma\left(\frac{n}{2}\right)}$$

$$\cdot\int_\sigma D(e^{it\eta_1},\cdots e^{it\eta_n})\phi_U(e^{it\eta_1},\cdots e^{it\eta_n})$$

$$\cdot D\left(\frac{\partial}{\partial\xi_1},\cdots,\frac{\partial}{\partial\xi_n}\right)\frac{1}{(1+R^2|\xi|^2)^{\frac{n+1}{2}}}\bigg|_{\xi=t\eta} d\sigma_\eta. \tag{5.4.5}$$

The Gauss-Sommerfeld mean (5.2.3) of the Fourier series of $u(U)$ can be expressed by

$$\frac{i^{\frac{n(n-1)}{2}} R^n \exp\left(\frac{(2n-1)n(n+1)}{6R^2}\right)}{\pi^{\frac{n}{2}}(n-1)!\cdots 2!1!2^n}$$

$$\int\cdots\int_{\infty>\xi_1\geqslant\cdots\geqslant\xi_n>-\infty} D\left(\frac{\partial}{\partial\xi_1},\cdots\frac{\partial}{\partial\xi_n}\right)e^{-\frac{|\xi|^2R}{4}}D(e^{i\xi_1},\cdots e^{i\xi_n})$$

$$\cdot\phi_U(e^{i\xi_1},\cdots,e^{i\xi_n})d\xi_1\cdots d\xi_n. \tag{5.4.6}$$

Since

$$\frac{\partial}{\partial\xi_i}e^{-\frac{R^2|\xi|^2}{4}} = -\frac{R^2\xi_i}{2} - e^{-\frac{R^2|\xi|^2}{4}},$$

$$\frac{\partial^2}{\partial\xi_i^2}e^{-\frac{R^2|\xi|^2}{4}} = \left(-\frac{R^2}{2}+\frac{R^4}{4}\xi_i^2\right)e^{-\frac{R^2|\xi|^2}{4}},$$

....

from the properties of determinants, it yields that

$$D\left(\frac{\partial}{\partial \xi_1}, \cdots \frac{\partial}{\partial \xi_n}\right)e^{-\frac{R^2|\xi|^2}{4}} = \left(-\frac{R^2}{2}\right)^{\frac{n(n-1)}{2}} D(\xi_1, \cdots \xi_n)e^{-\frac{R^2|\xi|}{4}}$$

Thus (5.2.3) can be put into

$$\frac{(-i)^{\frac{n(n-1)}{2}} R^{n^2} \exp\left(\frac{(2n-1)n(n-1)}{6R^2}\right)}{\pi^{\frac{n}{2}}(n-1)!\cdots 2!1!2^{\frac{n(n+1)}{2}}}$$

$$\cdot \int_{\infty > \xi_1 \geqslant \cdots \geqslant \xi_n > -\infty} \cdots \int e^{-\frac{R^2|\xi|^2}{4}} D(\xi_1, \cdots \xi_n) \cdot D(e^{i\xi_1}, \cdots e^{i\xi_n})$$

$$\cdot \psi_U(e^{i\xi_1}, \cdots e^{i\xi_n}) d\xi_1 \cdots d\xi_n, \tag{5.4.7}$$

which can also be expressed by (5.3.20) with

$$\frac{\exp\left(\frac{(2n-1)n(n-1)}{6R^2}\right)(-t)^{\frac{n(n-1)}{2}} R^{n^2-1}}{2^{\frac{n(n+1)}{2}} \pi^{\frac{n}{2}} e^{\frac{R^2t^2}{4}}}$$

$$\cdot \int_\sigma D(e^{it\eta_1}, \cdots e^{it\eta_n}) D(\eta_1 \cdots \eta_n) \psi_U(e^{it\eta_1} \cdots e^{it\eta_n}) d\sigma_\eta. \tag{5.4.8}$$

When $\delta > \frac{n^2-1}{2}$, the Riesz mean of order δ of the Fourier series of $u(U)$ can be written in the form

$$\frac{\left(1 - \frac{(2n-1)n(n+1)}{6R^2}\right)^{-\delta} \cdot R^n \cdot 2^{\delta - \frac{n}{2}} \Gamma(\delta+1)}{\pi^{\frac{n}{2}}(n-1)!\cdots 2!1!}$$

$$\cdot \int_{\infty > \xi_1 \geqslant \cdots \geqslant \xi_n > -\infty} \cdots \int D\left(\frac{\partial}{\partial \xi_1}, \cdots \frac{\partial}{\partial \xi_n}\right) V_{\delta + \frac{n}{2}}(|\xi| R)$$

$$\cdot D(e^{i\xi_1}, \cdots e^{i\xi_n}) \psi_U(e^{i\xi_1}, \cdots e^{i\xi_n}) d\xi_1 \cdots d\xi_n, \tag{5.4.9}$$

which can also be denoted by (5.3.20) with

$$h_U(t) = \frac{\left(1 - \frac{(2n-1)n(n+1)}{6R^2}\right)^{-\delta} \cdot 2^{\delta - \frac{n}{2}} \Gamma(\delta+1) R^{n-1}}{\pi^{\frac{n}{2}}}$$

$$\cdot \int_\sigma D(e^{it\eta_1}, \cdots e^{it\eta_n}) \psi_U(e^{it\eta_1}, \cdots e^{it\eta_n})$$

$$\cdot D\left(\frac{\partial}{\partial \xi_1}, \cdots \frac{\partial}{\partial \xi_n}\right) V_{\delta + \frac{n}{2}}(|\xi| R)|_{\xi = t\eta} d\sigma_\eta. \tag{5.4.10}$$

§ 5.5 Proof of Theorem 5.2.2

Theorem 5.5.1 *Let $\phi(t)$ be absolutely continuous on any finite interval and satisfy* (5.3.4) *and* (5.3.11). *Then*

$$\frac{\Gamma\left(\frac{n}{2}\right) i^{\frac{n(n-1)}{2}} \cdot R}{2\pi^{\frac{n}{2}}(n-1)!\cdots 2!1!} \int_{\infty>\xi_1\geqslant\cdots\geqslant\xi_n>-\infty}\cdots\int D\left(\frac{\partial}{\partial\xi_1},\cdots\frac{\partial}{\partial\xi_n}\right)\frac{H_\varphi(|\xi|R)}{|\xi|^{n-1}}$$

$$\cdot D(e^{i\xi_1},\cdots e^{i\xi_n})d\xi_1\cdots d\xi_n = 1, \tag{5.5.1}$$

for $R^2 \geqslant \dfrac{(2n-1)n(n+1)}{6}$.

Proof. The left side of (5.5.1) is

$$\frac{\Gamma\left(\frac{n}{2}\right)(-i)^{\frac{n(n-1)}{2}} \cdot R}{2\pi^{\frac{n}{2}}(n-1)!\cdots 2!1!}\int_{-\infty}^{\infty}\cdots\int_{-\infty}^{\infty}\frac{\partial^0}{\partial\xi_1^0}\frac{\partial^1}{\partial\xi_2^1}\cdots\frac{\partial^{n-1}}{\partial\xi_n^{n-1}}$$

$$\cdot\frac{H_\phi(|\xi|R)}{|\xi|^{n-1}}D(e^{i\xi_1},\cdots e^{i\xi_n})d\xi_1\cdots d\xi_n.$$

As H_φ satisfies (5.3.11), the preceding expression is

$$\frac{\Gamma\left(\frac{n}{2}\right) i^{\frac{n(n-1)}{2}} \cdot R}{2\pi^{\frac{n}{2}}(n-1)!\cdots 2!1!}\int_{-\infty}^{\infty}\cdots\int_{-\infty}^{\infty}\frac{H_\varphi(|\xi|R)}{|\xi|^{n-1}}$$

$$\cdot\frac{\partial^0}{\partial\xi_1^0}\cdots\frac{\partial^{n-1}}{\partial\xi_n^{n-1}}D(e^{i\xi_1},\cdots e^{i\xi_n})\cdot d\xi_1\cdots d\xi_n. \tag{5.5.2}$$

(5.5.2) can also be written as

$$R\int_0^\infty f_0(t)H_\varphi(tR)dt, \tag{5.5.3}$$

where

$$f_0(t) = \frac{i^{\frac{n(n-1)}{2}}\cdot\Gamma\left(\frac{n}{2}\right)}{(n-1)!\cdots 2!1!2\pi^{\frac{n}{2}}}\int_\sigma\frac{\partial^0}{\partial\xi_1^0}\frac{\partial^1}{\partial\xi_2^1}\cdots\frac{\partial^{n-1}}{\partial\xi_n^{n-1}}$$

$$\cdot D(e^{i\xi_1},\cdots e^{i\xi_n})|_{\xi=t\eta}d\sigma_\eta.$$

By $f(\theta_1,\cdots,\theta_n)$, we denote

$$\frac{i^{\frac{n(n-1)}{2}}}{(n-1)!\cdots 2!1!}\frac{\partial^0}{\partial\theta_1^0}\frac{\partial^1}{\partial\theta_2^1}\cdots\frac{\partial^{n-1}}{\partial\theta_n^{n-1}}D(e^{i\theta_1},\cdots e^{i\theta_n}),$$

whose multiple Fourier series is

$$\sum_{\nu_1,\cdots,\nu_n} b_{\nu_1,\cdots,\nu_n} e^{i(\nu_1\theta_1+\cdots+\nu_n\theta_n)}. \tag{5.5.4}$$

Evidently, if $\nu^2 \neq (2n-1)n(n+1)/6$, then $b_{\nu_1,\cdots,\nu_n}=0$. Since $\phi(t)$ satisfies (5.3.4), acturally (5.5.3) is the value of the spherical mean (in a sense of ϕ) of the multiple Fourier series (5.5.4) of $f(\theta_1, \cdots, \theta_n)$ at $\theta=0$ (cf. Bochner [1]). Namely, (5.5.3) is equal to the value of

$$\sum_{\substack{\nu=0\\ \nu^2=\nu_1^2+\cdots+\nu_n^2}}^{\infty} \sum_{\nu_1\cdots\nu_n} \phi\left(\frac{\nu}{R}\right) b_{\nu_1,\cdots\nu_n} e^{i(\nu_1\theta_1+\cdots+\nu_n\theta_n)} \tag{5.5.5}$$

at $\theta=0$. From

$$\phi(t)=\varphi(t)/\varphi\left(\sqrt{\frac{(2n-1)n(n-1)}{6R^2}}\right),$$

it follows that (5.5.1) holds for $R^2 \geqslant (2n-1)n(n+1)/6$.

we now proceed to prove Theorem 5.2.2. Set

$$\varphi_U(e^{-i\theta_1},\cdots,e^{-i\theta_n})=\frac{1}{\omega_n'}\int_{[W]}(u^*(WU)-u(V))[\dot{W}]. \tag{5.5.6}$$

From Theorem 5.5.1, we obtain the difference between $G_R(U)$, the Gauss-Sommerfeld mean (5.2.3) of the Fourier series of $u(U)$ and $u(U)$.

$$G_R(U)-u(U)=\frac{i^{\frac{n(n-1)}{2}}R}{n!(n-1)!\cdots 2!1!}\int_0^\infty h_U^{(2)}(t)t^{n-1}dt, \tag{5.5.7}$$

heve $h_U^{(2)}(t)$ is equal to

$$\exp\left(\frac{(2n-1)n(n-1)}{6R^2}\right)\frac{(-t)^{\frac{n(n-1)}{2}}R^{n^2-1}}{2^{\frac{n(n+1)}{2}}\pi^{\frac{n}{2}}e^{\frac{R^2t^2}{4}}}$$
$$\cdot\int_\sigma D(e^{it\eta_1},\cdots,e^{it\eta_n})D(\eta_1,\cdots\eta_n)$$
$$\cdot\,\phi_U(e^{it\eta_1},\cdots,e^{it\eta_n})d\sigma_\eta. \tag{5.5.8}$$

Partition the integral in (5.5.7) into three parts, i. e. $G_R(U)-u(U)$ is equal to

$$\frac{i^{\frac{n(n-1)}{2}}R}{n!(n-1)!\cdots 2!1!}\left(\int_0^{\frac{1}{R}}+\int_{\frac{1}{R}}^{\eta}+\int_\eta^\infty\right)=I_1+I_2+I_3,$$

where η is a constant greater than $1/R$.

First consider I_1. Let $Rt=u$. Then $h_U^{(2)}(t)$ is equal to

$$\exp\left(\frac{(2n-1)n(n+1)}{6R^2}\right)c_1u^{\frac{n(n+1)}{2}}\cdot R^{\frac{(n-1)(n+2)}{2}}\cdot e^{-\frac{u^2}{4}}$$

$$\cdot\int_\sigma D(e^{\frac{i\eta_1u}{R}},\cdots e^{\frac{i\eta_nu}{R}})\cdot D(\eta_1,\cdots\eta_n)\cdot\varphi_U(e^{\frac{i\eta_1u}{R}},\cdots e^{\frac{i\eta_nu}{R}})d\sigma_\eta,$$

where c_1 is an absolute constant. According to the definition of D and $0 \leqslant u \leqslant 1$ in I_1, we have

$$h_U^{(2)}(t) = o(u^{\frac{n(n-1)}{2}} \cdot R^{n-1} e^{-\frac{u}{4}}),$$

where the continuity of $u(U)$ is used. Thus

$$I_1 = o\left(\int_0^1 e^{-\frac{u^2}{4}} u^{\frac{(n+2)(n-1)}{2}} du\right) = o(1).$$

Now turn to I_2. Let

$$F(t) = \int_0^t |h_U^{(2)}(t)| t^{n-1} dt.$$

Then

$$F(t) \leqslant \frac{c_2 R^{n^2-1}}{e^{\frac{R^2 t^2}{4}}} \int \cdots \int_{\eta_1^2+\cdots+\eta_n^2 < t^2} |D(e^{i\eta_1}, \cdots, e^{i\eta_n}) D(\eta_1, \cdots \eta_n)$$

$$\cdot \varphi_U(e^{i\eta_1}, \cdots e^{i\eta_n})| d\eta_1 \cdots d\eta_n,$$

where c_2 is an absolute constant. From this, it follows that

$$F(t) = o(R^{n^2-1} \cdot e^{-\frac{R^2 t^2}{4}} t^{n^2}),$$

as $t \to 0$. Therefore

$$|I_2| \leqslant c_3 R \int_{\frac{1}{R}}^{\eta} dF(t) = c_3 R F(t) \Big|_{\frac{1}{R}}^{\eta}.$$

The right-hand side comes to

$$o(R^{n^2} e^{-\frac{R^2 \eta^2}{4}} \eta^{n^2}) + o(1).$$

Thus

$$I_2 = o(1).$$

Finally, for I_3 we have

$$|I_3| \leqslant c_4 \int_{\eta}^{\infty} R^{n^2} t^{\frac{n^2+n}{2}} e^{-\frac{R^2 t^2}{4}} dt = o(1),$$

where c_3 and c_4 are absolute constants. Thus Theorem 5.2.2 is proved. In fact, we do not use the properties of $u(U)$ in proving $I_3 = o(1)$. So we are led to the conclusion that the Gauss-Sommerfeld summability of the Fourier series of $u(U)$ is a local property.

§ 5.6 Proof of Theorem 5.2.3

Here we start from the definition of H_φ and (5.4.9). By the properties of the Bessel function (see Erdelyi [1] and Chandrasekharan-

Minakshisundaran [1]), we have

$$\left.\begin{aligned}
&\frac{\partial}{\partial\xi_j} V_{\delta+\frac{n}{2}}(|\xi|R) = -R^2\xi_j V_{\delta+\frac{n}{2}+1}(|\xi|R),\\
&\frac{\partial^2}{\partial\xi_j^2} V_{\delta+\frac{n}{2}}(|\xi|R) = -R^2 V_{\delta+\frac{n}{2}+1}(|\xi|R),\\
&\qquad + R^4\xi_j^2 V_{\delta+\frac{n}{2}+2}(|\xi|R),\\
&\frac{\partial^3}{\partial\xi_j^3} V_{\delta+\frac{n}{2}}(|\xi|R) = 3R^4\xi_j V_{\delta+\frac{n}{2}+2}(|\xi|R)\\
&\qquad -6R^6\xi_j^3\cdot V_{\delta+\frac{n}{2}+3}(|\xi|R),\\
&\frac{\partial^4}{\partial\xi_j^4} V_{\delta+\frac{n}{2}}(|\xi|R) = 3R^4 V_{\delta+\frac{n}{2}+2}(|\xi|R)\\
&\qquad - 6R^6\xi_j^2(|\xi|R) + R^8\xi_j^4 V_{\delta+\frac{n}{2}+4}(|\xi|R).
\end{aligned}\right\} \tag{5.6.1}$$

From the properties of determinants, it is readily proved that

$$D\left(\frac{\partial}{\partial\xi_1},\cdots\frac{\partial}{\partial\xi_n}\right)\frac{H_\phi(|\xi|R)}{|\xi|^{n-1}} = \frac{2^{\delta-\frac{n}{2}+1}\Gamma(\delta+1)}{\Gamma\left(\frac{n}{2}\right)}$$

$$\cdot(-1)^{\frac{n(n-1)}{2}}R^{n^2-1}\left(1-\frac{(2n-1)n(n+1)}{6R^2}\right)^{-\delta}$$

$$\cdot D(\xi_1,\cdots\xi_n)V_{\delta+\frac{n^2}{2}}(|\xi|R).$$

By Theorem 5.5.1 we obtain that the difference between the Riesz mean (5.2.4) (denoted by $s_R^\delta(U)$) of order δ of the Fourier series of $u(U)$ for $\delta > \dfrac{n^2-1}{2}$ and $u(U)$, i.e. $s_R(U) - u(U)$, is equal to

$$\frac{i^{\frac{n(n-1)}{2}}R}{n!(n-1)!\cdots 2!1!}\int_0^\infty h_U^{(3)}(t)t^{n-1}dt, \text{ where} \tag{5.6.2}$$

$$h_U^{(3)}(t) = (-1)^{\frac{1}{2}n(n-1)}\cdot 2^{\delta-\frac{n}{2}}\Gamma(\delta+1)\pi^{-\frac{n}{2}}\left(1-\frac{(2n-1)n(n+1)}{6R^2}\right)^{-\delta}$$

$$\cdot R^{n^2-1}t^{\frac{1}{2}n(n-1)}V_{\delta+\frac{n^2}{2}}(tR)\int_\sigma D(e^{it\eta_1},\cdots e^{it\eta_n})$$

$$\cdot\phi_U(e^{it\eta_1},\cdots e^{it\eta_n})D(\eta_1,\cdots\eta_n)d\sigma_\eta. \tag{5.6.3}$$

Partition the integral in (5.6.2) into three parts, i. e.

$$s_R^\delta(U) - u(U) = \frac{i^{\frac{1}{2}n(n-1)} \cdot R}{n!(n-1)!\cdots 2!1!}\left(\int_0^{\frac{1}{R}} + \int_{\frac{1}{R}}^{\eta} + \int_{\eta}^{\infty}\right) = I_1 + I_2 + I_3,$$

where η is a constant greater than $1/R$.

First consider I_1. Writing $Rt = u$, we get

$$I_1 = \frac{i^{\frac{1}{2}n(n-1)}}{n!(n-1)!\cdots 2!1!}\int_0^1 h_U^{(3)}(t)\frac{u^{n-1}}{R^{n-1}}\,du,$$

where $$h_U^{(3)}\left(\frac{u}{R}\right) = O(R^{-\frac{n(n-1)}{2}} R^{n-1} R^{\frac{n(n-1)}{2}} u^{\frac{n(n-1)}{2}}).$$

hence

$$I_1 = o(1).$$

Next, investigate I_2, in which $\frac{1}{R} \leqslant t \leqslant \eta$. By means of properties of the Bessel function, we have

$$V_{\delta+\frac{n^2}{2}}(tR) = O\left((Rt)^{-\delta-\frac{n^2}{2}-\frac{1}{2}}\right). \tag{5.6.4}$$

Thus

$$|h_U^{(3)}(t)| = \frac{\Gamma\left(\frac{n}{2}\right)}{2\pi^{\frac{n}{2}}}\int_\sigma |D(e^{it\eta_1},\cdots e^{it\eta_n})\varphi_U(e^{it\eta_1},\cdots e^{it\eta_n})|$$

$$\cdot\, d\sigma_\eta O(R^{-\delta+\frac{1}{2}n^2-\frac{1}{2}-1} t^{-\delta-\frac{n}{2}-\frac{1}{2}}). \tag{5.6.5}$$

Let

$$h_U^{(4)}(t) = \frac{\Gamma\left(\frac{n}{2}\right)}{2\pi^{\frac{n}{2}}}\int_\sigma |D(e^{it\eta_1},\cdots e^{it\eta_n})$$

$$\cdot\, \varphi_U(e^{it\eta_1},\cdots e^{it\eta_n})|\,d\sigma_\eta, \tag{5.6.6}$$

and

$$F(t) = \int_0^t h_U^{(4)}(t)t^{n-1}dt. \tag{5.6.7}$$

Then, when $t \to 0$, we have

$$F(t) = o(t^n t^{\frac{1}{2}n(n-1)}) = o(t^{\frac{1}{2}n(n+1)}). \tag{5.6.8}$$

From (5.6.2), (5.6.5), (5.6.6) and (5.6.7), it follows that

$$|I_2| = O\left(\int_{\frac{1}{R}}^{\eta} R^{-\delta+\frac{n^2}{2}-\frac{1}{2}} t^{-\delta-\frac{n}{2}-\frac{1}{2}}\,dF(t)\right).$$

Integrating by parts gives us

$$O\left(R^{-\delta-\frac{n^2}{2}-\frac{1}{2}} t^{-\delta-\frac{n}{2}-\frac{1}{2}} F(t) \Big|_{\frac{1}{R}}^{\eta} \right)$$

$$+ O\left(\int_{\frac{1}{R}}^{\eta} R^{-\delta+\frac{n}{2}-\frac{1}{2}} t^{-\delta+\frac{n^2}{2}-\frac{3}{2}} dt \right) = o(1).$$

Consequently

$$I_2 = o(1).$$

The estimate of I_3 can be done by the same way as in the case of I_2, but here we have

$$F(t) = O(t^n). \tag{5.6.9}$$

As a result

$$I_3 = O(R^{-\delta+\frac{n^2-1}{2}}).$$

Thus Theorem 5.2.3 is proved. Notice that in the proof of $I_3 = o(1)$, we replace (5.6.8) by(5.6.9). Therefore we proved that, for $\delta > \frac{n^2-1}{2}$, the Riesz summability of order δ of the Fourier series of $u(U)$ is a local property.

§ 5.7 A General Convergence Theorem

If $\Phi(t)$ is absolutely continuous on a finite interval and satisfies (5.3.4) and (5.3.11), then the spherical mean (in a sence of ϕ) of the Fourier series of $u(U)$ can be put into (5.3.18) or (5.3.19).

In fact, this mean can also be written as (5.3.20). Write $s_R^{\phi}(U)$ to be (5.2.1). Then, by Theorem 5.5.1, $s_R^{\phi}(U) - u(U)$ can be put into

$$\frac{i^{\frac{1}{2}n(n-1)}R}{n!(n-1)!\cdots 2!1!} \int_0^{\infty} h_U^{(0)}(t) t^{n-1} dt, \tag{5.7.1}$$

where

$$h_U^{(0)} = \frac{\Gamma\left(\frac{n}{2}\right)}{2\pi^{\frac{n}{2}}} \int_{\sigma} D(e^{it\eta_1}, \cdots e^{it\eta_n}) \varphi_U(e^{it\eta_1}, \cdots e^{it\eta_n})$$

$$\cdot D\left(\frac{\partial}{\partial \xi_1}, \cdots \frac{\partial}{\partial \xi_n}\right) \frac{H_{\phi}(|\xi| R)}{|\xi|^{n-1}} \Big|_{\xi = t\eta} d\sigma_{\eta}$$

and φ_U is defined by (5.5.6).

Divide the integral in (5.7.1) into three parts

$$\frac{i^{\frac{1}{2}n(n-1)} \cdot R}{n!(n-1)!\cdots 2!1!} \left(\int_0^{\frac{1}{R}} + \int_{\frac{1}{R}}^{\eta} + \int_{\eta}^{\infty} \right) = I_1 + I_2 + I_3,$$

where η is a fixed constant greater than $1/R$.

First we are ready to investigate I_1. Making substitution of $Rt = u$, we get

$$I_1 = \frac{i^{\frac{n(n-1)}{2}}}{n!(n-1)!\cdots 2!1!}\int_0^1 h_U^{(0)}\left(\frac{u}{R}\right)\frac{u^{n-1}}{R^{n-1}}du.$$

This would lead to $I_1 = o(1)$, if

$$h_U^{(0)}\left(\frac{u}{R}\right) = o(R^{n-1}). \tag{5.7.2}$$

However

$$h_U^{(0)}\left(\frac{u}{R}\right) = O\left(R^{-\frac{n(n-1)}{2}}D\left(\frac{\partial}{\partial\xi_1},\cdots\frac{\partial}{\partial\xi_n}\right)\right)\frac{H_\phi(|\xi|R)}{|\xi|^{n-1}}\bigg|_{\xi=\frac{u\eta}{R}}.$$

Thus (5.7.2) is valid provided

$$D\left(\frac{\partial}{\partial\xi_1}\cdots\frac{\partial}{\partial\xi_n}\right)\frac{H_\phi(|\xi|R)}{|\xi|^{n-1}}\bigg|_{\xi=\frac{u\eta}{R}}$$
$$= O(R^{\frac{1}{2}(n-1)(n+2)}). \tag{5.7.3}$$

Next we consider I_2. If

$$\left|D\left(\frac{\partial}{\partial\xi_1},\cdots\frac{\partial}{\partial\xi_n}\right)\frac{H_\phi(|\xi|R)}{(R|\xi|)^{n-1}}\right|\bigg|_{\xi=t\eta}$$
$$= O(R^{-p-n}t^{-p-\frac{1}{2}n(n+1)}), \tag{5.7.4}$$

then

$$|I_2| = O\left(\int_{\frac{1}{R}}^{\eta} R^{-p}t^{-p-\frac{1}{2}n(n+1)}dF(t)\right).$$

By partial integration, we obtain, for $p > 0$,

$$|I_2| = O\left(R^{-p}t^{-p-\frac{1}{2}n(n+1)}F(t)\bigg|_{\frac{1}{R}}^{\eta} + \left(p + \frac{n(n+1)}{2}\right)\right.$$
$$\left.\cdot\int_{\frac{1}{R}}^{\eta} R^{-p}t^{-p-\frac{1}{2}n(n+1)-1}F(t)dt\right) = o(1).$$

We can deduce the estimate of I_3 by running along the same method as that used in proving I_2, but here we have

$$F(t) = o(t^n).$$

Thus, if (5.7.4) still holds valid for $\eta \leqslant t < \infty$, then

$$I_3 = o(1)$$

for $p > 0$.

To sum up, we get

Theorem 5.7.1 *Let $u(U)$ be continuous on U_n, the spherical mean* (5.2.1) *(in a sense of ϕ) of its Fourier series be denoted by $s_R^\phi(U)$. Suppose that*

$\phi(t)$ is absolutely continuous over any finite interval and satisfies (5.3.4) and (5.3.11) and that H_ϕ determined by $\phi(t)$ (cf. (5.3.6)) satisfies the following two conditions

(1) *When* $0 \leqslant t \leqslant \frac{1}{R}$,

$$D\left(\frac{\partial}{\partial \xi_1}, \cdots, \frac{\partial}{\partial \xi_n}\right) \frac{H_\phi(|\xi| R)}{|\xi|^{n-1}} \bigg|_{\mu = t\eta} = O(R^{\frac{(n-1)(n-2)}{2}}); \tag{5.7.5}$$

(2) *When* $\frac{1}{R} \leqslant t < \infty$,

$$D\left(\frac{\partial}{\partial \xi_1}, \cdots \frac{\partial}{\partial \xi_n}\right) \frac{H_\phi(|\xi| R)}{|\xi|^{n-1}} \bigg|_{\xi = t\eta} = O(R^{-p-1} t^{-p+\frac{1}{2}n(n+1)}) \quad (\text{for } p > 0). \tag{5.7.6}$$

Then $s_R^\phi(U)$ converges to $u(U)$ as $R \to \infty$.

Thus it can be seen that, if $\phi(t)$ satisfies (5.3.4), (5.3.11), (5.7.5) and (5.7.6), the spherical summability (in a sense of ϕ) of the Fourier series is a local property.

Since

$$\frac{\partial}{\partial \eta_i} \frac{1}{(1 + R^2|\eta|^2)^{\frac{1}{2}(n+1)}} = \frac{-(n+1)R^2\eta_i}{(1 + R^2|\eta|^2)^{\frac{1}{2}(n+3)}},$$

$$\frac{\partial^2}{\partial \eta_i^2} \frac{1}{(1 + R^2|\eta|^2)^{\frac{1}{2}(n+1)}} = \frac{-(n+1) \cdot R^2}{(1 + R^2|\eta|^2)^{\frac{1}{2}(n+3)}} + \frac{(n+1)(n+3)R^4\eta_i^2}{(1 + R^2|\eta|^2)^{\frac{1}{2}(n+5)}},$$

$$\cdots\cdots\cdots\cdots\cdots,$$

theorem 5.2.1 immediately follows from Theorem 5.7.1. As a result, all conditions in Theorem 5.7.1 are satisfied and here we have $p = -\frac{n(n+1)}{2}$.

§ 5.8 A Tauber-Type Convergence Theorem

In the preceding chapter, by means of "cubical" summation, some approximation theorems are deduced. And obviously, the corresponding results are also valid for spherical summation. For example, we have (cf. Chapter 1 of this book)

Theorem 5.8.1 *If* $u(U) \in \mathrm{Lip}, \ p$, *then*

$$s_R^\delta(U) - u(U) = o(R^{-p}), \tag{5.8.1}$$

when $\delta > \frac{n^2 - 1}{2} + p$.

The theorem above can be easily deduced from the proof of Theorem 5.2.3.

On the basis of this result, the following convergence theorem can be proved.

Theorem 5.8.2 *If* $u(U) \in \mathrm{Lip}\ p$ *and*

$$N(f) \ \mathrm{tr}\ (C_{f_1,\cdots,f_n} \bar{C}'_{f_1,\cdots,f_n}) = o((l_1^2 + \cdots + l_n^2)^{-\alpha}), \tag{5.8.2}$$

where

$$\alpha = n - \frac{2p(1+\beta)}{n^2 - 1 - 2\beta + 2p + \varepsilon}, \ \beta \geqslant 0, \ \varepsilon > 0,$$

then

$$s_R^\beta(U) - u(U) = o(1). \tag{5.8.3}$$

Proof. By (5.8.1), we obtain that, when $\delta > \frac{n^2 - 1}{2} + p$,

$$s_R^\delta(U) - u(U) = O(R^{-p}).$$

when $t = O(R)$, we have

$$|S((R+t)^{\frac{1}{2}}) - S(R^{\frac{1}{2}})| \leqslant \sum_{R \leqslant m \leqslant R+t} |A_m(U)|,$$

where $s(R) \equiv s_R^0(U) = s_R(U)$,

$$A_m(U) = \sum_{\substack{f_1 \geqslant \cdots \geqslant f_n \\ m = l_1^2 + \cdots + l_n^2}} \mathrm{tr}\ (C_{f_1 \cdots f_n} \Phi'_{f_1 \cdots f_n}(U)).$$

By the Schwarz inequality, we get

$$|\mathrm{tr}(C_{f_1 \cdots f_n} \Phi'_{f_1 \cdots f_n}(U))|^2 \leqslant \mathrm{tr}(C_{f_1 \cdots f_n} \bar{C}'_{f_1 \cdots f_n})$$
$$\cdot\ \mathrm{tr}(\Phi_{f_1 \cdots f_n} \overline{\Phi_{f_1 \cdots f_n}(U)'}) = \frac{N(f)}{C} \mathrm{tr}(C_{f_1 \cdots f_n} \bar{C}'_{f_1 \cdots f_n}).$$

In virtue of (5.8.2), we have (cf. Chapter IV in Chandrasekharan & Minakshisundaran [1])

$$|s((R+t)^{\frac{1}{2}}) - s(R^{\frac{1}{2}})| = o(\Sigma(l_1^2 + \cdots + l_n^2)^{-\frac{\alpha}{2}})$$

$$= o\left(\sum_{R\leqslant m\leqslant R+t} r_n(m)m^{-\frac{\alpha}{2}}\right) = o\left(\left(\int_R^{R+t} x^{-\frac{\alpha}{2}} dR_n(x)\right)\right)$$

$$= o(tR^{\frac{n}{2}-\frac{\alpha}{2}-1}),$$

where $r_n(m)$ denotes the number of the integral points on the sphere $x_1^2 + \cdots + x_n^2 = m$. As we know (cf. Chandrasekharan-Minakshisundaran[1])

$$r_n(m) = O(m^{\frac{n}{2}-1+\eta}), \text{ for } \eta > 0,$$

and

$$R_n(t) = \sum_{m\leqslant x} r_n(m) = \pi^{\frac{n}{2}}\left\{\Gamma\left(\frac{n}{2}+1\right)\right\}^{-1} x^{\frac{n}{2}} + o(x^{\frac{n}{2}-\frac{n}{m+1}}),$$

applying Theorem 1.8.1 of Chandrasekharan-Minakshisundaran[1] leads to the conclusion

$$s_R^\delta(U) - u(U) = o(R^{\frac{1}{\delta+1}\left\{-\delta+\beta+(\delta-\beta)\left(\frac{n}{2}-\frac{\alpha}{2}\right)-p(1+\beta)\right\}}),$$

for $0 \leqslant \beta < \delta$. Taking

$$\alpha = n - \frac{p(1+\beta)}{\delta-\beta},$$

we get (5.8.3) for $\delta > (n^2-1)/2 + p$. Setting, in particular,

$$\delta = (n^2-1)/2 + p + \varepsilon, \quad (\varepsilon > 0)$$

we obtain the desired result.

From the above theorem, the following result follows immediately and may be regarded as a supplement to Theorem 5.2.3.

Theorem 5.8.3 *If* $u(U) \in \mathrm{Lip}\, p$ *and*

$$N(f) \text{ tr } (C_{f_1,\dots,f_n}\overline{C}'_{f_1,\dots,f_n}) = o(1), \tag{5.8.4}$$

then, when

$$\frac{n^2-1}{2} - \frac{p(n-1)^2}{2(n+p)} < \delta \leqslant \frac{n^2-1}{2},$$

$s_R^\delta(U)$ *converges to* $u(U)$ *as* $R \to \infty$.

Setting $\beta = 0$ in Theorem 5.8.2, we obtain a convergence Theorem.

Theorem 5.8.4 *If* $u(U) \in \mathrm{Lip}\, p$ *and*

$$N(f)\mathrm{tr}(C_{j_1,\dots,j_n}\overline{C}'_{f_1,\dots,f_n}) = o((l_1^2 + \cdots + l_n^2)^{-\alpha}), \tag{5.8.5}$$

then, when

$$\alpha = n - \frac{2p}{n^2-1+2p+\varepsilon}, \quad (\varepsilon > 0),$$

the Fourier series of $u(U)$ *converges to* $u(U)$ *itself.*

Certainly, the condition "$u(U)$ is continuous on U_n" in all theorems can be relaxed. Besides, based on the method used in this book, the theory of the spherical summation of Fourier series could continuously be developed. For example, there are no special difficulties at present either to establish the concept of absolute summability or to find the sufficient and necessary conditions of the summability. For all these, readers are referred to Bochner [1] or Chandrasekharan-Minakshisundaran[1].

PART II

HARMONIC ANALYSIS ON ROTATION GROUPS

Chapter 6. Abel Summation of Fourier Series on Rotation Groups

§ 6.1 Harmonic Analysis on Rotation Groups

It is well known that any complex representation of a compact Lie group is equivalent to a unitary representation. Therefore, any compact Lie group can be imbedded in a unitary group. Thus, the studies of harmonic analysis on unitary groups are of interest to the studies of general compact Lie groups. Similarly, any real representation of a compact Lie group is equivalent to an orthogonal representation, so any compact Lie group can be imbedded in an orthogonal group. Therefore, the studies of harmonic analysis on orthogonal groups are of equal interest to the studies of general compact Lie groups.

In Part I, harmonic analysis on unitary groups has been discussed. In this part, naturally, harmonic analysis on orthogonal groups is to be examined. An orthogonal group can be divided into two parts, the determinants of a part of elements are $+1$ and the determinants of the other part are -1. Besides, the first part forms a normal subgroup of orthogonal group, i. e. a rotation group. In this part, we will discuss harmonic analysis on rotation groups, i. e. on the group

$$SO(n)=O^{+}(n)=\{\Gamma \mid \Gamma\Gamma'=I,\ \det\Gamma=1\}.$$

The essential idea and method for studying harmonic analysis on rotation groups coincide with those used in Part I. In this chapter, we begin with the Poisson kernel on real classical domains, then introduce Abel summation of Fourier series on rotation groups and find out the coefficients for Abel summation explicitly. Here, as the method used in Chapter 1 is ineffective for rotation groups, we apply the generating function method adopted by Zhong Jiaqing instead (see Zhong Jiaqing[2]).

In the following seventh—ninth chapters, the main results are essentially obtained in the light of the methods used in Chapters 2, 3,4 of Part

I and by overcoming the difficulties caused by rotation groups. Much of the work was done by my graduate students Wang Shikun and Dong Daozhen [1][2][3].

As is well known, the representatives for the conjugate class of any element Γ in $SO(n)$ have two possible cases in accordance with the order of n.

1. When $n=2k$, we have

$$\Gamma \sim \begin{pmatrix} c(\theta_1) & & \\ & \ddots & \\ & & c(\theta_k) \end{pmatrix}, \quad c(\theta_i) = \begin{pmatrix} \cos\theta_i, & \sin\theta_i \\ -\sin\theta_i, & \cos\theta_i \end{pmatrix}.$$

2. When $n=2k+1$, we have

$$\Gamma \sim \begin{pmatrix} c(\theta_1) & & & \\ & \ddots & & \\ & & c(\theta_k) & \\ & & & 1 \end{pmatrix}.$$

Any single-valued irreducible representation of $SO(n)$ is characterized by k nonnegative integers

$$m = (m_1, \cdots, m_k),$$

with $m_1 \geqslant \cdots \geqslant m_k \geqslant 0$, and m is called its signature. By $N(m)$ we denote the dimension of the irreducible representation of which the signature is $m=(m_1,\cdots,m_k)$, and by $\sigma_m(\Gamma)$ the character of the representation. For them there are two possible cases in accordance with the order of n.

1. $n=2k$.

When $m_k=0$, $(\varphi_{ij}^m(\Gamma))_{1\leqslant i,j\leqslant N(m)}$ denotes the single-valued irreducible representation with signature $m=(m_1, m_2, \cdots, m_{k-1}, 0)$. Let

$$c(l_1,\cdots,l_k) = \begin{vmatrix} c_{l_1}(\theta_1), \cdots, c_{l_1}(\theta_k) \\ \cdots\cdots\cdots\cdots \\ c_{l_k}(\theta_1), \cdots, c_{l_k}(\theta_k) \end{vmatrix}, \tag{6.1.1}$$

where $l_1=m_1+k-1$, $l_2=m_2+k-2, \cdots, l_k=m_k$ and

$$c_q(\theta) = \begin{cases} 1, & \text{if } q=0, \\ 2\cos q\theta, & \text{if } q \geqslant 1. \end{cases}$$

Then the character $\sigma_m(\Gamma)$ of above representation is

$$[m] = \frac{c(l_1,\cdots,l_k)}{c(k-1,\cdots,1,0)}, \tag{6.1.2}$$

and its dimension $N(m)$ is

$$\lim_{\substack{\theta_1\to 0\\ \cdots\\ \theta_k\to 0}}[m] = \frac{2^{k-1}}{(2k-2)!\cdots 4!2!}\begin{vmatrix} l_1^{2k-2}, & \cdots, & l_k^{2k-2} \\ \cdots & \cdots & \cdots \\ l_1^2, & \cdots, & l_k^2 \\ 1, & \cdots, & 1 \end{vmatrix}. \tag{6.1.3}$$

When $m_k > 0$, the single-valued irreducible representation with signature $m = (m_1, \cdots, m_k)$ has two irreducible adjoint representations, denoted by $(\phi_{ij}^m(\Gamma))$ and $(\psi_{ij}^m(\Gamma))$ respectively, and the characters of them are denoted by $\sigma_m^\phi(\Gamma)$ and $\sigma_m^\psi(\Gamma)$ respectively. Then $\sigma_m^\phi(\Gamma)$ is equal to

$$[m]_+ = \frac{c(l_1, \cdots, l_k) + s(l_1, \cdots, l_k)}{2c(k-1, \cdots, 1, 0)}, \tag{6.1.4}$$

and $\sigma_m^\psi(\Gamma)$ is equal to

$$[m]_- = \frac{c(l_1, \cdots, l_k) - s(l_1, \cdots, l_k)}{2c(k-1, \cdots, 1, 0)}, \tag{6.1.5}$$

where $c(l_1, \cdots, l_k)$ is defined by (6.1.1),

$$s(l_1, \cdots, l_k) = \begin{vmatrix} s_{l_1}(\theta_1), & \cdots, & s_{l_1}(\theta_k) \\ \cdots & \cdots & \cdots \\ \cdots & \cdots & \cdots \\ s_{l_k}(\theta_1), & \cdots, & s_{l_k}(\theta_k) \end{vmatrix}, \tag{6.1.6}$$

$l_1 = m_1 + k - 1$, $l_2 = m_2 + k - 2, \cdots, l_k = m_k$ and $s_q(\theta) = 2i\sin q\theta$. Besides, these two irreducible adjoint representations have the same dimension which is defined by (6.1.3).

2. $n = 2k + 1$.

By $(\varphi_{ij}^m(\Gamma))_{1\leqslant i,j\leqslant N(m)}$, we denote the single-valued irreducible representation with signature $m = (m_1, \cdots, m_k)$. Its character $\sigma_m(\Gamma)$ is

$$[m] = \frac{s(l_1 + 1/2, \cdots, l_k + 1/2)}{s(k - 1/2, \cdots, 3/2, 1/2)}, \tag{6.1.7}$$

and its dimension $N(m)$ amounts to

$$\lim_{\substack{\theta_1\to 0\\ \cdots\cdots\\ \theta_k\to 0}}[m] = \frac{2^k}{(2k-1)!\cdots 3!1!} \cdot \begin{vmatrix} \left(l_1 + \frac{1}{2}\right)^{2k-1}, & \cdots, & \left(l_k + \frac{1}{2}\right)^{2k-1} \\ \cdots & \cdots & \cdots \\ l_1 + \frac{1}{2}, & \cdots, & l_k + \frac{1}{2} \end{vmatrix}. \tag{6.1.8}$$

We know that the volume c of $SO(n)$ is

$$(8\pi)^{\frac{1}{4}n(n-1)} \prod_{\nu=1}^{n} \frac{\Gamma\left(\frac{1}{2}(\nu-1)\right)}{\Gamma(\nu-1)}. \tag{6.1.9}$$

Let $u(\Gamma)$ be an integrable function on $SO(2k)$ and

$$a_{ij}^{m} = \frac{1}{c}\int_{SO(2k)} u(\Gamma)\phi_{ij}^{m}(\Gamma)\dot{\Gamma}, \tag{6.1.10}$$

if $m_1 \geqslant \cdots \geqslant m_k \geqslant 0$;

$$b_{ij}^{m} = \frac{1}{c}\int_{SO(2k)} u(\Gamma)\psi_{ij}^{m}(\Gamma)\dot{\Gamma}, \tag{6.1.11}$$

if $m_1 \geqslant \cdots \geqslant m_k > 0$. Write

$$\Phi_{m_1,\cdots,m_k}(\Gamma) = (\phi_{ij}^{m}(\Gamma))_{1\leqslant i,j\leqslant N(m)},$$

$$\Psi_{m_1,\cdots,m_k}(\Gamma) = (\psi_{ij}^{m}(\Gamma))_{1\leqslant i,j\leqslant N(m)},$$

$$A_{m_1,\cdots,m_k} = (a_{ij}^{m})_{1\leqslant i,j\leqslant N(m)},$$

$$B_{m_1,\cdots,m_k} = (b_{ij}^{m})_{1\leqslant i,j\leqslant N(m)}.$$

Then

$$\sum_{m_1\geqslant\cdots\geqslant m_k\geqslant 0} \sum_{i,j}^{N(m)} a_{ij}^{m}\phi_{ij}^{m}(\Gamma) + \sum_{m_1\geqslant\cdots\geqslant m_k> 0} \sum_{i,j}^{N(m)} b_{ij}^{m}\psi_{ij}^{m}(\Gamma)$$

$$= \sum_{m_1\geqslant\cdots\geqslant m_k\geqslant 0} \operatorname{tr}(A_{m_1,\cdots,m_k}\Phi'_{m_1,\cdots,m_k}(\Gamma))$$

$$+ \sum_{m_1\geqslant\cdots\geqslant m_k> 0} \operatorname{tr}(B_{m_1,\cdots,m_k}\Psi'_{m_1,\cdots,m_k}(\Gamma)) \tag{6.1.12}$$

is called the Fourier series of tunction $u(\Gamma)$ on $SO(2k)$, and $A_{m_1,\cdots,m_k}$ and $B_{m_1,\cdots,m_k}$ are called its Fourier coefficients.

Let $u(\Gamma)$ be an integrable function on $SO(2k+1)$ and

$$a_{ij}^{m} = \frac{1}{c}\int_{SO(2k+1)} u(\Gamma)\varphi_{ij}^{m}(\Gamma)\dot{\Gamma}. \tag{6.1.13}$$

Similarly, we write

$$A_{m_1,\cdots,m_k} = (a_{ij}^{m})_{1\leqslant i,j\leqslant N(m)},$$

and

$$\Phi_{m_1,\cdots,m_k}(\Gamma) = (\varphi_{ij}^{m}(\Gamma))_{1\leqslant i,j\leqslant N(m)}.$$

We name

$$\sum_{m_1\geqslant\cdots\geqslant m_k\geqslant 0} \sum_{i,j}^{N(m)} a_{ij}^{m}\varphi_{ij}^{m}(\Gamma) = \sum_{m_1\geqslant\cdots\geqslant m_k\geqslant 0} \operatorname{tr}(A_{m_1,\cdots,m_k}\Phi'_{m_1,\cdots,m_k}(\Gamma)) \tag{6.1.14}$$

the Fourier series of $u(\Gamma)$ and $A_{m_1,\cdots,m_k}$ its Fourier coefficients.

Now we study the set consisting of all orthogonal matrices G such that

$$G = \Gamma M \Gamma', \tag{6.1.15}$$

where $\Gamma \in SO(n)$ and

$$M = \begin{pmatrix} \cos\theta_1, & \sin\theta_1 \\ -\sin\theta_1, & \cos\theta_1 \end{pmatrix} \dot{+} \begin{pmatrix} \cos\theta_2, & \sin\theta_2 \\ -\sin\theta_2, & \cos\theta_2 \end{pmatrix} \dot{+} \cdots,$$

$$\pi \geqslant \theta_1 \geqslant \theta_2 \geqslant \cdots \geqslant \theta_k \geqslant 0. \tag{6.1.16}$$

When $n = 2k$, this direct sum takes

$$\begin{pmatrix} \cos\theta_k, & \sin\theta_k \\ -\sin\theta_k, & \cos\theta_k \end{pmatrix}$$

as its end, and when $n = 2k + 1$, it ends at

$$\begin{pmatrix} \cos\theta_k, & \sin\theta_k \\ -\sin\theta_k, & \cos\theta_k \end{pmatrix} \dot{+} 1.$$

In addition, when n is odd, (6.1.15) includes all matrices of $SO(n)$. When n is even, any matrix which belongs to $SO(n)$ can be expressed in the form of (6.1.15). As det Γ may be ± 1, this implies that $V(G) = C/2$. Thus any point on G can be expressed uniquely by a coset of $\Sigma(= \Gamma/\Delta)$ and an M (except several submanifolds of lower dimensions). Here

$$\Delta = \begin{pmatrix} \cos\delta_1, & \sin\delta_1 \\ -\sin\delta_1, & \cos\delta_1 \end{pmatrix} \dot{+} \begin{pmatrix} \cos\delta_2, & \sin\delta_2 \\ -\sin\delta_2, & \cos\delta_2 \end{pmatrix} \dot{+} \cdots$$

and the volume element on G is

$$\dot{\Gamma}_{2k} = 2^{2k(k-1)+\frac{1}{2}k} \prod_{1 \leqslant \alpha < \beta \leqslant k} (\cos\theta_\alpha - \cos\theta_\beta)^2 \dot{\Sigma}\, d\theta_1 \cdots d\theta_k \tag{6.1.17}$$

for $n = 2k$, and

$$\dot{\Gamma}_{2k+1} = 2^{2k^2+\frac{1}{2}k} \prod_{i=1}^{k} (1 - \cos\theta_i)$$

$$\cdot \prod_{1 \leqslant \alpha < \beta \leqslant k} (\cos\theta_\alpha - \cos\theta_\beta)^2 \dot{\Sigma}\, d\theta_1 \cdots d\theta_k \tag{6.1.18}$$

for $n = 2k + 1$, where $\dot{\Sigma}$ is the volume element of the coset. When $n = 2k$, the volume $V(\Sigma)$ of the coset is equal to

$$2^{2k^2-2k} \pi^{k^2-\frac{3}{2}k} \prod_{\nu=2}^{2k} \frac{\Gamma\left(\frac{1}{2}(\nu-1)\right)}{\Gamma(\nu-1)};$$

when $n = 2k + 1$, the volume is equal to

$$2^{2k^2}\pi^{k^2-\frac{k}{2}}\prod_{\nu=2}^{2k+1}\frac{\Gamma\left(\frac{1}{2}(\nu-1)\right)}{\Gamma(\nu-1)}.$$

The volume elements also can be put into

$$\dot{\Gamma}_{2k} = 2^{k^2-\frac{1}{2}k}c^2(k-1,\cdots,1,0)\dot{\Sigma}d\theta_1\cdots d\theta_k; \tag{6.1.19}$$

$$\dot{\Gamma}_{2k+1} = 2^{k^2+\frac{1}{2}k}(-1)^k s^2\left(k-\frac{1}{2},\cdots,\frac{3}{2},\frac{1}{2}\right)\dot{\Sigma}d\theta_1\cdots d\theta_k. \tag{6.1.20}$$

§ 6.2 Poisson Kernels on Real Classical Domains

Assume that $X = (X_{\alpha a})$ is an $m \times n$ matrix. Consider the domain $\mathscr{R}(m, n)(m \leqslant n)$ in the Euclidean space R^{mn} of dimension mn. The domain $\mathscr{R}(m, n)$ is defined by

$$I - XX' > 0,$$

and its characteristic manifold is $I = XX'$. The group of holomorphic automorphisms on $\mathscr{R}(m, n)$ is

$$W = (AX + B)(CX + D)^{-1}, \tag{6.2.1}$$

where

$$A'A - C'C = I,\ B'B - D'D = -I,\ A'B = C'D \tag{6.2.2}$$

and

$$\det\begin{pmatrix} A & B \\ C & D \end{pmatrix} = 1.$$

We consider only the case $m = n$ and denote $\mathscr{R}(n, n)$ by $\mathscr{R}_n$. Here, as in § 1.1, it can be proved that (6.2.1) maps $\mathscr{R}_n$, $O(n)$ and $SO(n)$ onto $\mathscr{R}_n$, $O(n)$ and $SO(n)$ respectively. The transformation

$$Y = A(X - X_0)(I - X_0'X)^{-1}D^{-1} \tag{6.2.3}$$

changes X_0 to 0 if A and D satisfy

$$A'A = (I - X_0X_0')^{-1},\ D'D = (I - X_0'X)^{-1}. \tag{6.2.4}$$

If $\Gamma \in SO(n)$, then (6.2.3) on $SO(n)$ takes the form

$$\Gamma_1 = A(\Gamma - X_0)(I - X_0'\Gamma)^{-1}D^{-1}. \tag{6.2.5}$$

Set $\delta\Gamma = d\Gamma\Gamma'$, $\delta\Gamma_1 = d\Gamma_1\Gamma_1'$. Let $\dot{\Gamma}$ and $\dot{\Gamma}_1$ be the volume elements spanned by the differential vectors of $\delta\Gamma$ and $\delta\Gamma_1$ respectively. Then, as in § 1.1, it can be shown that

$$\dot{\Gamma}_1 = \frac{\det(I - X_0X_0')^{\frac{n-1}{2}}}{\det(I - X_0\Gamma')^{n-1}}\dot{\Gamma}. \tag{6.2.6}$$

As in § 1.1,

$$P(X_0, \Gamma) = \frac{\det (I - X_0 X_0')^{\frac{n-2}{2}}}{\det (I - X_0)^{n-1}}$$

is called the Poisson kernel of domain $\mathscr{R}_n$. Thus we may deduce the Poisson integral

$$u(X) = \frac{1}{C} \int_{SO(n)} P(X, \Gamma) u(\Gamma) \dot{\Gamma} \tag{6.2.7}$$

on $\mathscr{R}_n$: $I - XX' > 0$.

As stated in § 0.3, to $\mathscr{R}_n$ there should be the corresponding Riemann metric and the corresponding Laplace-Beltrami operator. Here the corresponding Laplace-Beltrami operator turns out (cf. Hua Luogeng and Lu Qikeng[1])

$$\Delta = \sum_{i,j=1}^{n} \sum_{\alpha,\beta=1}^{n} \left(\delta_{ij} - \sum_{\sigma=1}^{n} x_{i\sigma} x_{j\sigma}\right)\left(\delta_{\alpha\beta} - \sum_{k=1}^{n} x_{k\alpha} x_{k\beta}\right)$$

$$\cdot \frac{\partial^2}{\partial x_{i\alpha} \partial x_{j\beta}} - 2 \sum_{k,p=1}^{n} \sum_{r=1}^{n} x_{pr}\left(\delta_{kp} - \sum_{\sigma=1}^{n} x_{k\sigma} x_{p\sigma}\right) \frac{\partial}{\partial x_{kr}}. \tag{6.2.8}$$

Let $u(x)$ be defined on $\mathscr{R}_n$ and have continuous partial derivatives of order 2. $u(x)$ is said to be harmonic if

$$\Delta u(x) = 0 \tag{6.2.9}$$

holds in $\mathscr{R}_n$. As in the case of $\mathscr{R}_I$, for the following Dirichlet problem there exists a solution which is given by (6.2.7): Given a continuous function on $SO(n)$, there exists a unique solution satisfying (6.2.9) and its values on $SO(n)$ are equal to those of the given function.

We can prove the following as we did theorem 1.1.1 in § 1.1.

Theorem 6.2.1 *If $u(\Gamma_0)$ is a continuous function on $SO(n)$, then*

$$u(\Gamma_0) = \lim_{r \to 1} \frac{1}{C} \int_{SO(n)} P(r\Gamma_0, \Gamma) u(\Gamma) \dot{\Gamma}. \tag{6.2.10}$$

The detailed proof can be easily completed by readers themselves, so it is omitted here. Indeed, this result can also be extended to the case where $u(\Gamma)$ is Lebesgue integrable.

§ 6.3 Expansion for Poisson Kernels

Our task now is to expand the Poisson kernel. As

$$P(r\Gamma_0, \Gamma) = \frac{(1 - r^2)^{\frac{n(n-2)}{2}}}{\det (I - r\Gamma_0 \Gamma')^{n-1}} = P(rI, \Gamma\Gamma_0'),$$

we only need to expand $P(rI, \Gamma)$.

Let $\sigma_m(\Gamma)$ be the character of the single-valued irreducible representation with signature $m=(m_1,\cdots,m_k)$, $m_1\geqslant m_2\geqslant\cdots\geqslant m_k\geqslant 0$, of the rotation group $SO(n)$ (n is denoted by $2k$ or $2k+1$ according as n is even or odd). Then we have the following basic identity (see Murneghan [1], p. 255).

Let $t_1,\cdots,t_k$ be k independent variables and set

$$f_i=f(t_i,\Gamma)=\det(I-t_i\Gamma).$$

Then

$$\frac{\prod_{i\leqslant j}^{k}(1-t_it_j)}{f_1\cdots f_k}=\sum_{m\geqslant 0}\sigma_m(\Gamma)\chi_m(t), \tag{6.3.1}$$

where $\chi_m(t)=A_{l_1,\cdots,l_k}(t_1,\cdots,t_k)/D_k(t_1,\cdots,t_k)$,

$$A_{l_1,\cdots,l_k}(t_1,\cdots,t_k)=\begin{vmatrix} t_1^{l_1},\cdots,t_k^{l_1}\\ \cdots\cdots\cdots\\ t_1^{l_k},\cdots,t_k^{l_k}\end{vmatrix},$$

$$D_k(t_1,\cdots,t_k)=\begin{vmatrix} t_1^{k-1},\cdots,t_k^{k-1}\\ t_1^{k-2},\cdots,t_k^{k-2}\\ \cdots\cdots\cdots\\ 1,\ \cdots,\ 1\end{vmatrix}=\prod_{i<j}^{k}(t_i-t_j),$$

$$l_i=m_i+k-i,\quad (i=1,\cdots,k).$$

When $\max|t_i|<r<1$, (6.3.1) converges uniformly with respect to Γ.

On $SO(n)$, when $n=2k+1$, $\sigma_m(\Gamma)$ is given by (6.1.7); when $n=2k$ and $m_1\geqslant\cdots\geqslant m_{k-1}\geqslant m_k=0$, $\sigma_m(\Gamma)$ is given by (6.1.2); when $n=2k$ and $m_1\geqslant\cdots\geqslant m_k>0$, (6.3.1) becomes

$$\frac{\prod_{i\leqslant j}^{k}(1-t_it_j)}{f_1\cdots f_k}=\sum_{m>0}\sigma_m^{\phi}(\Gamma)\chi_m(t)+\sum_{m>0}\sigma_m^{\psi}(\Gamma)\chi_m(t),$$

where $\sigma_m^{\phi}(\Gamma)=[m]_+$ is given by (6.1.4) and $\sigma_m^{\psi}(\Gamma)=[m]_-$ is given by (6.1.5). If we denote

$$\sigma_m(\Gamma)=\sigma_m^{\phi}(\Gamma)+\sigma_m^{\psi}(\Gamma), \tag{6.3.2}$$

then $\sigma_m(\Gamma)$ is just the value defined by (6.1.2).

Thus (6.3.1) holds valid on $SO(n)$. Besides, when $n=2k+1$, $\sigma_m(\Gamma)$ is defined by (6.1.7); when $n=2k$, $\sigma_m(\Gamma)$ is defined by (6.1.2). Henceforth, $\sigma_m(\Gamma)$ will be defined just like this.

We consider

$$\frac{\prod_{i\leq j}^{n-1}(1-t_it_j)}{f_1\cdots f_{n-1}}$$

instead of the left side of (6.3.1) and set all $t_i = r$. Then

$$\left.\frac{\prod_{i\leq j}^{n-1}(1-t_it_j)}{f_1\cdots f_{n-1}}\right|_{t_1=\cdots=t_{n-1}=r} = \frac{(1-r^2)^{\frac{n(n-1)}{2}}}{\det^{n-1}(I-r\Gamma)} = P(rI, \Gamma). \tag{6.3.3}$$

Therefore, if the basic formula (6.3.1) can be extended from k to $n-1$, the problem of expanding $P(rI, \Gamma)$ will be solved immediately.

Set

$$T_i = 1 + t_i^2,\ D_s(t) = \prod_{i<j}^{s}(t_i - t_j),\ L_s(t) = \prod_{i<j}^{s}(1 - t_it_j),$$

$$\bar{L}_s(t) = \prod_{i\leq j}^{s}(1 - t_it_j) = L_s(t)\prod_{i=1}^{s}(1 - t_i^2),$$

$$\langle g_1(t), \cdots, g_s(t)\rangle = \begin{vmatrix} g_1(t_1), & g_2(t_1), \cdots, & g_s(t_1) \\ g_1(t_2), & g_2(t_2), \cdots, & g_s(t_2) \\ \cdots & \cdots & \cdots \\ g_1(t_s), & g_2(t_s), \cdots, & g_s(t_s) \end{vmatrix},$$

where $g_i(t)$ is an arbitrary function of t. Thus we have

Lemma 6.3.1 *Let $s > 0$. Then we have the identity*

$$\begin{aligned} D_{s+1}(t)L_{s+1}(t) &= \langle t^s, t^{s-1}T, t^{s-2}T^2, \cdots, T^s\rangle \\ &= \langle t^s, t^{s-1} + t^{s+1}, t^{s-2} + t^{s+2}, \cdots, 1 + t^{2s}\rangle. \end{aligned} \tag{6.3.4}$$

Proof. Obviously,

$$\langle t^s, t^{s-1}T, \cdots, T^s\rangle = (T_1\cdots T_{s+1})^s\left\langle\left(\frac{t}{T}\right)^s, \left(\frac{t}{T}\right)^{s-1}, \cdots, \frac{t}{T}, 1\right\rangle$$

$$= (T_1\cdots T_{s+1})^s\prod_{i<j}^{s+1}\left(\frac{t_i}{T_i} - \frac{t_j}{T_j}\right) = \prod_{i<j}^{s+1}(t_iT_j - t_jT_i)$$

$$= \prod_{i<j}^{s+1}(t_i + t_it_j^2 - t_j - t_jt_i^2) = \prod_{i<j}^{s+1}(t_i - t_j)$$

$$\cdot (1 - t_it_j) = D_{s+1}(t)L_{s+1}(t).$$

As to $\langle t^s, t^{s-1}T, \cdots, T^s\rangle = \langle t^s, t^{s-1} + t^{s+1}, \cdots, 1 + t^{2s}\rangle$, it suffices to reduce each column of the left determinant by a proper linear combination of its preceding columns and then the right determinant is obtained.

Lemma 6.3.2 *Let* $s \geqslant k = \left[\frac{n}{2}\right]$ *in case*

$$\frac{D_s(t)\bar{L}_s(t)}{f_1\cdots f_s} = \sum_{m\geqslant 0} \sigma_m(\Gamma)\langle g_1(t), \cdots, g_s(t)\rangle, \tag{6.3.5}$$

where $f_i = \det(I - t_i\Gamma)$; *when* $n = 2k + 1$ *or* $n = 2k$ *and* $m_1 \geqslant \cdots \geqslant m_{k-1} \geqslant m_k = 0$, $\sigma_m(\Gamma)$ *is the character of the irreducible representation of* $SO(n)$ *with signature* $m = (m_1, \cdots, m_k)$; *when* $n = 2k$ *and* $m_1 \geqslant \cdots \geqslant m_k > 0$, $\sigma_m(\Gamma)$ *is the sum (6.3.2) of the characters* $\sigma_m^\phi(\Gamma)$ *and* $\sigma_m^\psi(\Gamma)$ *of the irreducible representation with signature* $m = (m_1, \cdots, m_k)$ *of* $SO(n)$. *Then we have*

$$\frac{D_{s+1}(t)\bar{L}_{s+1}(t)}{f_1\cdots f_{s+1}} = \sum_{m\geqslant 0} \sigma_m(\Gamma)\langle g_1(t)t, \cdots, g_s(t)t, d(t)(1 + t^2)^{s-k}\rangle, \tag{6.3.6}$$

where

$$d(t) = \begin{cases} 1 - t^2, & \text{when } \Gamma \in SO(2k); \\ 1 + t, & \text{when } \Gamma \in SO(2k+1). \end{cases}$$

Proof. 1° Assume that $\Gamma \in SO(2k)$. By

$$\Gamma \sim \begin{pmatrix} c(\theta_1) & & \\ & \ddots & \\ & & c(\theta_k) \end{pmatrix}, \quad c(\theta_i) = \begin{pmatrix} \cos\theta_i, & \sin\theta_i \\ -\sin\theta_i, & \cos\theta_i \end{pmatrix},$$

we have

$$f = \det(I - t\Gamma) = \prod_{i=1}^{k} (1 - 2t\cos\theta_i + t^2)$$

$$= \prod_{i=1}^{k} (1 - 2t\cos\theta_i + t^2) = \prod_{i=1}^{k} (T - 2t\cos\theta_i) = T^k + \cdots,$$

where the missing term is a linear combination of t^iT^{k-i}, $(i = 1, \cdots, k)$. Thus

$$T^s = T^kT^{s-k} = T^{s-k}(f + \cdots) = T^{s-k}f + \cdots, \tag{6.3.7}$$

where the missing term is a linear combination of t^iT^{s-i}, $(i=1,\cdots,k)$. From Lemma 6.3.1 and (6.3.7), it follows that

$$D_{s+1}(t)\bar{L}_{s+1}(t) = \langle t^s, t^{s-1}T, \cdots, T^s\rangle$$

$$= \langle t^s, t^{s-1}T, \cdots, tT^{s-1}, T^{s-k}f\rangle = \sum_{i=1}^{s+1} (-1)^{s+1-i}T_i^{s-k}$$

$$\cdot f_i \cdot t_1\cdots\hat{t}_i\cdots t_{s+1} \begin{vmatrix} t_1^{s-1}, & t_1^{s-2}T_1, & \cdots\cdots, & t_1T_1^{s-2}, & T_1^{s-1} \\ \cdots & \cdots & \cdots & \cdots & \cdots \\ \hat{t}_i^{s-1}, & \hat{t}_i^{s-2}\hat{T}_i, & \cdots\cdots, & \hat{t}_i\hat{T}_i^{s-2}, & \hat{T}_i^{s-1} \\ \cdots & \cdots & \cdots & \cdots & \cdots \\ \cdots & \cdots & \cdots & \cdots & \cdots \\ t_{s+1}^{s-1}, & t_{s+1}^{s-2}T_{s+1}, & \cdots\cdots, & t_{s+1}T_{s+1}^{s-2}, & T_{s+1}^{s-1} \end{vmatrix}$$

$$= \sum_{i=1}^{s+1} (-1)^{s+1-i} T_i^{s-k} f_i t_1 \cdots \hat{t}_i \cdots t_{s+1} D_s^{(i)}(t) L_s^{(i)}(t), \tag{6.3.8}$$

where $\hat{t}_i$ means that there is no t_i term; $D_s^{(i)}(t)$ expresses the difference product of $(t_1, \cdots, \hat{t}_i, \cdots, t_{s+1})$ according to $D_s(t)$; $L_s^{(i)}(t)$ denotes the product of $(t_1, \cdots, \hat{t}_i, \cdots, t_{s+1})$ according to $L_s(t)$. Multiplying the two sides of the preceding formula by $\prod_{i=1}^{s+1} (1 - t_i^2)/f_1 \cdots f_{s+1}$ and noticing the relation between $L_s(t)$ and $\bar{L}_s(t)$, we get

$$\frac{D_{s+1}(t)\bar{L}_{s+1}}{f_1 \cdots f_{s+1}} = \sum_{i=1}^{s+1} (-1)^{s+1-i} T_i^{s-k} f_i t_1 \cdots \hat{t}_i \cdots t_{s+1}$$

$$\cdot (1 - t_i^2) \frac{D_s^{(i)}(t)\bar{L}_s^{(i)}(t)}{f_1 f_2 \cdots f_{s+1}} = \sum_{i=1}^{s+1} (-1)^{s+1-i} (1 - t_i^2)$$

$$\cdot T_i^{s-k} t_1 \cdots \hat{t}_i \cdots t_{s+1} \frac{D^{(i)}(t)\bar{L}_s^{(i)}(t)}{f_1 \cdots \hat{f}_i \cdots f_{s+1}}.$$

By the assumption of the lemma,

$$\frac{D_s^{(i)}(t)\bar{L}_s^{(i)}(t)}{f_1 \cdots \hat{f}_i \cdots f_{s+1}} = \sum_{m \geq 0} \sigma_m(\Gamma) \begin{vmatrix} g_1(t_1), \cdots, g_s(t_1) \\ \cdots\cdots\cdots\cdots \\ g_1(t_{i-1}), \cdots, g_s(t_{i-1}) \\ g_1(t_{i+1}), \cdots, g_s(t_{i+1}) \\ \cdots\cdots\cdots\cdots \\ g_1(t_{s+1}), \cdots, g_s(t_{s+1}) \end{vmatrix},$$

so we have

$$\frac{D_{s+1}(t)\bar{L}_{s+1}(t)}{f_1 f_2 \cdots f_{s+1}} = \sum_{m \geq 0} \sigma_m(\Gamma) \sum_{i=1}^{s+1} (-1)^{s+1-i} T_i^{s-k} (1 - t_i^2)$$

$$\cdot t_1 \cdots \hat{t}_i \cdots t_{s+1} \begin{vmatrix} g_1(t_1), \cdots, g_s(t_1) \\ \cdots\cdots\cdots\cdots \\ g_1(t_{i-1}), \cdots, g_s(t_{i-1}) \\ g_1(t_{i+1}), \cdots, g_s(t_{i+1}) \\ \cdots\cdots\cdots\cdots \\ g_1(t_{s+1}), \cdots, g_s(t_{s+1}) \end{vmatrix}$$

$$= \sum_{m \geq 0} \sigma_m(\Gamma) \langle g_1(t) \cdot t, g_2(t) \cdot t, \cdots, g_s(t) \cdot t, (1 - t^2) T^{s-k} \rangle,$$

which proves that our lemma is valid for $\Gamma \in SO(2k)$.

2° Assume that $\Gamma \in SO(2k+1)$. Because of

$$\Gamma \sim \begin{pmatrix} c(\theta_1) & & & \\ & \ddots & & \\ & & c(\theta_k) & \\ & & & 1 \end{pmatrix},$$

we have

$$f = \det(I - t\Gamma) = (1-t)\prod_{i=1}^{k}(1 - 2t\cos\theta_i + t^2)$$

$$= (1-t)h,$$

where

$$h = \prod_{i=1}^{k}(T - 2t\cos\theta_i) = T^k + \cdots.$$

Therefore $T^s = T^{s-k}T^k = T^{s-k}h + \cdots$. The rest of the proof is all alike to that of 1°. Namely, replace T^s by $T^{s-k}h$ in

$$D_{s+1}(t)L_{s+1}(t) = \langle t^s, t^{s-1}T, \cdots tT^{s-1}, T^s\rangle$$

and expand it. By a similitude we get

$$\frac{D_{s+1}(t)\bar{L}_{s+1}(t)}{f_1\cdots f_{s+1}} = \sum_{i=1}^{s+1}(-1)^{s+1-i}T_i^{s-k}(1-t_i^2)$$

$$\cdot\, h_i t_1\cdots\hat{t}_i\cdots t_{s+1}\frac{D_s^{(i)}(t)\bar{L}_s^{(i)}(t)}{f_1\cdots f_{s+1}}.$$

From

$$(1-t_i^2)h_i = (1+t_i)(1-t_i)h_i = (1+t_i)f_i,$$

it follows that

$$\frac{D_{s+1}(t)\bar{L}_{s+1}(t)}{f_1\cdots f_{s+1}} = \sum_{i=1}^{s+1}(-1)^{s+1-i}(1+t_i)$$

$$\cdot\, T_i^{s-k}t_1\cdots\hat{t}_i\cdots t_{s+1}\frac{D_s^{(i)}(t)\bar{L}_s^{(i)}(t)}{f_1\cdots\hat{f}_i\cdots f_{s+1}}.$$

By the assumption of the lemma, finally, we obtain

$$\frac{D_{s+1}(t)\bar{L}_{s+1}(t)}{f_1\cdots f_{s+1}}$$

$$= \sum_{m\geqslant 0}\sigma_m(\Gamma)\langle g_1(t)\cdot t, \cdots, g_s(t)t, (1+t)\Gamma^{s-k}\rangle.$$

Thus the proof is completed.

Take (6.3.1) as our starting point and rewrite it as

$$\frac{D_k(t)\bar{L}_k(t)}{f_1\cdots f_k} = \sum_{m\geqslant 0}\sigma_m(\Gamma)\langle t^{l_1}, \cdots, t^{l_k}\rangle. \tag{6.3.9}$$

For the coefficients of $\sigma_m(\Gamma)$, using Lemma 6.3.2 repeatedly leads to the following sequence

$$\langle t^{l_1}, \cdots, t^{l_k}\rangle \xrightarrow{k+1} \langle t^{l_1+1}, \cdots, t^{l_k+1}, d(t)\rangle \xrightarrow{k+2} \langle t^{l_1+2}, \cdots,$$
$$t^{l_k+2}, d(t)t, d(t)(1+t^2)\rangle \xrightarrow{k+3} \cdots \xrightarrow{n-1} \langle t^{l_1+n-1-k}, \cdots, t^{l_k+n-1-k},$$
$$d(t)t^{n-k-2}, d(t)t^{n-k-3}(1+t^2), \cdots, d(t)(1+t^2)^{n-k-2}\rangle. \tag{6.3.10}$$

By $c_{n-1}(t_1, \cdots, t_{n-1})$, we denote the last term of (6.3.10). Thus the above discussion can be formulated into the following theorem.

Theorem 6.3.1 *The notations are the same as those in Lemma 6.3.1. Then the following identity*

$$\frac{\prod_{i<j}^{n-1}(t_i - t_j)\prod_{i\leq j}^{n-1}(1 - t_i t_j)}{f_1 f_2 \cdots f_{n-1}} = \sum_{m\geq 0} \sigma_m(\Gamma) c_{n-1}(t_1, \cdots, t_{n-1}) \tag{6.3.11}$$

holds, where $f_i = \det(I - t_i\Gamma)$, $\Gamma \in SO(n)$, $c_{n-1}(t_1, \cdots, t_{n-1})$ is expressed by (6.3.10). And, when $n = 2k+1$ or $n = 2k$ and $m_1 \geq \cdots \geq m_{k-1} \geq m_k = 0$, $\sigma_m(\Gamma)$ is the character of the irreducible representation of $SO(n)$ with the signature $m = (m_1, \cdots, m_k)$; when $n = 2k$ and $m_1 \geq \cdots \geq m_k > 0$, $\sigma_m(\Gamma)$ is the sum (6.3.2) of the characters $\sigma_m^{\phi}(\Gamma)$ and $\sigma_m^{\psi}(\Gamma)$ of the irreducible representation with the signature $m = (m_1, \cdots, m_k)$ of $SO(n)$.

By (6.3.11) we have

$$\frac{\prod_{i\leq j}^{n-1}(1 - t_i t_j)}{f_1 \cdots f_{n-1}} = \sum_{m\geq 0} \sigma_m(\Gamma) \frac{c_{n-1}(t_1, \cdots, t_{n-1})}{D_{n-1}(t_1, \cdots, t_{n-1})}. \tag{6.3.12}$$

Set $t_1 = t_2 = \cdots = t_{n-1} = r$. From (6.3.3), it follows that

$$P(rI, \Gamma) = \frac{(1 - r^2)^{\frac{n(n-1)}{2}}}{\det^{n-1}(I - r\Gamma)} = \sum_{m\geq 0} \rho_m(r)\sigma_m(\Gamma), \tag{6.3.13}$$

where

$$\rho_m(r) = \left.\frac{c_{n-1}(t_1, \cdots, t_{n-1})}{D_{n-1}(t_1, \cdots, t_{n-1})}\right|_{t_1 = \cdots = t_{n-1} = r}. \tag{6.3.14}$$

Thus we have

Corollary 6.3.1 *The Poisson kernel $P(rI, \Gamma)$ of $SO(n)$ can be expanded to a linear combination of the characters of its irreducible representations.*

When $n = 2k+1$, (6.3.13) is the expansion of $P(rI, \Gamma)$, in which $\sigma_m(\Gamma)$ and $\rho_m(r)$ are defined by (6.1.7) and (6.3.14), respectively.

When $n=2k$, the expansion of $P(rI,\Gamma)$ is given by

$$P(rI,\Gamma)=\sum_{m_1\geqslant\cdots\geqslant m_{k-1}\geqslant m_k=0}\rho_m(r)\sigma_m(\Gamma)+\sum_{m>0}\rho_m(r)(\sigma_m^{\phi}(\Gamma)+\sigma_m^{\psi}(\Gamma)),\tag{6.3.15}$$

where $\sigma_m^{\phi}(\Gamma)$, $\sigma_m^{\psi}(\Gamma)$, $\sigma_m(\Gamma)$ and $\rho_m(r)$ are defined by (6.1.4), (6.1.5), (6.1.2) and (6.3.14), respectively.

§ 6.4 Abel Summation

Having given the expansion of Poisson kernels, we are able to define the Abel mean of the Fourier series (6.1.12) and (6.1.14) of any integrable function $u(\Gamma)(\Gamma\in SO(n))$ on $SO(n)$.

When $n=2k+1$, since

$$P(rI,\Gamma\Gamma_0')=\sum_{m\geqslant 0}\rho_m(r)\sigma_m(\Gamma\Gamma_0')=\sum_{m\geqslant 0}\rho_m(r)\sum_{i,j}^{N(m)}\varphi_{ij}^m(\Gamma)\varphi_{ij}^m(\Gamma_0),\tag{6.4.1}$$

by (6.2.7), we get

$$u(r\Gamma_0)=\sum_{m\geqslant 0}\rho_m(r)\sum_{i,j}^{N(m)}\frac{a_{ij}^m}{N(m)}\varphi_{ij}^m(\Gamma_0)=\sum_{m\geqslant 0}\frac{\rho_m(r)}{N(m)}\sum_{i,j}^{N(m)}a_{ij}^m\varphi_{ij}^m(\Gamma_0).\tag{6.4.2}$$

The series on the right side of (6.4.2) is defined as the Abel mean of the Fourier series (6.1.14).

When $n=2k$, we denote

$$P(rI,\Gamma\Gamma_0)=\sum_{m\geqslant 0}\rho_m(r)\sum_{i,j}^{N(m)}\phi_{ij}^m(\Gamma)\phi_{ij}^m(\Gamma_0)$$

$$+\sum_{m>0}\rho_m(r)\sum_{i,j}^{N(m)}\psi_{ij}^m(\Gamma)\psi_{ij}^m(\Gamma_0).\tag{6.4.3}$$

By (6.2.7) we have

$$u(r\Gamma_0)=\sum_{m\geqslant 0}\rho_m(r)\sum_{i,j}^{N(m)}\frac{a_{ij}^m}{N(m)}\phi_{ij}^m(\Gamma_0)$$

$$+\sum_{m>0}\rho_m(r)\sum_{i,j}^{N(m)}\frac{b_{ij}^m}{N(m)}\psi_{ij}^m(\Gamma_0)$$

$$=\sum_{m\geqslant 0}\frac{\rho_m(r)}{N(m)}\sum_{i,j}^{N(m)}a_{ij}^m\phi_{ij}^m(\Gamma_0)+\sum_{m>0}\frac{\rho_m(r)}{N(m)}\sum_{i,j}^{N(m)}b_{ij}^m\psi_{ij}^m(\Gamma_0).\tag{6.4.4}$$

The series on the right side of (6.4.4) is defined as the Abel mean of the Fourier series (6.1.12).

Thus Theorem 6.2.1 can be equivalently stated as follows:

Theorem 6.4.1 *If $u(\Gamma_0)$ is a continuous function on $SO(n)$, then the Fourier series of $u(\Gamma_0)$ is Abel-summable to itself.*

When $n = 2k + 1$, taking $u(\Gamma) = \sigma_\nu(\Gamma)$ and $\Gamma_0 = I$ in (6.2.7), we have

$$u(rI) = \frac{1}{C}\int_{SO(n)} P(rI,\Gamma)\sigma_m(\Gamma)\dot{\Gamma}$$

$$= \sum_{m\geqslant 0} \rho_m(r)\frac{1}{C}\int_{SO(n)} \sigma_m(\Gamma)\sigma_m(\Gamma)\dot{\Gamma} = \rho_m(r),$$

so $\rho_m(r) \to \sigma_m(I) = N(m)$ as $r \to 1$.

When $n = 2k$, take $u(\Gamma) = \sigma_\nu(\Gamma)$, $\Gamma_0 = I$ and $u(\Gamma) = \sigma_\nu^\phi(\Gamma)$, $\Gamma_0 = I$, $u(\Gamma) = \sigma_\nu^\psi(\Gamma)$, $\Gamma_0 = I$ according as $m_1 \geqslant \cdots \geqslant m_{k-1} \geqslant m_k = 0$ and $m_1 \geqslant \cdots \geqslant m_{k-1} \geqslant m_k > 0$, respectively. Then, in the same way, we obtain that $\rho_\nu(r)$ tends to $\sigma_\nu(I) = N(\nu)$, $\sigma_\nu^\phi(I) = N(\nu)$ and $\sigma_\nu^\psi(I) = N(\nu)$, respectively. In other words, for any signature $m = (m_1, \cdots, m_k)$ we have

$$\rho_m(r) \to N(m),$$

as $r \to 1$. Next, we will give a concrete expression for $\rho_\nu(r)$.

By (6.3.10), $c_{n-1}(t_1, \cdots, t_{n-1})$ can be put into

$$c_{n-1}(t_1, \cdots, t_{n-1}) = \langle t^{l_1+n-1-k}, \cdots, t^{l_k+n-1-k}, d(t)t^{n-k-2},$$
$$d(t)(t^{n-k-3} + t^{n-k-1}), d(t)(t^{n-k-4} + t^{n-k}), \cdots, d(t)(1 + t^{(2n-2k-4)})\rangle,$$

where

$$d(t) = \begin{cases} 1 + t, & \text{if } n = 2k + 1, \\ 1 - t^2, & \text{if } n = 2k. \end{cases}$$

In the following, the discussion will be covered by odd state and even state

1° $n = 2k + 1$.

As

$$d(t)t^{n-k-2} = (1 + t)^{n-k-2} = t^{n-k-2} + t^{n-k-1},$$

$$d(t)(t^{n-k-3} + t^{n-k-1}) = (1 + t)(t^{n-k-3} + t^{n-k-1}) = t^{n-k-3} + t^{n-k}$$
$$+ (t^{n-k-2} + t^{n-k-1}),$$

$$d(t)(t^{n-k-4} + t^{n-k}) = (1 + t)(t^{n-k-4} + t^{n-k}) = t^{n-k-4} + t^{n-k+1}$$
$$+ (t^{n-k-3} + t^{n-k}),$$

....

$$d(t)(1 + t^{2n-2k-4}) = (1 + t)(1 + t^{2n-2k-4}) = 1 + t^{2n-2k-3}$$
$$+ (t + t^{2n-2k-4}),$$

we have, by the properties of determinant, that

$$c_{n-1}(t_1,\cdots,t_{n-1}) = \langle t^{l_1+n-1-k},\cdots,t^{l_k+n-1-k}, d(t)t^{n-k-2},$$
$$d(t)(t^{n-k-3}+t^{n-k-1}),\cdots,d(t)(1+t^{2n-2k-4})\rangle = \langle t^{l_1+n-1-k},\cdots,$$
$$t^{l_k+n-1-k}, (t^{n-k-2}+t^{n-k-1}), (t^{n-k-3}+t^{n-k}),\cdots,(1+t^{2n-2k-3})\rangle$$
$$= \sum_{(p_1,\cdots,p_{n-k-1})\in D} (-1)^{p_1+\cdots+p_s-sk}\langle t^{l_1+n-1-k},\cdots,t^{l_k+n-1-k},t^{p_1},\cdots,$$
$$t^{p_{n-k-1}}\rangle, \tag{6.4.5}$$

where the set D of indices satisfies the following conditions

(1) $2n-2k-3\geqslant p_1>p_2>\cdots>p_s\geqslant n-k-1>p_{s+1}>\cdots>p_{n-k-1}\geqslant 0$;

(2) $p_i+p_j\neq 2n-2k-3$ for any $i\neq j$.

Inserting them into (6.3.14), we get

$$\rho_m(r) = \frac{c_{n-1}(t_1,\cdots,t_{n-1})}{D_{n-1}(t_1,\cdots,t_{n-1})}$$
$$= \sum_{(p_1,\cdots,p_{n-k-1})\in D} (-1)^{p_1+\cdots+p_s-sk}\cdot\frac{1}{D_{n-1}(t_1,\cdots,t_{n-1})}$$
$$\cdot\langle t^{l_1+n-k-1},\cdots,t^{l_k+n-k-1},t^{p_1},\cdots,t^{p_{n-k-1}}\rangle\big|_{t_1=\cdots=t_n=r}. \tag{6.4.6}$$

Let $q_1=p_1-n+2+k$, $q_2=p_2-n+3+k,\cdots$, $q_{n-k-1}=p_{n-k-1}$.

We have already known that

$$l_1=m_1+k-1,\ l_2=m_2+k-2,\cdots,\ l_k=m_k.$$

Thus

$$\langle t^{l_1+n-1-k},\cdots,\ t^{l_k+n-1-k},t^{p_1},\cdots,\ t^{p_{n-k-1}}\rangle$$
$$=\langle t^{m_1+n-2},\ t^{m_2+n-3},\cdots,\ t^{m_k+n-k-1},t^{q_1+n-k-2},\cdots,\ t^{q_{n-k-1}}\rangle.$$

Inserting it into (6.4.6), we get

$$\rho_m(r) = \sum_{(q_1,\cdots,q_{n-k-1})\in D^0} (-1)^{q_1+\cdots+q_s-\frac{s(s+1)}{2}}\cdot\frac{1}{D_{n-1}(t_1,\cdots,t_{n-1})}$$
$$\cdot\langle t^{m_1+n-2},t^{m_2+n-3},\cdots,t^{m_k+n-k-1},t^{q_1+n-k-2},\cdots,t^{q_{n-k-1}}\rangle\big|_{t_1=\cdots=t_{n-1}=r}$$
$$= \sum_{(q_1,\cdots,q_{n-k-1})\in D^0} (-1)^{q_1+\cdots+q_s-\frac{s(s+1)}{2}} r^{\sum_1^k m_i+\sum_1^{n-k-1} q_j}$$
$$\cdot N(m_1,\cdots,m_k,q_1,\cdots,q_{n-k-1}), \tag{6.4.7}$$

where the set D^0 of indices is transformed from D. We have

$$(q_1,\cdots,q_{n-k-1})\in D^0$$

if and only if $(p_1,\cdots,q_{n-k-1})\in D$.

It is easy to check that the conditions for determining D^0 are as follows

(1) $n-k-1 \geqslant q_1 \geqslant \cdots \geqslant q_s \geqslant s \geqslant q_{s+1} \geqslant \cdots \geqslant q_{n-k-1} \geqslant 0$;

(2) $q_i + q_j \neq i + j - 1$ for any $i \neq j$.

Moreover, $N(m_1, \cdots, m_k, q_1, \cdots, q_{n-k-1})$ is defined as in § 1.2 with a change that the order of magnitude among the numbers $m_1, \cdots, m_k, q_1, \cdots, q_{n-k-1}$ is not required any more. In addition, (6.4.7) is just the expression of $\rho_n(r)$ for $SO(n)$ when $n = 2k+1$.

2° $n = 2k$.

Similarly, by the expression of $c_{n-1}(t_1, \cdots, t_{n-1})$. since

$$d(t)t^{n-k-2} = (1-t^2)t^{n-k-2} = t^{n-k-2} - t^{n-k},$$

$$d(t)(t^{n-k-2} + t^{n-k-1}) = (1-t^2)(t^{n-k-3} + t^{n-k-1})$$

$$= t^{n-k-3} - t^{n-k+1}, \cdots\cdots\cdots\cdots$$

$$d(t)(1 + t^{2n-2k-4}) = (1-t^2)(1 + t^{2n-2k-4})$$

$$= 1 - t^{2n-2k-2} - (t^2 - t^{2n-2k-4}),$$

we have by the properties of determinant that

$$\begin{aligned} c_{n-1}(t_1, \cdots, t_{n-1}) &= \langle t^{l_1+n-1-k}, \cdots, t^{l_k+n-1-k}, \\ & d(t)t^{n-k-2}, \cdots, d(t)(1 + t^{2n-2k-4})\rangle = \langle t^{l_1+n-1-k}, \cdots, \\ & t^{l_k+n-1-k}, (t^{n-k-2} - t^{n-k}), (t^{n-k-3} - t^{n-k-1}), \cdots, \\ & (1 - t^{2n-2k-2})\rangle = \sum_{(p_1, \cdots, p_{n-k-1}) \in E} (-1)^{p_1+\cdots+p_s - s(k+1)} \\ & \cdot \langle t^{l_1+n-1-k}, \cdots, t^{l_k+n-1-k}, t^{p_1}, \cdots, t^{p_{n-k-1}}\rangle, \end{aligned}$$

where the set E of indices is determined by the following conditions

(1) $2n-2k-2 \geqslant p_1 > \cdots > p_s \geqslant n-k$
$\geqslant p_{s+1} > \cdots > p_{n-k-1} \geqslant 0$;

(2) $p_i \neq n-k-1$ for any i;

(3) $p_i + p_j \neq 2n-2k-2$ for any $i \neq j$.

Therefore, $\rho_m(r)$ is equal to

$$\left.\frac{c_{n-1}(t_1, \cdots, t_{n-1})}{D_{n-1}(t_1, \cdots, t_{n-1})}\right|_{t_1=\cdots=t_{n-1}=r} = \sum_{(p_1, \cdots, p_{n-k-1}) \in E} (-1)^{p_1+\cdots+p_s - s(k+1)}$$

$$\cdot \left.\frac{\langle t^{l_1+n-1-k}, \cdots, t^{l_k+n-1-k}, t^{p_1}, \cdots, t^{p_{n-k-1}}\rangle}{D_{n-1}(t_1, \cdots, t_{n-1})}\right|_{t_1=\cdots=t_{n-1}=r}.$$

Similarly, set

$$q_1 = p_1 - n + 2 + k, \; q_2 = p_2 - n + 3 + k, \cdots, \; q_{n-k-1} = p_{n-k-1}.$$

We have already known

$$l_1 = m_1 + k - 1, \cdots, l_k = m_k,$$

thus $\langle t^{l_1+n-1-k}, \cdots, t^{l_k+n-1-k}, t^{p_1}, \cdots, t^{p_{n-k-1}} \rangle$

$$= \langle t^{m_1+n-2}, t^{m_2+n-3}, \cdots, t^{m_k+n-k-1}, t^{q_1+n-k-2}, \cdots, t^{q_{n-k-1}} \rangle,$$

which is inserted into the above expression, and then we get

$$\rho_m(r) = \sum_{(q_1,\cdots,q_{n-k-1})\in \mathring{E}} (-1)^{q_1+\cdots+q_s-\frac{s(s+1)}{2}}$$

$$\cdot \left. \frac{\langle t^{m_1+n-2}, \cdots, t^{m_k+n-k-1}, t^{q_1+n-k-2}, t^{q_{n-k-1}} \rangle}{D_{n-1}(t_1, \cdots, t_{n-1})} \right|_{t_1=\cdots=t_{n-1}=r}$$

$$= \sum_{(q_1,\cdots,q_{n-k-1})\in \mathring{E}} (-1)^{q_1+\cdots+q_s-\frac{s(s+1)}{2}}$$

$$N(m_1, \cdots, m_k, q_1, \cdots, q_{n-k-1}) \cdot r^{\sum_1^k m_i + \sum_1^{n-k-1} q_j} . \tag{6.4.8}$$

Here, the set $\mathring{E}$ of indices is transformed from E. Thus we have

$$(q_1, \cdots, q_{n-k-1}) \in \mathring{E},$$

if and only if $(p_1, \cdots\cdots, p_{n-k-1}) \in E$. It can be easily checked that the conditions for determining $\mathring{E}$ are the following

(1) $n - k \geqslant q_1 \geqslant \cdots \geqslant q_s \geqslant s + 1 \geqslant q_{s+1} \geqslant \cdots \geqslant q_{n-k-1} \geqslant 0$;

(2) $q_i \neq i$ for any i;

(3) $q_i + q_j \neq i + j$ for any $i \neq j$.

Besides, when $n=2k$, (6.4.8) is just the expression for $\rho_m(r)$ of $SO(n)$.

Summing up the above discussion, it leads to the following theorem (see Zhong Jiaqing [2])

Theorem 6.4.2 *The Poisson kernel $P(rI, \Gamma)$ of rotation group $SO(n)$ has (6.3.13) as its expansion and the coefficient $\rho_m(r)$ is given by*

$$\rho_m(r) = \sum_{(q_1,\cdots,q_{n-k-1})\in \mathring{D} \text{ or } \mathring{E}} (-1)^{q_1+\cdots+q_s-\frac{s(s+1)}{2}}$$

$$\cdot N(m_1, \cdots, m_k, q_1, \cdots, q_{n-k-1}) r^{\sum_1^k m_i + \sum_1^{n-k-1} q_j},$$

where the set of indices is $\mathring{D}$ or $\mathring{E}$ according as n is odd or even, and $N(m_1, \cdots, m_k, q_1, \cdots, q_{n-k-1})$ is defined as in § 1.2 with a change that the order of the magnitude among the numbers $m_1, \cdots, m_k, q_1, \cdots, q_{n-k-1}$ is not required any more.

We now come to point out that in addition to the expression given by its expansion, $\rho_m(r)$ has an expression made by a determinant. To this end, recall (6.3.14)

$$\rho_m(r) = \left.\frac{c_{n-1}(t_1, \cdots, t_{n-1})}{D_{n-1}(t_1, \cdots, t_{n-1})}\right|_{t_1=\cdots=t_{n-1}=r}.$$

Using the well-known identity (see Theorem 1.2.4 in Hua Luogeng [1])

$$\lim_{\substack{x_1 \to x \\ \cdots \\ x_n \to x}} \frac{\begin{vmatrix} f_1(x_1), \cdots, f_n(x_1) \\ \cdots\cdots\cdots \\ \cdots\cdots\cdots \\ f_1(x_n), \cdots, f_n(x_n) \end{vmatrix}}{D_n(x_1, \cdots, x_n)}$$

$$= \frac{(-1)^{\frac{n(n-1)}{2}}}{1!2!\cdots(n-1)!} \cdot \begin{vmatrix} f_1(x), \cdots, f_n(x) \\ f_1'(x), \cdots, f_n'(x) \\ \cdots\cdots\cdots \\ f_1^{(n-1)}(x), \cdots, f_n^{(n-1)}(x) \end{vmatrix},$$

we obtain the expression for $\rho_m(r)$

$$\rho_m(r) = \frac{(-1)^{\frac{(n-1)(n-2)}{2}}}{1!2!\cdots(n-2)!} \cdot \begin{vmatrix} \xi_1(r), \cdots, \xi_{n-1}(r) \\ \xi_1'(r), \cdots, \xi_{n-1}'(r) \\ \cdots\cdots\cdots \\ \xi_1^{(n-2)}(r), \cdots, \xi_{n-1}^{(n-2)}(r) \end{vmatrix}, \quad (6.4.9)$$

where

$$\xi_1(r) = r^{m_1+n-2}, \ \xi_2(r) = r^{m_2+n-3}, \cdots, \ \xi_k(r) = r^{m_k+n-k-1},$$

$$\xi_{k+1}(r) = \begin{cases} r^{n-k-2} + r^{n-k-1}, & \text{if } n = 2k+1; \\ r^{n-k-2} - r^{n-k}, & \text{if } n = 2k; \end{cases}$$

$$\xi_{k+2}(r) = \begin{cases} r^{n-k-3} + r^{n-k}, & \text{if } n = 2k+1; \\ r^{n-k-3} - r^{n-k-1}, & \text{if } n = 2k; \end{cases}$$

....

$$\xi_{n-1}(r) = \begin{cases} 1 + r^{2n-2k-3}, & \text{if } n = 2k+1; \\ 1 - r^{2n-2k-4}, & \text{if } n = 2k. \end{cases}$$

Now, we write down the simplest case, i. e. the rotation group of dimension 3. Take $n = 3$ and $k = 1$ in (6.4.9). The corresponding irreducible representation is characterized by a nonnegative integer $m \geqslant 0$. And we have

$$\xi_1(r) = r^{m+1}, \quad \xi_2(r) = 1 + r,$$

$$\rho_m(r) = (-1)\begin{vmatrix} r^{m+1}, & 1+r \\ (m+1)r^m, & 1 \end{vmatrix} = (m+1)r^m + mr^{m+1}.$$

Zhong Jiaqing([2]) studied the Abel summation of the Fourier series on rotation groups and, in the same paper, he also gave some other results on group representation.

Chapter 7. Cesàro Summation of Fourier Series on Rotation Groups

§ 7.1 Definition and Kernel of Cesàro Summation

Just as in Chapter 2, here we consider Cesàro summation of Fourier series on rotation groups.

If $u(\Gamma)$ is an integrable function on $SO(n)$, $\Gamma \in SO(n)$. The Cesàro (c,α) mean of its Fourier series (6.1.14) and (6.1.12) are defined by

$$\sum_{(n-1)N \geq m_1 \geq \cdots \geq m_k \geq 0} B_m^\alpha \sum_{i,j}^{N(m)} a_{ij}^m \varphi_{ij}^m(\Gamma) \tag{7.1.1}$$

and

$$\sum_{(n-1)N \geq m_1 \geq \cdots \geq m_k \geq 0} B_m^\alpha \sum_{i,j}^{N(m)} a_{ij}^m \phi_{ij}^m(\Gamma) + \sum_{(n-1)N \geq m_1 \geq \cdots \geq m_k \geq 0} B_m^\alpha \sum_{i,j}^{N(m)} b_{ij}^m \phi_{ij}^m(\Gamma), \tag{7.1.2}$$

respectively, where

$$B_m^\alpha = \frac{1}{N(m)C} \int_{SO(n)} \sigma_m(\Gamma) K_N^\alpha(\Gamma) \dot{\Gamma}, \tag{7.1.3}$$

in which $\sigma_m(\Gamma)$ is defined as in § 6.3, C represents the volume of $SO(n)$, and $N(m)$ stands for the dimension of the irreducible representation of $SO(n)$ with signature $m = (m_1, \cdots, m_k)$. $K_N^\alpha(\Gamma)$ is called the kernel of Cesàro (c, α)-summation and is equal to

$$\frac{1}{B_N^\alpha} \det^{\frac{n-1}{2}} \left(\frac{A_N^\alpha I + \sum_{j=1}^{N} (\Gamma^j + \Gamma'^j) \sum_{v=0}^{N-j} A_v^{\alpha-1}}{A_N^\alpha} \right), \tag{7.1.4}$$

where

$$B_N^\alpha = \frac{1}{C} \int_{SO(n)} \det^{\frac{n-1}{2}} \left(\frac{A_N^\alpha I + \sum_{j=1}^{N} (\Gamma^j + \Gamma'^j) \sum_{v=0}^{N-j} A_v^{\alpha-1}}{A_N^\alpha} \right) \dot{\Gamma}, \tag{7.1.5}$$

and

$$A_N^\alpha = C_N^{N+\alpha} = \frac{(\alpha+1)(\alpha+2)\cdots(\alpha+N)}{N!} \quad (\text{for } \alpha > -1).$$

When $n = 2k$, (7.1.4), with the exception of a manifold of lower dimension, can be rewritten as

$$\frac{\det^{\frac{n-1}{2}}\left[(I-\Gamma)^{-1}\left(\sum_{v=0}^{N} A_{N-v}^{v-1}\Gamma'^v(I-\Gamma^{2v+1})\right)\right]}{B_N^\alpha(2A_N^\alpha)^{\frac{n(n-1)}{4}}},$$

here (7.1.5) appears in the form

$$B_N^\alpha = \frac{1}{C}\int_{SO(n)} \frac{\det^{\frac{n-1}{2}}\left[(I-\Gamma)^{-1}\left(\sum_{v=0}^{N} A_{N-v}^{\alpha-1}\Gamma'^v(I-\Gamma^{2v+1})\right)\right]}{(2A_N^\alpha)^{\frac{n(n-1)}{2}}}\dot{\Gamma},$$

Theorem 7.1.1 *Both* (7.1.1) *and* (7.1.2) *can be expressed as*

$$\frac{1}{C}\int_{SO(n)} u(W\Gamma)K_N^\alpha(W)\dot{W}. \tag{7.1.6}$$

Proof. We prove only the case of $n = 2k$.

If $\Gamma \in SO(n)$ and $\Gamma \sim c(\theta_1)\dotplus\cdots\dotplus c(\theta_k)$, then we can easily show that

$$\sum_{(n-1)N \geqslant m_1 \geqslant m_k \geqslant 0} B_m^\alpha N(m)\sigma_m(\Gamma) = K_N^\alpha(\Gamma). \tag{7.1.7}$$

To prove this, we multiply the two sides of the above expression by $c(k-1,\cdots,1,0)$. Thus both sides are multiple trigonometric polynomials. By (7.1.2), we get

$$B_m^\alpha N(m) = \frac{1}{(2\pi)^k}\int_0^{2\pi}\cdots\int_0^{2\pi} \cos l_1\theta_1\cdots\cos l_k\theta_k K_N^\alpha(\Gamma)$$
$$\cdot\, c(k-1,\cdots,1,0)d\theta_1\cdots d\theta_k.$$

More precisely, the coefficients of the two trigonometric polynomials are identical, which proves (7.1.7).

From Chapter 6, it follows that (7.1.1) and (7.1.2) exactly turn out

$$\frac{1}{C}\int_{SO(n)} u(\dot{W}\Gamma)\left(\sum_{(n-1)N \geqslant m_1 \geqslant \cdots \geqslant m_k \geqslant 0} B_m^\alpha N(m)\sigma_m(W)\right)\dot{W},$$

which, by (7.1.7), is just (7.1.6).

It can be seen that, as α tends to infinity, the Cesàro (c,α) sum thus defined tends to the Abel sum defined in Chapter 6 and (c,α) kernel becomes the Poisson kernel.

The Fourier series of $u(\Gamma)$ is called (C,α)-summable if the limits of (7.1.1) and (7.1.2) exist for $N \to \infty$.

§ 7.2 Semi-Positivity of Cesàro Kernel

We now turn to prove (Wang Shikun & Dong Daozheng [2]).

Theorem 7.2.1 *If $u(\Gamma)$ is a continuous function on $SO(n)$, $\Gamma \in SO(n)$, then its Fourier series is (c,α) summable to itself for $\alpha > \frac{n-2}{n-1}$.*

To this end, we first show that the Cesàro kernel, when $\alpha > \frac{n-2}{n-1}$, is "semi-positive".

Theorem 7.2.2 *The (c, α) kernel $K_N(\Gamma)$ is semi-positive. Namely, if $\alpha > \frac{n-2}{n-1}$, then*

$$\int_{SO(n)} |K_N^\alpha(\Gamma)| \dot{\Gamma} \leqslant M, \tag{7.2.1}$$

where M is dependent only on n and α, and is independent of N.

Proof. When $n = 2k+1, K_N^\alpha(\Gamma)$ is nonnegative. Thus (7.2.1) is obviously valid.

When $n = 2k$, for $k = 1$, (7.2.1) is self-evident. Assume that (7.2.1) holds valid for all natural numbers less than k. We should show (7.2.1) still holds for k.

B_N does not tend to zero as $N \to \infty$. For proving the result, we only need to prove that

$$I = \int \cdots \int_{\pi \geqslant \theta_1 \geqslant \cdots \geqslant \theta_k \geqslant 0} |\sigma_N^\alpha(\theta_1) \cdots \sigma_N^\alpha(\theta_k)|^{2k-1} \cdot D^2(\cos\theta_1, \cdots, \cos\theta_k) d\theta_1 \cdots d\theta_k \tag{7.2.2}$$

is bounded, where

$$\sigma_N^\alpha(\theta) = \frac{1}{A_N^\alpha \sin\frac{\theta}{2}} I_m \sum_{\nu=0}^{N} A_{N-\nu}^{\alpha-1} e^{i(\nu+\frac{1}{2})\theta}.$$

Take $\delta \geqslant \frac{1}{N}$ and divide the integral domain into

R_1: $\delta \geqslant \theta_1 \geqslant \cdots \geqslant \theta_k \geqslant 0$,

R_2: $\pi \geqslant \theta_1 \geqslant \delta \geqslant \theta_2 \geqslant \cdots \geqslant \theta_k \geqslant 0$,

. . . .

R_k: $\pi \geqslant \theta_1 \geqslant \cdots \geqslant \theta_{k-1} \geqslant \delta \geqslant \theta_k \geqslant 0$,

R_{k+1}: $\pi \geqslant \theta_1 \geqslant \cdots \geqslant \theta_k \geqslant \delta$.

By I_j, we denote the part of the integral in (7.2.2) which corresponds to a sub-domain R_j. Therefore

$$I = I_1 + \cdots + I_{k+1}.$$

First we consider I_2. We have

$$I_2 = \int_{R_2} |\sigma_N^\alpha(\theta_1)\cdots\sigma_N^\alpha(\theta_k)|^{2k-1} D^2(\cos\theta_1, \cdots, \cos\theta_k)\, d\theta_1\cdots d\theta_k$$

$$= \underset{\delta \geqslant \theta_2 \geqslant \cdots \geqslant \theta_k \geqslant 0}{\int\cdots\int} |\sigma_N^\alpha(\theta_2)\cdots\sigma_N^\alpha(\theta_k)|^{2k-1}$$

$$\cdot\, D^2(\cos\theta_2, \cdots, \cos\theta_k)\, d\theta_2\cdots d\theta_k$$

$$\cdot \int_\delta^\pi |\sigma_N^\alpha(\theta_1)|^{2k-1} \prod_{j=2}^{k} (\cos\theta_1 - \cos\theta_j)^2 d\theta_1.$$

As

$$\prod_{j=2}^{k} (\cos\theta_1 - \cos\theta_j)^2 = \prod_{j=2}^{k} ((1 - \cos\theta_1) - (1 - \cos\theta_j))^2$$

$$= \sum_{s_1+\cdots+s_k = 2(k-1)} a_{s_1\cdots s_k}(1 - \cos\theta_1)^{s_1}(1 - \cos\theta_2)^{s_2}\cdots(1 - \cos\theta_k)^{s_k},$$

where $a_{s_1\cdots s_k}$ are absolute constants depending only on $s_1, \cdots, s_k$, thus

$$I_2 \leqslant \sum_{s_1+\cdots+s_k = 2(k-1)} |a_{s_1\cdots s_k}| \underset{\delta \geqslant \theta_2 \geqslant \cdots \geqslant \theta_k \geqslant 0}{\int\cdots\int} (1 - \cos\theta_2)^{s_2}\cdots$$

$$\cdot\, (1 - \cos\theta_k)^{s_k} |\sigma_N^\alpha(\theta_2)\cdots\sigma_N^\alpha(\theta_k)|^{2k-1}$$

$$\cdot\, D^2(\cos\theta_2, \cdots, \cos\theta_k)\, d\theta_2\cdots d\theta_k \int_\delta^\pi (1 - \cos\theta_1)^{s_1}$$

$$\cdot\, |\sigma_N^\alpha(\theta_1)|^{2k-1} d\theta_1.$$

Take an arbitrary integral on the right side of the above expression and denote it by

$$I_{s_1\cdots s_k} = \int_\delta^\pi (1 - \cos\theta_1)^{s_1} |\sigma_N^\alpha(\theta_1)|^{2k-1} d\theta_1$$

$$\cdot \underset{\delta \geqslant \theta_2 \geqslant \cdots \geqslant \theta_k \geqslant 0}{\int\cdots\int} (1 - \cos\theta_2)^{s_2}\cdots(1 - \cos\theta_k)^{s_k} |\sigma_N^\alpha(\theta_2)\cdots\sigma_N^\alpha(\theta_k)|^{2k-1}$$

$$\cdot\, D^2(\cos\theta_2, \cdots, \cos\theta_k)\, d\theta_2\cdots d\theta_k.$$

On account of $\delta \geqslant \dfrac{1}{N}$, we have

$$|\sigma_N^\alpha(\theta)| = O(N^{-\alpha}|\theta|^{-\alpha-1}).$$

Thus

$$I_{s_1\cdots s_k} = O\Big(\int_\delta^\pi N^{-\alpha(2k-1)}\theta^{-\alpha(2k-1)-(2k-1)+2s_1} d\theta N^{2(k-1)} \delta^{2(s_1+\cdots+s_k)}\Big),$$

here we resort to the assumption for induction. Because of

$$\alpha > \frac{n-2}{n-1} = \frac{2k-2}{2k-1}, \quad 0 \leqslant s_1 \leqslant 2(k-1),$$

and

$$s_1 + \cdots + s_k = 2(k-1),$$

we have

$$I_{s_1 \cdots s_k} = O((N\delta)^{-\alpha(2k-1)+(2k-2)}).$$

Consequently,

$$I_2 = O((N\delta)^{-\alpha(2k-1)+2k-2}).$$

We now consider I_3 and replace R_3 by a larger domain $R_3^* = (\pi \geqslant \theta_1 \geqslant \delta) \cdot (\pi \geqslant \theta_2 \geqslant \delta) \cdot \delta \geqslant \theta_3 \geqslant \cdots \geqslant \theta_k \geqslant 0$. By I_3, we still denote the part of the integral in (7.2.2) corresponding to the domain R_3^*. Then

$$I_3 = \frac{1}{(2\pi)^k} \int \cdots \int_{\delta \geqslant \theta_3 \geqslant \cdots \geqslant \theta_k \geqslant 0} |\sigma_N^\alpha(\theta_3) \cdots \sigma_N^\alpha(\theta_k)|^{2k-1}$$

$$D^2(\cos\theta_1, \cdots, \cos\theta_k) d\theta_3 \cdots d\theta_k \cdot \int_\delta^\pi |\sigma_N^\alpha(\theta_2)|^{2k-1}$$

$$\cdot \prod_{i=3}^k (\cos\theta_2 - \cos\theta_i)^2 d\theta_2 \int_\delta^\pi |\sigma_N^\alpha(\theta_1)|^{2k-1} \prod_{j=2}^k (\cos\theta_1 - \cos\theta_j)^2 d\theta_1.$$

Similarly,

$$\prod_{j=2}^k (\cos\theta_1 - \cos\theta_j)^2 \prod_{i=3}^k (\cos\theta_2 - \cos\theta_i)^2$$

$$= \prod_{j=2}^k ((1-\cos\theta_1) - (1-\cos\theta_j))^2 \prod_{l=3}^k ((1-\cos\theta_2) - (1-\cos\theta_l))^2$$

$$= \sum_{t_1+\cdots+t_k = 4k-6} b_{t_1 \cdots t_k} (1-\cos\theta_1)^{t_1} (1-\cos\theta_2)^{t_2} \cdots (1-\cos\theta_k)^{t_k},$$

where $b_{t_1, \cdots, t_k}$ are absolute constants depending only on $t_1, \cdots t_k$, and

$$\int_\delta^\pi (1-\cos\theta_1)^{t_1} |\sigma_N^\alpha(\theta_1)|^{2k-1} d\theta_1 \int_\delta^\pi (1-\cos\theta_2)^{t_2} \cdot |\sigma_N^\alpha(\theta_2)|^{2k-1} d\theta_2$$

$$\cdot \int \cdots \int_{\delta \geqslant \theta_3 \geqslant \cdots \geqslant \theta_k \geqslant 0} (1-\cos\theta_3)^{t_3} \cdots (1-\cos\theta_k)^{t_k} |\sigma_N^\alpha(\theta_3) \cdots \sigma_N^\alpha(\theta_k)|^{2k-1}$$

$$\cdot D^2(\cos\theta_3, \cdots, \cos\theta_k) d\theta_3 \cdots d\theta_k$$

$$= O\left(\left(\int_\delta^\pi N^{-\alpha(2k-1)} \cdot \theta_1^{-\alpha(2k-1)-(2k-1)+2t_1} d\theta_1 \int_\delta^\pi N^{-\alpha(2k-1)} \theta_2^{-\alpha(2k-1)-(2k-1)+2t_2} d\theta_2\right.\right.$$

$$\left.\left.\cdot N^{4(k-2)} \delta^{2(4k-6-t_1-t_2)}\right)\right)$$

$$= O((N\delta)^{2(-2(2k-1)\alpha+2k-4)}),$$

where the assumption for induction is used. Therefore we get

$$I_3 = O((N\delta)^{2(-(2k-1)\alpha+2k-4)}.$$

The same method leads to

$$I_j = O((N\delta)^{(j-1)(-(2k-1)\alpha+2k-2j+2)}),$$

for $j = 4, \cdots, k, k+1$.

Finally we consider I_1. We have

$$I_1 = \int_{\delta\geq\theta_1\geq\cdots\geq\theta_k\geq 0} |\sigma_N^\alpha(\theta_1)\cdots\sigma_N^\alpha(\theta_k)|^{2k-1} D^2(\cos\theta_1, \cdots, \cos\theta_k)\, d\theta_1\cdots d\theta_k$$

$$\leq \int\cdots\int_{\delta\geq\theta_2\geq\cdots\geq\theta_k\geq 0} |\sigma_N^\alpha(\theta_2)\cdots\sigma_N^\alpha(\theta_k)|^{2k-1} D^2(\cos\theta_2, \cdots, \cos\theta_k) d\theta_2\cdots d\theta_k$$

$$\cdot \int_{\theta_2}^{\delta} |\sigma_N^\alpha(\theta_1)|^{2k-1} \prod_{j=2}^{k} (\cos\theta_1 - \cos\theta_j)^2 d\theta_1$$

$$= O(N^{2(k-1)}N^{2(k-1)}\int_{\theta_2}^{\delta} |\sigma_N^\alpha(\theta_1)|\, d\theta_1 \delta^{4(k-1)}$$

$$\cdot \int\cdots\int_{\delta\geq\theta_2\geq\cdots\geq\theta_k\geq 0} |\sigma_N^\alpha(\theta_2)\cdots\sigma_N^\alpha(\theta_k)|^{2k-3} D^2(\cos\theta_2, \cdots, \cos\theta_k) d\theta_2\cdots d\theta_k)$$

$$= O\Bigg((N\delta)^{4(k-1)} \int\cdots\int_{\delta\geq\theta_2\geq\cdots\geq\theta_k\geq 0} |\sigma_N^\alpha(\theta_2)\cdots\sigma_N^\alpha(\theta_k)|^{2k-3}$$

$$\cdot D^2(\cos\theta_2, \cdots, \cos\theta_k) d\theta_2\cdots d\theta_k\Bigg).$$

In the light of the assumption for induction, the right-hand side in the above expression is equal to $O((N\delta)^{4(k-1)})$.

Summing up, we conclude

$$I = O((N\delta)^{4(k-1)}) + O((N\delta)^{-(2k-1)+2k-2}) + \cdots + O((N\delta)^{-k(2k-1)\alpha}).$$

Taking $\delta = 1/N$ leads to

$$I = O(1),$$

which proves Theorem 7.2.2.

§ 7.3 Proof of Riesz-Type Theorem

Now we turn to prove Theorem 7.2.1. From

$$\frac{1}{C}\int_{SO(n)} K_N^\alpha(\Gamma)\dot{\Gamma} = 1$$

and (7.1.6), it follows that

$$\sum_N^\alpha (\Gamma) - u(\Gamma) = \frac{1}{C}\int_{SO(n)} (u(W\Gamma) - u(\Gamma))K_N^\alpha(W)\dot{W}, \quad (7.3.1)$$

where $\sum_N^\alpha (\Gamma)$ denotes the (c,α) sum of the Fourier series of $u(\Gamma)$. What we want to show is that the right side of (7.3.1) tends to zero as $N\to\infty$. Below we prove only the case for $n=2k$ and the case for odd n can be proved in the same way. Divide the integral on the right-hand side of (7.3.1) into two parts: one is over G (for the definition of G see § 6.1 in Chapter 6), the other is over the complement of G. First we consider the part over G. It suffices to prove that, when $N\to\infty$

$$I = \frac{1}{(2\pi)^k}\int\cdots\int_{\pi\geqslant\theta_1\geqslant\cdots\geqslant\theta_k\geqslant 0} \varphi(\theta_1,\cdots,\theta_k)(\sigma_N^\alpha(\theta_1)\cdots\sigma_N^\alpha(\theta_k))^{2k-1}$$
$$\cdot D^2(\cos\theta_1,\cdots,\cos\theta_k)d\theta_1\cdots d\theta_k \quad (7.3.2)$$

tends to zero, where

$$\varphi(\theta_1,\cdots,\theta_k) = \int_{[\Sigma]}(u(W\Gamma) - u(\Gamma))\dot{\Sigma}.$$

As in the proof of Theorem 7.2.2, we divide the integral domain

$$\pi\geqslant\theta_1\geqslant\cdots\geqslant\theta_k\geqslant 0$$

into $R_1, R_2, \cdots, R_{k+1}$.

By Q_j we denote the part of the integral in (7.3.2) corresponding to the sub-domain R_j. Since $u(\Gamma)$ is continuous on $SO(n)$, the following inequality is valid

$$|\varphi(\theta_1,\cdots,\theta_k)| \leqslant L,$$

where L is an absolute constant. Take $\delta\geqslant 1/N$. Then

$$|Q_2| \leqslant \frac{L}{(2\pi)^k}\int\cdots\int_{\delta\geqslant\theta_2\geqslant\cdots\geqslant\theta_k\geqslant 0} |\sigma_N^\alpha(\theta_2)\cdots\sigma_N^\alpha(\theta_k)|^{2k-1}\cdot D^2(\cos\theta_2,\cdots,\cos\theta_k)d\theta_2\cdots d\theta_k$$
$$\cdot\int_\delta^\pi |\sigma_N^\alpha(\theta_1)|^{2k-1}\prod_{j=2}^k(\cos\theta_1 - \cos\theta_j)^2 d\theta_1$$
$$= \frac{L}{(2\pi)^k}|I_2|,$$

where I_2 is just I_2 appearing in the proof of Theorem 7.2.1.

Thus

$$|Q_2| = O((N\delta)^{-(2k-1)\alpha+2k-2}).$$

The same argument gives us

$$|Q_j| = O((N\delta)^{(j-1)(-(2k-1)\alpha+2k-2j+2)}),$$

for $j = 3,\cdots,k+1$. Therefore

$$\sum_{j=2}^{k+1} |Q_j| = \sum_{j=2}^{k+1} O((N\delta)^{(j-1)(-(2k-1)\alpha+2k-2j+2)}).$$

Finally we consider Q_1. We have

$$Q_1 = \frac{1}{(2\pi)^k} \int\cdots\int_{\delta\geqslant\theta_1\geqslant\cdots\geqslant\theta_k\geqslant 0} \varphi(\theta_1,\cdots,\theta_k)(\sigma_N^\alpha(\theta_1)\cdots\sigma_N^\alpha(\theta_k))^{2k-1}$$
$$\cdot D^2(\cos\theta_1,\cdots,\cos\theta_k)d\theta_1\cdots d\theta_k.$$

As $u(\Gamma)$ is continuous on $SO(n)$, for any given $\eta > 0$, we can choose δ sufficiently small such that, when $\delta \geqslant \theta_1 \geqslant \cdots \geqslant \theta_k \geqslant 0$, we have

$$|\varphi(\theta_1,\cdots,\theta_k)| < \eta.$$

By Theorem 7.2.1, when $\alpha > (n-2)/(n-1)$, we have

$$\frac{1}{C}\int_{SO(n)} |K_N^\alpha(\Gamma)|\dot{\Gamma} \leqslant M.$$

Thus, for any $\varepsilon > 0$, we can choose δ sufficiently small such that

$$|Q_1| < \varepsilon/2.$$

Having chosen $\delta = \delta(\varepsilon)$, because of $\alpha(n-1)-(n-2) > 0$, we can choose N sufficiently large such that

$$|Q_2| + \cdots + |Q_{k+1}| < \varepsilon/2.$$

Therefore, for any given $\varepsilon > 0$, we have

$$|I| < \varepsilon.$$

The rest part of the integral in (7.3.1) can be treated similarly. This proves Theorem 7.2.1.

From the proof of the theorem, it can be seen that the condition that "$u(\Gamma)$ is continuous on $SO(n)$" can be weakened as "$u(\Gamma)$ is bounded and integrable on $SO(n)$", under which $\sum_N^\alpha(\Gamma)$, as $N\to\infty$, tends to $u(\Gamma)$ at those points where $u(\Gamma)$ is continuous.

As to the condition $\alpha > \frac{n-2}{n-1}$ in Theorem 7.2.1, it is the best possible one when $n=2$, and $\frac{n-2}{n-1} = 1/2$ when $n=3$. In the latter case it is easily shown that Theorem 7.2.1 holds valid for $\alpha > 0$. However, whether α can be further improved for $n \geqslant 4$ is an open problem.

§ 7.4 Fejér Summation

For $\alpha = 1$, the Cesàro summation is just the Fejér summation, and, in this case, the Fejér kernel takes a much simpler form than the general (c,α) kernel. In the meantime, the Riesz-type theorem becomes the Fejér -type theorem: If $u(\Gamma)$ is continuous on $SO(n)$, then its Fourier series is Fejér summable to itself.

The Fejér kernel can be denoted as

$$K_N(\Gamma) = K'_N(\Gamma) = \frac{1}{B_N}\left|\det\left(\frac{\sum_{j=0}^{N}\Gamma^{j(N-j+1)}}{N+1}\right)\right|^{n-1}, \tag{7.4.1}$$

where

$$B_N = B_N^1 = \frac{1}{C}\int_{SO(n)}\left|\det\left(\frac{\sum_{j=0}^{N}\Gamma^{j(N-j+1)}}{N+1}\right)\right|^{n-1}\dot{\Gamma}. \tag{7.4.2}$$

When $n = 2k$, (7.4.1), except a manifold of lower dimension, becomes

$$\frac{1}{B_N}\cdot\frac{1}{(N+1)^{\frac{n(n-1)}{2}}}\cdot\left|\frac{\det(I-\Gamma^{N\pm1})}{\det(I-\Gamma)}\right|^{n-1},$$

where

$$B_N = \frac{1}{(N+1)^{\frac{n(n-1)}{2}}C}\int_{SO(n)}\left|\frac{\det(I-\Gamma^{N+1})}{\det(I-\Gamma)}\right|^{n-1}\dot{\Gamma}.$$

Obviously the Fejér kernel is positive definite. Thus, for $\alpha > 1$, all (c,α) kernels are positive definite. Naturally, we can give a direct proof for the Fejér-type theorem by using the definite positivity of the Fejér kernel.

The coefficients in Fejér summation are given by

$$B_m = \frac{1}{N(m)B_N C}\int_{SO(n)}\sigma_m(\Gamma)\left|\det\left(\frac{\sum_{j=0}^{N}\Gamma^{j(N+1-j)}}{N+1}\right)\right|^{n-1}\dot{\Gamma}, \tag{7.4.3}$$

and the Fejér sum of the Fourier series of $u(\Gamma)$ becomes

$$\frac{1}{CB_N}\int_{SO(n)}u(W\Gamma)\left|\det\left(\frac{\sum_{j=0}^{N}W^{j(N+1-j)}}{N+1}\right)\right|^{n-1}\dot{W}.$$

§7.5 Explicit Expression for Coefficients

In this section we will derive the explicit expressions for B_m and B_N, from which it could be seen how to find out the explicit expressions for general B_m^a and B_N^a.

By (7.4.3), $B_mN(m)$ is equal to

$$\frac{2^{\frac{k^2-k}{2}}}{B_N(N+1)^{k(2k-1)}}\det(a_{ij})_{1\leqslant i,j\leqslant k}, \tag{7.5.1}$$

for $n=2k$, where

$$a_{ij}=\frac{1}{2\pi}\int_0^{2\pi}\frac{[1-\cos(N+1)\theta]^{2k-1}}{(1-\cos\theta)^{2k-i}}\cos(l_j\theta)d\theta.$$

When $n=2k+1$, $B_mN(m)$ is equal to

$$\frac{2^{\frac{k^2-k}{2}}}{B_N(N+1)^{k(2k-1)}}\det(b_{ij})_{1\leqslant i,j\leqslant k}, \tag{7.5.2}$$

where

$$b_{ij}=\frac{1}{\pi}\int_0^{\pi}\frac{[1-\cos(N+1)\theta]}{(1-\cos\theta)^{2k-i}}\sin\left(l_j+\frac{1}{2}\right)\theta\sin\frac{\theta}{2}\,d\theta.$$

Let $b_{jt(j,t=1,\cdots,k)}$ denote the elements of the determinant in (7.5.1). Since

$$\frac{1}{2\pi}\int_0^{2\pi}\frac{(1-\cos(N+1)\theta)^{2k-1}}{(1-\cos\theta)^{2k-i+1}}\sin l_t\theta d\theta=0,$$

b_{jt} can be expressed by

$$\frac{1}{2\pi}\int_0^{2\pi}\frac{(1-\cos(N+1)\theta)^{2k-1}}{(1-\cos\theta)^{2k-j}}e^{il_t\theta}d\theta,$$

which means that

$$\frac{(-2)^{1-j}}{2\pi}\int_0^{2\pi}\frac{(1-e^{i(N+1)\theta})^{4k-2}e^{il_t\theta}d\theta}{(1-e^{i\theta})^{4k-2j}e^{i((2k-1)N+j-1)\theta}}$$

$$=\lim_{r\to1}\frac{(-2)^{1-j}}{2\pi i}\int_r\frac{(1-z^{N+1})^{4k-2}dz}{(1-z)^{4k-2j}z^{(2k-1)N+j-l_t}},$$

where $|z|=r$, $z=re^{i\theta}$ and $0<r<1$. However, we have

$$\frac{(1-z^{N+1})^{4k-2}}{(1-z)^{4k-2j}z^{(2k-1)N+j-l_t}}$$

$$=\sum_{e=0}^{\infty}\sum_{s=0}^{4k-2}\frac{(-1)^s(4k-2)!(4k-2j+e-1)!}{(4k-2-s)!(4k-2j-1)!s!e!}z^{e+(N+1)s-(2k-1)Nj+l_t}.$$

Thus

$$b_{jt}=\sum_{\substack{s=0\\ e\geq 0}}^{4k-2}\frac{(-1)^s(4k-2)!(4k-2j+e-1)!(-2)^{1-j}}{(4k-2j-1)!(4k-2-s)!s!e!},$$

where $e=(2k-1)N-(N+1)s+j-l_t-1(t=1,2,\cdots,k)$. Therefore, $B_mN(m)$ is equal to

$$\frac{[(4k-2)!]^k(-1)^{\frac{k(k-1)}{2}}}{(2k-1)!(2k+1)!\cdots(4k-3)!B_N(N+1)^{k(2k-1)}}\det(c_{ij})_{1\leq i,j\leq k},\quad(7.5.3)$$

where

$$c_{tj}=\sum_{\substack{s_j=0\\ e_j\geq 1-t}}^{4k-2}\frac{(-1)^{s_j}(4k-2-t+e_j)!}{s_j!(4k-2-s_j)!(e_j-1+t)!},$$

and

$$e_t=(2k-1)N-(N+1)s_t-l_t,\quad t=1,2,\cdots,k.$$

Let $a_{tj}(j,t=1,2,\cdots,k)$ denote the elements of the determinant in (7.5.2). Applying similar calculation, we find

$$a_{tj}=\sum_{\substack{s=0\\ e\geq 0}}^{4k}\frac{(-1)^s(4k)!(4k-2j+e)!(-2)^{1-j}}{(4k-s)!s!(4k-2j)!e!2},$$

where $e=2kN-(N+1)s-l_t+j-1$ $(t,j=1,2,\cdots,k)$. Inserting it into (7.5.2), we obtain that $B_mN(m)$ is equal to

$$\frac{(-1)^{\frac{1}{2}k(k-1)}((4k)!)^k}{(2k)!(2k+2)!\cdots(4k-2)!B_N(N+1)^{k(2k-1)}}\det(d_{ij})_{1\leq i,j\leq k},\quad(7.5.4)$$

for $n=2k+1$, where

$$d_{ij}=\sum_{\substack{s_j=0\\ e_j\geq 1-t}}^{4k}\frac{(-1)^{s_j}(4k-1-t+e_j)!}{(4k-s_j)!s_j!(e_j+t-1)}$$

and

$$e_t=2kN-(N+1)s_t-l_t\ (t=1,2,\cdots k).$$

Let the convention be that $\frac{1}{(-Q)!}=0$, for $Q>0$. Then we obtain that (7.5.3) is equal to

$$\frac{((2k-1)!)^k(-1)^{\frac{k(k-1)}{2}}(N+1)^{-k(2k-1)}}{(2k-1)!(2k+1)!\cdots(4k-3)!B_N}$$

$$\cdot\sum_{\substack{s_1=0\\ l_1\geq 1-k}}^{4k-2}\cdots\sum_{\substack{s_k=0\\ l_k\geq 1-k}}^{4k-2}(-1)^{s_1+\cdots+s_k}C_{s_1}^{4k-2}\cdots C_{s_k}^{4k-2}C_{2k-1}^{3k-2+e_1}\cdots C_{3k+1}^{3k-2+e_k}A_{e_1,\cdots,e_k}^{s_1,\cdots,s_k}$$

for $n=2k$, and (7.5.4) is equal to

$$\frac{((2k)!)^k(-1)^{\frac{k(k-1)}{2}}(N+1)^{-k(2k-1)}}{(2k)!(2k+2)!\cdots(4k-2)!B_N}$$

$$\cdot\sum_{\substack{s_1=0\\ l_1\geqslant 1-k}}^{4k}\cdots\sum_{\substack{s_k=0\\ l_k\geqslant 1-k}}^{4k}(-1)^{s_1+\cdots+s_k}C_{s_1}^{4k}\cdots C_{s_k}^{4k}C_{2k}^{3k-1+e_1}\cdots C_{2k}^{3k-1+e_k}\tilde{\Lambda}_{e_1,\cdots,e_k}^{s_1,\cdots,s_k}$$

for $n=2k+1$, where $e_t=(n-1)N-(N+1)s_t-l_t(t=1,2,\cdots,k)$, and

$$\Lambda_{e_1,\cdots,e_k}^{s_1,\cdots,s_k}=\det(f_{ij})_{1\leqslant i,j\leqslant k},\tag{7.5.5}$$

$$f_{ij}=\begin{cases}(4k-2-i+e_j)\cdots(3k-1+e_j)(e_j+i)\cdots(e_j+k-1), & \text{if } i\neq k;\\ 1, & \text{if } i=k,\end{cases}$$

$$\tilde{\Lambda}_{e_1,\cdots,e_k}^{s_1,\cdots,s_k}=\det(g_{ij})_{1\leqslant i,j\leqslant k},\tag{7.5.6}$$

$$g_{ij}=\begin{cases}(4k-1-i+e_j)\cdots(3k+e_j)(e_j+i)\cdots(e_j+k-1), & \text{if } i\neq k;\\ 1, & \text{if } i=k.\end{cases}$$

Since

$$\begin{aligned}(3k-1+e_t)(e_t+k-1)&=((2k-1+e_t)+k)\cdot((2k-1+e_t)-k)\\&=(2k-1+e_t)^2-k^2,\end{aligned}$$

$$\begin{aligned}(3k+e_t)(3k-1+e_t)(e_t+k-1)(e_t+k-2)&\\=[(2k-1+e_t)^2-k^2][(2k-1+e_t)^2-(k+1)^2],&\end{aligned}$$

$$\cdots\cdots\cdots\cdots\cdots\cdots$$

$$\begin{aligned}(4k-3+e_t)\cdots(3k-1+e_t)(e_t+k-1)\cdots(e_t+1)&\\=[(2k-1+e_t)^2-(2k-2)^2][(2k-1+e_t)^2&\\-(2k-3)^2]\cdots[(2k-1+e_t)^2-k^2],&\end{aligned}$$

the determinant (7.5.5) can be simplified to

$$\Lambda_{e_1,\cdots,e_k}^{s_1,\cdots,s_k}=\begin{vmatrix}(2k-1+e_1)^{2(k-1)}, & \cdots, & (2k-1+e_k)^{2(k-1)}\\ & \cdots\cdots\cdots & \\ (2k-1+e_1)^4, & \cdots, & (2k-1+e_k)^4\\ (2k-1+e_1)^2, & \cdots, & (2k-1+e_k)^2\\ 1, & \cdots\cdots, & 1\end{vmatrix}$$

$$=\prod_{0\leqslant j<l\leqslant k-1}(j^2-l^2)$$

$$\cdot N((n-1-s_1)(N+1)-m_1,\cdots,(N-1-s_k)(N+1)-m_k).$$

Similarly, (7.5.6) can be simplified to

$$\tilde{A}^{s_1\cdots s_k}_{e_1\cdots e_k} = \frac{1}{\prod_{t=1}^{k}\left((n-1-s_t)(N+1)-l_t-\frac{1}{2}\right)}$$

$$\cdot \prod_{0\leqslant j<l\leqslant k-1}\left(\left(j+\frac{1}{2}\right)^2-\left(l+\frac{1}{2}\right)^2\right)\frac{1}{2}\cdot\frac{3}{2}\cdots\left(k-\frac{1}{2}\right)$$

$$\cdot N((n-1-s_1)(N+1)-m_1,\cdots,(n-1-s_k)(N+1)-m_k).$$

Theorem 7.5.1 *When $n=2k$, it implies*

$$B_mN(m)=\frac{((2k-1)!)^k\prod_{0\leqslant j<l\leqslant k-1}(l^2-j^2)}{(2k-1)!(2k+1)!\cdots(4k-3)!B_N(N+1)^{k(2k-1)}}$$

$$\cdot\sum_{\substack{s_1=0\\ e_1\geqslant 1-k}}^{4k-2}\cdots\sum_{\substack{s_k=0\\ e_k\geqslant 1-k}}^{4k-2}(-1)^{s_1+\cdots+s_k}C_{s_1}^{4k-2}\cdots C_{s_k}^{4k-2}C_{2k-1}^{3k-2+e_1}\cdots C_{2k-1}^{3k-2+e_k}$$

$$\cdot N((n-1-s_1)(N+1)-m_1,\cdots,(n-1-s_k)(N+1)$$
$$-m_k).$$

When $n=2k+1$, it implies

$$B_mN(m)=\frac{((2k)!)^k\frac{1}{2}\cdot\frac{3}{2}\cdots\left(k-\frac{1}{2}\right)\prod_{0\leqslant j<l\leqslant k-1}\left(\left(l+\frac{1}{2}\right)^2-\left(j+\frac{1}{2}\right)^2\right)}{(2k)!(2k+2)!\cdots(4k-2)!B_N(N+1)^{k(2k-1)}}$$

$$\cdot\sum_{\substack{s_1=0\\ e_1\geqslant 1-k}}^{4k}\cdots\sum_{\substack{s_k=0\\ e_k\geqslant 1-k}}^{4k}(-1)^{s_1+\cdots+s_k}C_{s_1}^{4k}\cdots C_{s_k}^{4k}C_{2k}^{3k-1+e_1}\cdots C_{2k}^{3k-1+e_k}$$

$$\cdot\prod_{t=1}^{k}\left((n-1-s_t)(N+1)-l_t-\frac{1}{2}\right)^{-1}N((n-1-s_1)(N+1)$$
$$-m_1,\cdots,(n-1-s_k)(N+1)-m_k),$$

where $e_t=(n-1)N-(N+1)s_t-l_t,(t=1,2,\cdots k)$.

When $m=(0,\cdots,0)$ in particular, there are $B_m=1$ and $N(m)=1$.

Corollary 7.5.1 *When $n=2k$, we admit*

$$B_N=\frac{((2k-1)!)^k\prod_{0\leqslant j<l\leqslant k-1}(l^2-j^2)}{(2k-1)!(2k+1)!\cdots(4k-3)!(N+1)^{k(2k-1)}}$$

$$\cdot\sum_{\substack{s_1=0\\ e_1\geqslant 1-k}}^{2k-2}\cdots\sum_{\substack{s_k=0\\ e_k\geqslant 1-k}}^{2k-2}(-1)^{s_1+\cdots+s_k}\cdot C_{s_1}^{4k-2}\cdots C_{s_k}^{4k-2}C_{2k-1}^{3k-2+e_1}\cdots C_{2k-1}^{3k-2+e_k}$$

$$\cdot N((n-1-s_1)(N+1),\cdots,(n-1-s_k)(N+1)).$$

When $n = 2k + 1$, *we admit*

$$B_N = \frac{((2k)!)^k \frac{1}{2} \cdot \frac{3}{2} \cdots \left(k - \frac{1}{2}\right) \prod_{0 \leqslant j < l \leqslant k-1} \left(\left(l + \frac{1}{2}\right)^2 - \left(j + \frac{1}{2}\right)^2\right)}{(2k)!(2k+2)! \cdots (4k-2)!(N+1)^{k(2k-1)}}$$

$$\cdot \sum_{\substack{s_1=0 \\ e_1 \geqslant 1-k}}^{2k-1} \cdots \sum_{\substack{s_k=0 \\ e_k \geqslant 1-k}}^{2k-1} (-1)^{s_1+\cdots+s_k} C_{s_1}^{4k} \cdots C_{s_k}^{4k} C_{2k}^{3k-1+e_1} \cdots C_{2k}^{3k-1+e_k}$$

$$\cdot \prod_{t=1}^{k} \left((n-1-s_t)(N+1) - k + t - \frac{1}{2}\right)^{-1}$$

$$\cdot N((n-1-s_t)(N+1), \cdots, (n-1-s_k)(N+1)),$$

where $e_t = (n-1)N - (N+1)s_t - k + t (t = 1, 2, \cdots, k)$.

§ 7.6 Approximation by Cesàro Means

Let Γ and W be two points on $SO(n)$, $\Gamma = (r_{ij})_{1 \leqslant i,j \leqslant n}$, and $W = (w_{ij})_{1 \leqslant i,j \leqslant n}$. Thus the square of the Euclidean distance $d(\Gamma, W)$ between Γ and W is equal to

$$\sum_{i,j}^{n} |r_{ij} - w_{ij}|^2 = \operatorname{tr}((\Gamma - W)(\Gamma - W)')$$

$$= \operatorname{tr}(2I - W\Gamma' - \Gamma W').$$

For $n = 2k$, it is inferred that $W\Gamma' \sim C(\theta_1) + \cdots + C(\theta_k)$ and, for $n = 2k + 1$, it is inferred that $W\Gamma' \sim C(\theta_1) + \cdots + C(\theta_k) + 1$. That is

$$(d(\Gamma, W))^2 = 2 \sum_{j=1}^{k} (1 - \cos\theta_j).$$

Let $u(\Gamma)$ be a continuous function on $SO(n)$. Then

$$w(\Gamma, \delta) = \max_{d(\Gamma, w) \leqslant \delta} |u(\Gamma) - u(w)|$$

is called the modulus of continuity of $u(\Gamma)$. If $w(\Gamma, \delta) = O(\delta^p)$, then $u(\Gamma)$ is referred to as "satisfying Lipschitz condition", and denoted by $u(\Gamma) \in \operatorname{Lip} p$.

Imitating the method used in Chapter 4, we can prove the following approximation theorem by Cesàro means.

Theorem 7.6.1 *If* $u(\Gamma)$ *is a continuous function on rotation group* $SO(n)$ *and* $u(\Gamma) \in \operatorname{Lip} p (0 < p < 1)$, *then the Nth term* $\sum_N^{a} (\Gamma)$ *of* (c, α) *means of its Fourier series satisfies*

$$\text{(i)} \quad \left| u(\Gamma) - \sum_N^{\alpha} (\Gamma) \right| = O(N^{-p}),$$

if $\alpha(n-1) - n + 2 > p$;

$$\text{(ii)} \quad \left| u(\Gamma) - \sum_N^{\alpha} (\Gamma) \right| = O(N^{-p} \ln N),$$

if $\alpha(n-1) - n + 2 = p$;

$$\text{(iii)} \quad \left| u(\Gamma) - \sum_N^{\alpha} (\Gamma) \right| = O(N^{-\alpha(n-1)+n-2}),$$

if $\alpha(n-1) - n + 2 < p$.

Chapter 8. Partial Sum of Fourier Series on Rotation Groups

§ 8.1 Dirichlet Kernels

Let $u(\Gamma)$ be an integrable function on $SO(n)$, $\Gamma \in SO(n)$, its Fourier series be (6.1.12) and (6.1.14), and $l_1 = m_1 + k - 1, \cdots, l_k = m_k$. Then

$$s_N(\Gamma) = \sum_{N \geqslant l_1 > \cdots > l_k \geqslant 0} \operatorname{tr}(A_{f_1 \cdots f_n} \phi'_{f_1 \cdots f_n}(\Gamma)), \tag{8.1.1}$$

for $n = 2k + 1$, and

$$s_N(\Gamma) = \sum_{N \geqslant l_1 > \cdots > l_k \geqslant 0} \operatorname{tr}(A_{f_1 \cdots f_n} \phi'_{f_1 \cdots f_n}(\Gamma))$$

$$+ \sum_{N \geqslant l_1 > \cdots > l_k > 0} \operatorname{tr}(B_{f_1 \cdots f_n} \psi'_{f_1 \cdots f_n}), \tag{8.1.2}$$

for $n = 2k$. They are called the partial sum of the Fourier series of $u(\Gamma)$.

Now we try to find out its Dirichlet kernel. Obviously,

$$s_N(\Gamma_0) = \frac{1}{C} \int_{SO(n)} u(\Gamma \Gamma_0) \sum_{N \geqslant l_1 > \cdots > l_k \geqslant 0} N(m) \sigma_m(\Gamma) \dot{\Gamma},$$

where $\sigma_m(\Gamma)$ is given by (6.1.7) for $n = 2k + 1$; and

$$\sigma_m(\Gamma) = \sigma_m^{\phi}(\Gamma) + \sigma_m^{\psi}(\Gamma) \quad \text{for } n = 2k,$$

where σ_m^{ϕ} and σ_m^{ψ} are given by (6.1.4) and (6.1.5), respectively, for $m_1 \geqslant m_2 \geqslant \cdots \geqslant m_k > 0$; and $\sigma_m(\Gamma)$ is given by (6.1.2) for $m_1 \geqslant \cdots \geqslant m_{k-1} \geqslant m_k = 0$.

$$\mathscr{D}_N(\Gamma) = \sum_{N \geqslant l_1 > \cdots > l_n \geqslant 0} N(m) \sigma_m(\Gamma) \tag{8.1.3}$$

is referred to as the Dirichlet kernel of the Fourier series on $SO(n)$.

We will prove the following result (see Sheng Kung (Gong Sheng) [6]).

Theorem 8.1.1 *If $n=2k$, then $\mathscr{D}_N(\Gamma)$, defined by (8.1.3), is expressed as*

$$\mathscr{D}_N(\Gamma)=\frac{\det(d_N^{(2i-2)}(\theta_j))_{1\leqslant i,j\leqslant k}}{\det(d_{k-1}^{(2i-2)}(\theta_j))_{1\leqslant i,j\leqslant k}}; \tag{8.1.4}$$

if $n=2k+1$, then

$$\mathscr{D}_N(\Gamma)=\frac{\det(e_N^{(2i-1)}(\theta_j))_{1\leqslant i,j\leqslant k}}{\det(e_{k-1}^{(2i-1)}(\theta_j))_{1\leqslant i,j\leqslant k}}, \tag{8.1.5}$$

where $d_\nu(\theta)$ is the Dirichlet kernel $\sin(\nu+1/2)\theta/\sin\frac{\theta}{2}$ of one variable, and

$$e_\nu(\theta)=\frac{\sin(\nu+1)\theta}{\sin(\theta/2)}. \tag{8.1.6}$$

(8.1.4) can be rewritten as

$$\mathscr{D}_N(\Gamma)=\frac{\det(d_N^{(2i-2)}(\theta_j))_{1\leqslant i,j\leqslant k}}{a_{2k}c(k-1,\cdots,1,0)} \tag{8.1.7}$$

and (8.1.5) can be rewritten as

$$\mathscr{D}_N(\Gamma)=\frac{\det(e_N^{(2i-1)}(\theta_j))_{1\leqslant i,j\leqslant k}}{a_{2k+1}s\left(k-\frac{1}{2},\cdots,\frac{3}{2},\frac{1}{2}\right)}, \tag{8.1.8}$$

where

$$a_{2k}=\frac{(2k-2)!\cdots4!2!}{2^{k-1}},$$

$$a_{2k+1}=\frac{(2k-1)!\cdots3!1!}{(-2i)^k},$$

$c(p_1,\cdots,p_k)$ and $s(p_1,\cdots,p_k)$ are defined by (6.1.1) and (6.1.6), respectively.

First we show an identity.

Lemma 8.1.1 *Let $l_1,\cdots,l_k$ be integers with $l_1>l_2>\cdots>l_k\geqslant 0$, $p_j(l_i)$ be a function depending only on l_i $(l_i=1,\cdots,k)$, N be a positive integer, a and b be arbitrary real numbers. Then*

$$\sum_{N\geqslant l_1>l_1>\cdots>l_k\geqslant 0}\begin{vmatrix} l_1^{a-b}, & l_2^{a-b}, & \cdots, & l_k^{a-b}\\ l_1^{2a-b}, & l_2^{2a-b}, & \cdots, & l_k^{2a-b}\\ \cdots & \cdots & \cdots & \cdots\\ l_1^{ka-b}, & l_2^{ka-b}, & \cdots, & l_k^{ka-b}\end{vmatrix}$$

$$\cdot\begin{vmatrix} p_1(l_1) & p_2(l_1), & \cdots & p_k(l_1)\\ p_1(l_2), & p_2(l_2), & \cdots & p_k(l_2)\\ \cdots & \cdots & \cdots & \cdots\\ p_1(l_k), & p_2(l_k), & \cdots & p_k(l_k)\end{vmatrix}=\det\left(\sum_{l=0}^{N} l^{ai-b}p_j(l)\right)_{1\leqslant i,j\leqslant k}. \tag{8.1.9}$$

Proof. The left side of the preceding formula is equal to

$$\frac{1}{k!}\sum_{l_1=0}^{N}\cdots\sum_{l_k=0}^{N}\det\left(\sum_{\mu=1}^{k} l_\mu^{ia-b}p_j(l_\mu)\right)_{1\leqslant i,j\leqslant k}.$$

$\det\left(\sum_{k=1}^{k} l_\mu^{ia-b}p_j(l_\mu)\right)_{1\leqslant i,j\leqslant k}$, however, can be separated into k^k determinants, each being

$$\begin{vmatrix} l_{i_1}^{a-b}p_1(l_{i_1}), & l_{i_1}^{a-b}p_2(l_{i_1}), & \cdots, & l_{i_1}^{a-b}p_k(l_{i_1}) \\ l_{i_2}^{2a-b}p_1(l_{i_2}), & l_{i_2}^{2a-b}p_2(l_{i_2}), & \cdots, & l_{i_2}^{2a-b}p_k(l_{i_2}) \\ \cdots & \cdots & \cdots & \cdots \\ l_{i_k}^{ka-b}p_1(l_{i_k}), & l_{i_k}^{ka-b}p_2(l_{i_k}), & \cdots, & l_{i_k}^{ka-b}p_k(l_{i_k}) \end{vmatrix}.$$

Obviously, except that $(i_1, i_2, \cdots, i_k)$ is a permutation of $(1, 2, \cdots, k)$, all the others vanish. Thus

$$\frac{1}{k!}\sum_{l_1=0}^{N}\cdots\sum_{l_k=0}^{N}\det\left(\sum_{\mu=1}^{k} l_\mu^{ia-b}p_j(l_\mu)\right)$$

is equal to

$$\sum_{l_1=0}^{N}\cdots\sum_{l_k=0}^{N}\det(l_i^{ia-b}p_j(l_i))_{1\leqslant i,j\leqslant k}.$$

Namely

$$\det\left(\sum_{l=0}^{N} l^{ia-b}p_j(l)\right)_{1\leqslant i,j\leqslant k}.$$

§ 8.2 Proof for Dirichlet Kernels

Now let us prove Theorem 8.1.1.

First consider the case $n = 2k + 1$.

The character $\sigma_m(\Gamma)$ of the irreducible representation with signature $m = (m_1, \cdots, m_k)$ is given by

$$[m] = \frac{s\left(l + \frac{1}{2}\right)}{s\left(k - \frac{1}{2}, \cdots, \frac{3}{2}, \frac{1}{2}\right)},$$

(see Murnaghan [1]), where $l = (l_1, \cdots, l_k)$, $l_1 = m_1 + k - 1, \cdots, l_{k-1} = m_{k-1} + 1$, $l_k = m_k$, $s(l + 1/2)$ is defined by (6.1.6) and the dimension of the representation is

$$N(m) = \lim_{\substack{\theta_1 \to 0 \\ \cdots \\ \theta_k \to 0}} [m],$$

which is just

$$l_1'\cdots l_k'\begin{vmatrix} 1, & l_1'^2, & \cdots, & l_1'^{2k-2} \\ 1, & l_2'^2, & \cdots, & l_2'^{2k-2} \\ \cdots & \cdots & \cdots & \cdots \\ \cdots & \cdots & \cdots & \cdots \\ 1, & l_k'^2, & \cdots, & l_k'^{2k-2} \end{vmatrix} \Big/ \left(k-\frac{1}{2}\right)\cdots\frac{1}{2}D,$$

where

$$D=\begin{vmatrix} 1, & \left(k-\frac{1}{2}\right)^2, & \cdots, & \left(k-\frac{1}{2}\right)^{2k-2} \\ 1, & \left(k-\frac{3}{2}\right)^2, & \cdots, & \left(k-\frac{3}{2}\right)^{2k-2} \\ \cdots & \cdots & \cdots & \cdots \\ \cdots & \cdots & \cdots & \cdots \\ 1, & \left(\frac{1}{2}\right)^2, & \cdots, & \left(\frac{1}{2}\right)^{2k-2} \end{vmatrix}$$

and

$$l_j' = l_j + \frac{1}{2}.$$

Thus the Dirichlet kernel

$$\mathscr{D}_N(\Gamma) = \sum_{N\geq l_1>\cdots>l_k\geq 0} N(m)[m] = \frac{P}{Q},$$

where P is equal to

$$\sum_{N\geq l_1>\cdots>l_k\geq 0}\begin{vmatrix} l_1', & l_1'^3, & \cdots, & l_1'^{2k-1} \\ l_2', & l_2'^3, & \cdots, & l_2'^{2k-1} \\ \cdots & \cdots & \cdots & \cdots \\ l_k', & l_k'^3, & \cdots, & l_k'^{2k-1} \end{vmatrix}\begin{vmatrix} \sin l_1'\theta_1, & \cdots, & \sin l_1'\theta_k \\ \sin l_2'\theta_1, & \cdots, & \sin l_2'\theta_k \\ \cdots & \cdots & \cdots \\ \sin l_k'\theta_1, & \cdots, & \sin l_k'\theta_k \end{vmatrix}$$

and Q is equal to $A\cdot B$.

$$A=\begin{vmatrix} \left(k-\frac{1}{2}\right), & \left(k-\frac{1}{2}\right)^3, & \cdots, & \left(k-\frac{1}{2}\right)^{2k-1} \\ \left(k-\frac{3}{2}\right), & \left(k-\frac{3}{2}\right)^3, & \cdots, & \left(k-\frac{3}{2}\right)^{2k-1} \\ \cdots & \cdots & \cdots & \cdots \\ \frac{1}{2}, & \left(\frac{1}{2}\right)^3, & \cdots, & \left(\frac{1}{2}\right)^{2k-1} \end{vmatrix},$$

$$B=\begin{vmatrix} \sin\left(k-\frac{1}{2}\right)\theta_1, & \cdots, & \sin\left(k-\frac{1}{2}\right)\theta_k \\ \sin\left(k-\frac{3}{2}\right)\theta_1, & \cdots, & \sin\left(k-\frac{3}{2}\right)\theta_k \\ \cdots & \cdots & \cdots \\ \sin\frac{1}{2}\theta_1, & \cdots, & \sin\frac{1}{2}\theta_k \end{vmatrix}.$$

To calculate P, we apply Lemma 8.1.1. Taking $a=2$, $b=1$ and $p_i(l) = \sin l\theta_j$, we get

$$P = \det\left(\sum_{l=0}^{N} l'^{2i-1} \sin l'\theta_j\right)_{1 \leqslant i,j \leqslant k}.$$

It is obvious that

$$\sum_{l=0}^{N} l'^{2i-1} \sin l'\theta = (-1)^i \frac{d^{2i-1}}{d\theta^{2i-1}}\left(\frac{\sin(N+1)\theta}{2\sin\theta/2}\right).$$

Therefore

$$P = (-1)^{\frac{k\cdot(k+1)}{2}} \det\left(\left(\frac{\sin(N+1)\theta_j}{2\sin\dfrac{\theta_j}{2}}\right)^{(2i-1)}\right)_{1 \leqslant i,j \leqslant k}.$$

Similarly

$$Q = (-1)^{\frac{k(k+1)}{2}} \det\left(\left(\frac{\sin k\theta_i}{2\sin\dfrac{\theta_i}{2}}\right)^{(2i-1)}\right)_{1 \leqslant i,j \leqslant k}.$$

Thus the conclusion for $n = 2k+1$ has been justified.

Next we consider the case $n = 2k$.

When $l_k = 0$, the character $\sigma_m(\Gamma)$ of the irreducible representation with signature $m = (m_1, \cdots, m_{k-1}, 0)$ reads

$$[m] = \frac{c(l_1, \cdots, l_k)}{c(k-1, \cdots, 1, 0)}$$

(see Murnaghan [1]), where $l = (l_1, \cdots, l_k)$, $l_1 = m_1 + k - 1, \cdots, l_{k-1} = m_{k-1} + 1$, $l_k = m_k$, $c(l)$ is defined by (6.1.1) and the dimension of the representation is written as

$$N(m) = \lim_{\substack{\theta_1 \to 0 \\ \cdots \\ \theta_k \to 0}} [m],$$

which is just

$$\left|\begin{matrix} 1, & l_1^2, & \cdots, & l_1^{2k-2} \\ 1, & l_2^2, & \cdots, & l_2^{2k-2} \\ \cdots & \cdots & \cdots & \cdots \\ 1, & l_k^2, & \cdots, & l_k^{2k-2} \end{matrix}\right| \Bigg/ \left|\begin{matrix} 1, & (k-1)^2, & \cdots, & (k-1)^{2k-2} \\ 1, & (k-2)^2, & \cdots, & (k-2)^{2k-2} \\ \cdots & \cdots & \cdots & \cdots \\ 1, & 1, & \cdots, & 1 \\ 1, & 0, & \cdots, & 0 \end{matrix}\right|. \tag{8.2.1}$$

When $l_k > 0$, the characters $\sigma_m^{\phi}(\Gamma)$ and $\sigma_m^{\psi}(\Gamma)$ of the irreducible representation with signature $m = (m_1, \cdots, m_k)$ are reduced to

$$[m]_+ = \frac{c(l) + s(l)}{2c(k-1, \cdots, 1, 0)}$$

and

$$[m]_- = \frac{c(l) - s(l)}{2c(k-1,\cdots,1,0)}.$$

Both of their dimensions are $N(m)$ defined by (8.2.1) (see Murnaghan [1]).

Therefore we have the Dirichlet kernel

$$\mathscr{D}_N(\Gamma) = \sum_{N\geq l_1>\cdots>l_k\geq 0} N(m)[m] = \frac{P_1}{Q_1},$$

where P_1 is equal to

$$\sum_{N\geq l_1\cdots l_k\geq 0}\begin{vmatrix} 1, & l_1^2, & \cdots, & l_1^{2k-2} \\ 1, & l_2^2, & \cdots, & l_2^{2k-2} \\ & \cdots & \cdots & \\ 1, & l_k^2, & \cdots, & l_k^{2k-2}\end{vmatrix}\begin{vmatrix} c_{l_1}(\theta_1), & \cdots, & c_{l_1}(\theta_k) \\ c_{l_2}(\theta_1), & \cdots, & c_{l_2}(\theta_k) \\ & \cdots\cdots & \\ c_{l_k}(\theta_1), & \cdots, & c_{l_k}(\theta_k)\end{vmatrix},$$

and Q_1 is equal to

$$\begin{vmatrix} 1, & (k-1)^2, & \cdots, & (k-1)^{2k-2} \\ 1, & (k-2)^2, & \cdots, & (k-2)^{2k-2} \\ & \cdots & \cdots & \\ 1, & 1, & \cdots, & 1 \\ 1, & 0, & \cdots, & 0 \end{vmatrix}\begin{vmatrix} c_{k-1}(\theta_1), & \cdots, & c_{k-1}(\theta_k) \\ & \cdots\cdots & \\ & \cdots\cdots & \\ c_1(\theta_1), & \cdots, & c_1(\theta_k) \\ c_0(\theta_1), & \cdots, & c_0(\theta_k)\end{vmatrix}.$$

To calculate P_1, we apply Lemma 8.1.1. Taking $a = 2$, $b = 2$ and $p_j(l) = c_l(\theta_j)$, we have

$$P_1 = \det\left(\sum_{l=0}^{N} l^{2s-2}c_l(\theta_t)\right)_{1\leq s,t\leq k}.$$

It is obviously that, when $s > 1$,

$$\sum_{l=0}^{N} l^{2s-2}c_l(\theta) = 2\sum_{l=0}^{N} l^{2s-2}\cos l\theta = (-1)^{s-1}\left(\frac{\sin\left(N+\frac{1}{2}\right)\theta}{\sin\frac{\theta}{2}}\right)^{(2s-2)},$$

and, when $s = 1$,

$$\sum_{l=0}^{N} l^{2s-2}c_l(\theta) = \sum_{l=0}^{N} c_l(\theta) = \frac{\sin\left(N+\frac{1}{2}\right)\theta}{\sin\frac{\theta}{2}} = d_N(\theta).$$

Consequently

$$P_1 = (-1)^{\frac{k(k-1)}{2}}\det\left(\left(\frac{\sin\left(N+\frac{1}{2}\right)\theta_j}{\sin\frac{\theta_j}{2}}\right)^{(2i-2)}\right)_{1\leq i,j\leq k}.$$

Similarly

$$Q_1 = (-1)^{\frac{k(k-1)}{2}} \det\left(\left(\frac{\sin\left(N-\frac{1}{2}\right)\theta_j}{\sin\frac{\theta_j}{2}}\right)^{(2i-2)}\right)_{1\leqslant i,j\leqslant k}.$$

Thus the conclusion for $n = 2k$ has been justified.

§ 8.3 Partial Sum of Fourier Series

As mentioned above, if $u(\Gamma)$ is an integrable function on $SO(n)(\Gamma \in SO(n))$, then the partial sums (8.1.1) and (8.1.2) of its Fourier series (6.1.12) and (6.1.14) can be expressed as

$$s_N(\Gamma_0) = \frac{1}{C}\int_{SO(n)} u(\Gamma\Gamma_0)\mathscr{D}_N(\Gamma)\dot{\Gamma}, \tag{8.3.1}$$

where $\mathscr{D}_N(\Gamma)$ is the Dirichlet kernel, and when n is odd or even, it is expressed by (8.1.5) or (8.1.4) respectively. Having found out the Dirichlet kernel, we are apt to obtain a convergence criterion of the Fourier series. We can easily show.

Lemma 8.3.1 *If $\mathscr{D}_N(\Gamma)$ is the Dirichlet kernel of the Fourier series on $SO(n)$, then*

$$\frac{1}{C}\int_{SO(n)} \mathscr{D}_N(\Gamma)\dot{\Gamma} = 1. \tag{8.3.2}$$

This lemma is immediate from the orthogonality between irreducible representations. Here we present another proof not relying on the theory of group representations.

Proof. First we consider the case $n = 2k+1$. From (8.1.8) and (6.1.20), it follows that

$$\frac{1}{C}\int_{SO(n)} \mathscr{D}_N(\Gamma)\dot{\Gamma} = \frac{(+i)^k}{(2k-1)!\cdots 3!1!\pi^k}$$
$$\cdot \int\cdots\int_{\pi\geqslant\theta_1>\cdots>\theta_k\geqslant 0} \det(e_N^{(2i-1)}(\theta_j))_{1\leqslant i,j\leqslant k}$$
$$\cdot\, s\left(k-\frac{1}{2},\cdots,\frac{3}{2},\frac{1}{2}\right) d\theta_1\cdots d\theta_k.$$

The right side is obviously equal to

$$\frac{(+i)^k}{(2k-1)!\cdots 3!1!\pi^k}\int_0^\pi\cdots\int_0^\pi s\left(k-\frac{1}{2},\cdots,\frac{3}{2},\frac{1}{2}\right)$$
$$\cdot\, e_N'(\theta_1)e_N''(\theta_2)\cdots e_N^{(2k-1)}(\theta_k)d\theta_1\cdots d\theta_k,$$

which is equivalent to

$$\frac{(+i)^k}{(2k-1)!\cdots 3!1!\pi^k}\int_0^\pi\cdots\int_0^\pi s\left(k-\frac{1}{2},\cdots,\frac{3}{2},\frac{1}{2}\right)$$

$$\cdot\left\{e'_{k-1}(\theta_1)-\left(2\left(k+\frac{1}{2}\right)\sin\left(k+\frac{1}{2}\right)\theta_1+\cdots\right.\right.$$

$$\left.\left.+2\left(N+\frac{1}{2}\right)\sin\left(N+\frac{1}{2}\right)\theta_1\right)\right\}\cdots\left\{e_{k-1}^{(2k-1)}(\theta_k)+(-1)^k\right.$$

$$\cdot\left(2\left(k+\frac{1}{2}\right)^{2k-1}\sin\left(k+\frac{1}{2}\right)\theta_k+\cdots+2\right.$$

$$\left.\left.\cdot\left(N+\frac{1}{2}\right)^{2k-1}\sin\left(N+\frac{1}{2}\right)\theta_k\right)\right\}d\theta_1\cdots d\theta_k$$

$$=\frac{(-i)^k}{(2k-1)!3!1!\pi^k}\int_{\pi>\theta_1>\cdots>\theta_k>0}\cdots\int s\left(k-\frac{1}{2},\cdots,\frac{3}{2},\frac{1}{2}\right)$$

$$\cdot\det(e_{k-1}^{(2i-1)}(\theta_j))_{1\leq i,j\leq k}d\theta_1\cdots d\theta_k=1.$$

The case $n=2k$ can be treated in the same way. It should be noticed that because of

$$\frac{1}{C}\int_{SO(n)}\mathscr{D}_N(\Gamma)\dot{\Gamma}=\frac{2}{C}\int_G\mathscr{D}_N(\Gamma)\dot{\Gamma},$$

we need to separate G before we enter into discussion.

Any $\Gamma\in SO(n)$ can be expressed as

$$\Gamma=K(c_1\dot{+}c_2\dot{+}\cdots)K',\quad K\in SO(n).$$

Take $K_1\in O(n)$ such that

$$K_1(c_1\dot{+}c_2\dot{+}\cdots)K_1'=c_{r_1}\dot{+}c_{r_2}\dot{+}\cdots,$$

where $(r_1,r_2,\cdots,r_k)$ is a permutation of $(1,2,\cdots,k)$. Let

$$\Gamma^*=KK_1(c_1\dot{+}c_2\dot{+}\cdots)K_1'K'=H\Gamma H',$$

where $H=KK_1K'\in SO(n)$. Set

$$u^*(\Gamma\Gamma_0)=\begin{cases}\dfrac{1}{k!}\displaystyle\sum_{(r_1,\cdots,r_k)}u(\Gamma^*\Gamma_0), & n=2k+1;\\[2ex] \dfrac{1}{2k!}\displaystyle\sum_{(r_1,\cdots,r_k)}(u(K_1\Gamma^*K_1'\Gamma_0)+u(P^*P_0)), & n=2k,\end{cases}$$

where $\Gamma\in G$, K_1 is a fixed orthogonal square matrix with the determinant -1. By the invariance of integrals on compact groups, we obtain

$$s_N(\Gamma_0)=\frac{1}{C}\int_{SO(n)}u^*(\Gamma\Gamma_0)\mathscr{D}_N(\Gamma)\dot{\Gamma}.$$

By (8.3.2) we get

$$s_N(\Gamma_0) - u(\Gamma_0) = \frac{1}{C}\int_{SO(n)} (u^*(\Gamma\Gamma_0) - u(\Gamma_0))\mathscr{D}_N(\Gamma)\dot{\Gamma}. \tag{8.3.3}$$

Write

$$g_{\Gamma_0}(\theta_1, \cdots \theta_k) = \frac{1}{V(\Sigma)}\int_\Sigma (u^*(\Gamma\Gamma_0) - u(\Gamma_0))\dot{\Sigma},$$

which, by the definition of u^*, is a symmetric function of $\theta_1, \cdots, \theta_k$ obviously. It is easily seen that (8.3.3) can be expressed as

$$s_N(\Gamma_0) - u(\Gamma_0) = C_0\int_0^\pi \cdots \int_0^\pi g_{\Gamma_0}(\theta_1, \cdots, \theta_k)d\tau, \tag{8.3.4}$$

where

$$C_0 = \begin{cases} \dfrac{(+i)^k}{(2k-1)!\cdots 3!1!\pi^k}, & n = 2K+1; \\ \dfrac{1}{(2k-2)!\cdots 4!2!\pi^k}, & n = 2K, \end{cases}$$

and

$$d\tau = \begin{cases} s\left(k-\frac{1}{2}, \cdots, \frac{3}{2}, \frac{1}{2}\right) e'_N(\theta_1)\cdots e_N^{(2k-1)}(\theta_k)d\theta_1\cdots d\theta_k, & n = 2k+1; \\ c(k-1, \cdots, 1, 0)d_N(\theta_1)\cdots d_N^{(2k-2)}(\theta_k)d\theta_1\cdots d\theta_k, & n = 2k. \end{cases}$$

Consider the integral

$$C_0\int_\pi^{2\pi}\int_0^\pi \cdots \int_0^\pi g_{\Gamma_0}(\theta_1, \cdots, \theta_k)d\tau,$$

which, under the substitution $\theta_1 \to 2\pi - \theta_1$, becomes

$$C_0\int_0^\pi \cdots \int_0^\pi g_{\Gamma_0}(-\theta_1, \theta_2, \cdots, \theta_k)d\tau = \frac{1}{C}$$

$$\cdot \int_{SO(n)} \{u^*(K(C_1(-\theta_1) \dotplus C_2 \dotplus \cdots)K'\Gamma_0) - u(\Gamma_0)\}\mathscr{D}_N(\Gamma)\dot{\Gamma}.$$

Take

$$K_1 = \begin{cases} \begin{pmatrix} 1 & & & & \\ & -1 & & & 0 \\ & & \ddots & & \\ 0 & & & 1 & \\ & & & & -1 \end{pmatrix}, & n = 2k+1; \\ \begin{pmatrix} 1 & & & & \\ & -1 & & & 0 \\ 0 & & 1 & & \\ & & & \ddots & \\ & & & & 1 \end{pmatrix}, & n = 2k. \end{cases}$$

Then

$$K(c_1(-\theta_1)\dotplus c_2\dotplus\cdots)K' = KK_1(c_1\dotplus c_2\dotplus\cdots)K_1'K'$$
$$= KK_1K'\Gamma KK_1'K'.$$

By the invariance of integral, the right side of (8.3.4) is just

$$\frac{C_0}{2^k}\int_0^{2\pi}\cdots\int_0^{2\pi} g_{\Gamma_0}(\theta_1,\theta_2,\cdots,\theta_k)d\tau. \tag{8.3.5}$$

§ 8.4 A Convergence Theorem of Fourier Series

First consider the case $n = 2k+1$

Let $u(\Gamma)\in c^{k^2}+P(0<P<1)$ and apply integration by parts to (8.3.5). Since $g_{\Gamma_0}(\theta_1,\cdots,\theta_k)$ is periodic, we get

$$s_N(\Gamma_0)-u(\Gamma_0) = \frac{(-i)^k}{(2k-1)!\cdots 3!1!}\int_0^{2\pi}\cdots\int_0^{2\pi}\frac{\partial}{\partial\theta_1}\cdots\frac{\partial^{2k-1}}{\partial\theta_k^{2k-1}}$$
$$\cdot\left[g_{\Gamma_0}(\theta_1,\cdots,\theta_k)s\left(k-\frac{1}{2},\ \cdots\frac{3}{2},\frac{1}{2}\right)\right]$$
$$\cdot\, e_N(\theta_1)\cdots e_N(\theta_k)d\theta_1\cdots d\theta_k. \tag{8.4.1}$$

Let $I_{\nu_1,\cdots,\nu_k}$ denote the integral

$$\frac{(-i)^k}{(2k-1)!\cdots 3!1!}\int_0^{2\pi}\cdots\int_0^{2\pi}\left(\frac{\partial^{\mu_1}}{\partial\theta_1^{\mu_1}}\cdots\frac{\partial^{\mu_k}}{\partial\theta_k^{\mu_k}}g_{\Gamma_0}(\theta_1,\cdots,\theta_k)\right)$$
$$\cdot\left(\frac{\partial^{\nu_1}}{\partial\theta_1^{\nu_1}}\cdots\frac{\partial^{\nu_k}}{\partial\theta_k^{\nu_k}}s\left(k-\frac{1}{2},\cdots,\frac{3}{2},\frac{1}{2}\right)\right)$$
$$\cdot\, e_N(\theta_1)\cdots e_N(\theta_k)d\theta_1\cdots d\theta_k,$$

where both μ_l and ν_l are nonnegative integers, and

$$\mu_l+\nu_l = 2l-1,\quad l = 1,\cdots,k.$$

Thus

$$s_N(\Gamma_0)-u(\Gamma_0) = \sum_{\nu_1\cdots\nu_k} I_{\nu_1\cdots\nu_k}. \tag{8.4.2}$$

First we treat the case that at least one out of $\nu_1,\nu_2,\cdots,\nu_k$, say ν_i, is even. In this case

$$\frac{\partial^{\nu_1}}{\partial\theta_1^{\nu_1}}\cdots\frac{\partial^{\nu_k}}{\partial\theta_k^{\nu_k}}s\left(k-\frac{1}{2},\cdots,\frac{3}{2},\frac{1}{2}\right),$$

at any rate, takes $\sin\frac{\theta_i}{2}$ as its factor. If $f(\theta)\in \mathrm{Lip}p,\ 0<p<1$, then

$$\int_0^{2\pi} f(\theta)\sin N\theta d\theta = O\left(\frac{1}{N^p}\right).$$

However

$$\int_0^{2\pi}|e_N(\theta)|d\theta \leqslant 2\int_0^{2\pi}|d_{2N}(\theta)|d\theta + 2\int_0^{2\pi}|d_N(\theta)|d\theta$$
$$= O(\ln N).$$

Consequently, from $u(\Gamma)\in c^{k^2}$, it follows that

$$I_{\nu_1,\cdots,\nu_k} = O\left(\frac{\ln^{k-1}N}{N^p}\right).$$

When all of $\nu_1,\nu_2,\cdots,\nu_k$ are odd and $\mu_1=\mu_2=\cdots=\mu_k=0$,

$$I_{1,3\cdots,2k-1} = \frac{(-i)^k}{(2k-1)!\cdots 3!1!\pi^k}\int_0^\pi\cdots\int_0^\pi g_{\Gamma_0}(\theta_1,\theta_2,\cdots\theta_k)$$
$$\cdot\left(\frac{\partial}{\partial\theta_1}\cdots\frac{\partial^{2k-1}}{\partial\theta_k^{2k-1}}s\left(k-\frac{1}{2},\cdots,\frac{3}{2},\frac{1}{2}\right)\right)$$
$$\cdot\, e_N(\theta_1)\cdots e_N(\theta_k)d\theta_1\cdots d\theta_k.$$

Separate the integral domain into

R_1: $\delta\geqslant\theta_1,\cdots,\ \theta_k\geqslant 0$;

R_2: $\pi\geqslant\theta_{i1}\geqslant\delta\geqslant\theta_1,\cdots,\hat{\theta}_{i_1},\cdots,\theta_k\geqslant 0,\ i_1=1,2,\cdots k$;

R_3: $\pi\geqslant\theta_{i_1},\theta_{i_2}\geqslant\delta\geqslant\theta_1,\cdots\hat{\theta}_{i_1},\cdots,\hat{\theta}_{i_2},\cdots\theta_k\geqslant 0$,
$i_1,i_2=1,2,\cdots,k;\ i_1\neq i_2$,

....

R_{k+1}: $\pi\geqslant\theta_1,\theta_2,\cdots,\theta_k\geqslant\delta$.

Let I_i donote

$$\frac{(-i)^k}{(2k-1)!\cdots 3!1!\pi^k}\int_{R_i} g_{\Gamma_0}(\theta_1,\cdots,\theta_k)$$
$$\cdot\left(\frac{\partial}{\partial\theta_1}\cdots\frac{\partial^{2k-1}}{\partial\theta_k^{2k-1}}s\left(k-\frac{1}{2},\cdots,\frac{3}{2},\frac{1}{2}\right)\right)$$
$$\cdot\, e_N(\theta_1)\cdots e_N(\theta_k)d\theta_1,\cdots d\theta_k.$$

First consider I_1. Since $\delta\geqslant\theta_1,\theta_2,\cdots,\theta_k\geqslant 0$, we have

$$|g_{\Gamma_0}(\theta_1,\theta_2,\cdots,\theta_k)| = O(\delta).$$

Thus

$$I_1 = O(\delta\ln^k N).$$

Next consider I_2. As $f(\theta)$ is continuously differentiable, we have

$$\int_\delta^\pi f(\theta)e_N(\theta)d\theta = \int_\delta^\pi f(\theta)d_N(\theta)d\theta + \int_\delta^\pi f(\theta)\cos\left(N+\frac{1}{2}\right)\theta d\theta$$

$$= O\left(\frac{1}{\delta}\omega\left(\frac{1}{N}\right)\right) + O\left(\frac{1}{N}\right) = O\left(\frac{1}{\delta N}\right),$$

where $\omega(x)$ denotes the modulus of continuity of f. Thus

$$I_2 = O\left(\frac{\ln^{k-1}N}{\delta N}\right).$$

Similarly

$$I_3 = O\left(\frac{\ln^{k-2}N}{\delta^2 N}\right), \cdots, I_{k+1} = O\left(\frac{1}{\delta^k N}\right).$$

Taking $\delta = (N\ln^k N)^{\frac{-1}{k+1}}$, we have

$$I_{1,3,\cdots 2k-1} = O\left(\left(\frac{\ln^{k_2}N}{N}\right)^{\frac{1}{k+1}}\right).$$

If all of $\nu_1, \nu_2, \cdots, \nu_k$ are odd and at least two of them, say ν_i and ν_j, are equal, then

$$\frac{\partial^{\nu_1}}{\partial\theta_1^{\nu_1}}\cdots\frac{\partial^{\nu_k}}{\partial\theta_k^{\nu_k}} s\left(k-\frac{1}{2}, \cdots, \frac{3}{2}, \frac{1}{2}\right)$$

must have a factor $\cos\frac{\theta_i}{2} - \cos\frac{\theta_j}{2}$. From this it follows that

$$I_{\nu_1,\cdots\nu_k} = \frac{(-i)^k}{(2k-1)!\cdots 3!1!\pi^k}$$

$$\cdot \int_0^\pi\cdots\int_0^\pi\left(\frac{\partial^{\mu_1}}{\partial\theta^{\mu_1}}\cdots\frac{\partial^{\mu_k}}{\partial\theta^{\mu_k}} gr_0(\theta_1, \cdots, \theta_k)\right)$$

$$\cdot\left(\left(1-\cos\frac{\theta_j}{2}\right) - \left(1-\cos\frac{\theta_i}{2}\right)\right)\varphi(\theta_1, \cdots, \theta_k)$$

$$\cdot e_N(\theta_1)\cdots e_N(\theta_k)d\theta_1\cdots d\theta_k,$$

where

$$\frac{\partial^{\nu_1}}{\partial\theta_1^{\nu_1}}\frac{\partial^{\nu_2}}{\partial\theta_2^{\nu_2}}\cdots\frac{\partial^{\nu_k}}{\partial\theta_k^{\nu_k}} s\left(k-\frac{1}{2}, \cdots, \frac{3}{2}, \frac{1}{2}\right)$$

$$= \left(\cos\frac{\theta_i}{2} - \cos\frac{\theta_j}{2}\right)\varphi(\theta_1, \theta_2, \cdots, \theta_k)$$

and φ is a differentiable function. To sum up, we arrive at

$$\sum_{\nu_1,\cdots,\nu_k} I_{\nu_1,\cdots,\nu_k} = O\left(\left(\frac{\ln^{k^2}N}{N}\right)^{\frac{1}{k+1}}\right).$$

We now treat the case $n = 2k$. In this case, (8.3.5) becomes

$$\frac{1}{(2k-2)!\cdots 4!2!}\frac{1}{(2\pi)^k}\int_0^{2\pi}\cdots\int_0^{2\pi} g_{\Gamma_0}(\theta_1,\cdots,\theta_k)$$
$$\cdot c(k-1,\cdots,1,0)d_N(\theta_1)d_N''(\theta_2)\cdots d_N^{(2k-2)}(\theta_k)d\theta_1\cdots d\theta_k.$$

When $u(\Gamma)\in C^{k(k-1)+p}(0<p<1)$, using integration by parts as before, we denote the integral

$$\frac{1}{(2k-2)!\cdots 4!2!}\frac{1}{(2\pi)^k}$$
$$\cdot\int_0^{2\pi}\cdots\int_0^{2\pi}\left(\frac{\partial^{\mu_1}}{\partial\theta^{\mu_1}}\cdots\frac{\partial^{\mu_k}}{\partial\theta_k^{\mu_k}}\cdot g_{\Gamma_0}(\theta_1,\cdots,\theta_k)\right)\left(\frac{\partial^{\nu_1}}{\partial\theta_1^{\nu_1}}\cdots\frac{\partial^{\nu_k}}{\partial\theta_k^{\nu_k}}\right)$$
$$\cdot c(k-1,\cdots,1,0)d\theta_1\cdots d\theta_k,$$

by $I_{\nu_1,\cdots,\nu_k}$.

If at least one of $\nu_1,\cdots,\nu_k$ is odd, we have

$$I_{\nu_1,\cdots,\nu_k} = O\left(\frac{\ln^{k-1}N}{N}\right).$$

If all of $\nu_1,\cdots,\nu_k$ are even and $\mu_1=\mu_2=\cdots=\mu_k=0$, we have

$$I_{0,2,\cdots,2k-2} = O\left(\left(\frac{\ln^{k^2}N}{N}\right)^{\frac{1}{k+1}}\right).$$

Otherwise, at least two out of $\nu_1,\cdots,\nu_k$, say ν_i and ν_j, are equal. Thus

$$\frac{\partial^{\nu_1}}{\partial\theta_1^{\nu_1}}\cdots\frac{\partial^{\nu_k}}{\partial\theta_k^{\nu_k}}c(k-1,\cdots,1,0)$$

must contain a factor $\cos\theta_i-\cos\theta_j$. Therefore

$$I_{\nu_1,\cdots,\nu_k} = O\left(\frac{\ln^{k-1}N}{N^p}\right).$$

To sum up, we arrive at the following theorem (see Wang Shikun and Dong Daozheng [1]).

Theorem 8.4.1 *Let $u(\Gamma)$ be defined on $SO(n)$, $\Gamma\in SO(n)$. Let $u(\Gamma)\in C^{k^2+p}$ for $n=2k+1$ and $u(\Gamma)\in C^{k(k-1)+p}$ for $n=2k(p>0)$. Then the Fourier series (6.1.12) and (6.1.14) of $u(\Gamma)$ are convergent. Furthermore, if $s_N(\Gamma)$ is the partial sum defined by (8.1.1) and (8.1.2), then*

$$|s_N(\Gamma)-u(\Gamma)|\leqslant A\max\left(\left(\frac{\ln^{k^2}N}{N}\right)^{\frac{1}{k+1}},\ \frac{\ln^{k-1}N}{N^p}\right),$$

where A is an absolute constant.

§ 8.5 Absolute Convergence of Fourier Series

Let $u(\Gamma)$ be integrable on $SO(n)$ and its Fourier series be (6.1.12) and (6.1.14). The Fourier series of $u(\Gamma)$ is said to be absolutely convergent, if the series

$$\sum_{m_1 \geqslant \cdots \geqslant m_n \geqslant 0} \sum_{i,j}^{N(m)} |a_{ij}^m| \cdot |\varphi_{ij}^m(\Gamma)|$$

is convergent for $n = 2k + 1$ and the series

$$\sum_{m_1 \geqslant \cdots \geqslant m_n \geqslant 0} \sum_{i,j}^{N(m)} |a_{ij}^m| \cdot |\phi_{ij}^m(\Gamma)| + \sum_{m_1 \geqslant \cdots \geqslant m_n > 0} \sum_{i,j}^{N(m)} |b_{ij}^m| \, |\varphi_{ij}^m(\Gamma)|$$

is convergent for $n = 2k$.

In virtue of the order, it can be divided into two cases.

1. When $n = 2k + 1$, we have

$$A_m A_m' = \frac{N^2(m)}{C^2} \int_{SO(n)} \int_{SO(n)} u(\Gamma) u(K) \Phi_m(\Gamma) \Phi_m(K') \dot{\Gamma} \dot{K},$$

thus
$$\operatorname{tr}(A_m A_m') = \frac{N^2(m)}{C^2} \int_{SO(n)} \int_{SO(n)} u(\Gamma K) u(K) \sigma_m(\Gamma) \dot{\Gamma} \dot{K}.$$

Write
$$g(\Gamma) = \int_{SO(n)} u(\Gamma K) u(K) \dot{K},$$

$$h(\theta_1, \cdots, \theta_k) = \frac{1}{\nu(\Sigma)} \int_{\Sigma} g(\Gamma) \dot{\Sigma},$$

$$h^*(\theta_1, \cdots, \theta_k) = \frac{1}{k!} \sum_{(r_1, \cdots, r_k)} h(\theta_{r_1}, \cdots, \theta_{r_k}),$$

where the summation index refers to all permutations of $(1, 2, \cdots, k)$. By the invariance of integrals, $\operatorname{tr}(A_m A_m')$ can be written as

$$\frac{(-1)^k N^2(m)}{c} \cdot \frac{1}{(2\pi)^k} \int \cdots \int_{\pi \geqslant \theta_1 > \cdots > \theta_k \geqslant 0} h^*(\theta_1, \cdots, \theta_k)$$

$$\cdot \sigma_m(\Gamma) s^2 \left(k - \frac{1}{2}, \cdots, \frac{3}{2}, \frac{1}{2}\right) d\theta_1, \cdots, d\theta_k. \tag{8.5.1}$$

Substituting the value of $\sigma_m(\Gamma)$ into (8.5.1), we get

$$\operatorname{tr}(A_m A_m') = \frac{(-i)^k N^2(m)}{c} \frac{1}{(2\pi)^k} \int_0^{2\pi} \cdots \int_0^{2\pi} H(\theta_1, \cdots, \theta_k)$$

$$\cdot \sin\left(l_1 + \frac{1}{2}\right) \theta_1 \cdots \sin\left(l_k + \frac{1}{2}\right) \theta_k d\theta_1 \cdots d\theta_k, \tag{8.5.2}$$

where

$$H(\theta_1,\cdots,\theta_k) = h^*(\theta_1,\cdots,\theta_k)s\left(k-\frac{1}{2},\cdots,\frac{3}{2},\frac{1}{2}\right). \tag{8.5.3}$$

Consider the integral

$$I = \int_0^{2\pi}\cdots\int_0^{2\pi}\left[\frac{\partial}{\partial\theta_1}\cdots\frac{\partial}{\partial\theta_k}D\left(\frac{\partial^2}{\partial\theta_1^2},\cdots,\frac{\partial^2}{\partial\theta_k^2}\right)H(\theta_1,\cdots,\theta_k)\right]$$

$$\cdot\sin\left(l_1+\frac{1}{2}\right)\theta_1\cdots\sin\left(l_k+\frac{1}{2}\right)\theta_k d\theta_1\cdots d\theta_k. \tag{8.5.4}$$

It is obvious that

$$\left(\frac{\partial^{\mu_1}}{\partial\theta_1^{\mu_1}}\cdots\frac{\partial^{\mu_k}}{\partial\theta_k^{\mu_k}}H(\theta_1,\cdots,\theta_k)\right)$$

$$\cdot\left(\frac{\partial^{\nu_1}}{\partial\theta_1^{\nu_1}}\cdots\frac{\partial^{\nu_k}}{\partial\theta_k^{\nu_k}}\sin\left(l_1+\frac{1}{2}\right)\theta_1\cdots\sin\left(l_k+\frac{1}{2}\right)\theta_k\right)$$

takes $\sin\frac{\theta_i}{2}$ or $\sin\left(l_i+\frac{1}{2}\right)\theta_i$ as its factor, otherwise it takes

$$\cos\frac{\theta_i}{2}\cos\left(l_i+\frac{1}{2}\right)\theta_i = \frac{1}{2}\left(\cos l_i\theta_i + \cos(l_i+1)\theta_i\right)$$

as its factor. Applying integration by parts to (8.5.4), we find

$$I = (-1)^{k^2}\int_0^{2\pi}\cdots\int_0^{2\pi}H(\theta_1,\cdots,\theta_k)$$

$$\cdot\left\{\frac{\partial}{\partial\theta_1}\cdots\frac{\partial}{\partial\theta_k}\cdot D\left(\frac{\partial^2}{\partial\theta_1^2},\cdots,\frac{\partial^2}{\partial\theta_k^2}\right)\sin\left(l_1+\frac{1}{2}\right)\theta_1\cdots\right.$$

$$\left.\cdot\sin\left(l_k+\frac{1}{2}\right)\theta_k\right\}d\theta_1\cdots d\theta_k. \tag{8.5.5}$$

Thus

$$\mathrm{tr}(A_m A_m') = (-1)^{\frac{k(k-1)}{2}}\frac{2^k N(m)}{C(2n-1)!\cdots 3!1!}\left(\frac{-i}{2\pi}\right)^k$$

$$\cdot\int_0^{2\pi}\cdots\int_0^{2\pi}\left\{\frac{\partial}{\partial\theta_1}\cdots\frac{\partial}{\partial\theta_k}D\left(\frac{\partial^2}{\partial\theta_1^2},\cdots,\frac{\partial^2}{\partial\theta_k^2}\right)H(\theta_1,\cdots,\theta_k)\right\}$$

$$\cdot\cos\left(l_1+\frac{1}{2}\right)\theta_1\cdots\cos\left(l_k+\frac{1}{2}\right)\theta_k\cdot d\theta_1\cdots d\theta_k. \tag{8.5.6}$$

Similarly

$$\mathrm{tr}(A_m A_m')N^3(m) = (-1)^{\frac{k(k-1)}{2}} \frac{1}{C}\left(\frac{2^k}{(2k-1)!\cdots 3!1!}\right)^5\left(\frac{-i}{2\pi}\right)^k$$

$$\cdot \int_0^{2\pi}\cdots\int_0^{2\pi}\left\{\left[\frac{\partial}{\partial\theta_1}\cdots\frac{\partial}{\partial\theta_k} D\left(\frac{\partial^2}{\partial\theta_1^2},\cdots,\frac{\partial^2}{\partial\theta_k^2}\right)\right]^5\right.$$

$$\left.\cdot H(\theta_1,\cdots,\theta_k)\right\}\cos\left(l_1+\frac{1}{2}\right)\theta_1\cdots\cos\left(l_k+\frac{1}{2}\right)\theta_k d\theta_1\cdots d\theta_k$$

$$= (-1)^{\frac{k(k-1)}{2}} \frac{1}{C}\left(\frac{2^k}{(2k-1)!\cdots 3!1!}\right)^5\left(\frac{-i}{2}\right)^k\left\{c_{-l_1,-l_2,\cdots,-l_k}\right.$$

$$+\sum_{1\leqslant i_1\leqslant k} c_{-l_1,\cdots,l_{i_1}+1,\cdots,-l_k}+\sum_{1\leqslant i_1<i_2\leqslant k}\left(c_{-l_1,\cdots,l_{i_1}+1,\cdots,l_{i_2}+1,\cdots,-l_k}\right.$$

$$\left.\left.+\cdots+c_{l_1+1,\cdots,l_k+1}\right)\right\}, \tag{8.5.7}$$

where $c_{\alpha_1,\cdots,\alpha_k}$ are coefficients of the multiple Fourier series of

$$\left\{\left(\frac{\partial}{\partial\theta_1}\cdots\frac{\partial}{\partial\theta_k} D\left(\frac{\partial^2}{\partial\theta_1^2},\cdots,\frac{\partial^2}{\partial\theta_k^2}\right)\right)^5 H(\theta_1,\cdots,\theta_k)\right\} e^{\frac{1}{2}i(\theta_1+\cdots+\theta_k)}.$$

Consequently

$$\left|\sum_{N\geqslant l_1>\cdots>l_k\geqslant 0} N^3(m)\mathrm{tr}(A_m A_m')\right| \leqslant B_1 \sum_{N+1\geqslant l_1>\cdots>l_k\geqslant -(N+1)} |c_{l_1,\cdots,l_k}|, \tag{8.5.8}$$

where B_1 stands for an absolute constant.

2. When $n=2k$, we can obtain as in case 1 that

$$\mathrm{tr}(B_m B_m') = \frac{2N^2(m)}{C}\cdot\frac{1}{(2\pi)^k}\cdot\int_{\pi\geqslant\theta_1>\cdots>\theta_k\geqslant 0}\cdots\int h^*(\theta_1,\cdots,\theta_k)$$

$$\cdot\,\sigma_m^{\psi}(\Gamma)c^2(k-1,\cdots,1,0)d\theta_1\cdots d\theta_k. \tag{8.5.9}$$

Substituting $\sigma_m^{\psi}(\Gamma)$ into (8.5.9), we find

$$\mathrm{tr}(B_m B_m') = \frac{N^2(m)}{C}\frac{1}{(2\pi)^k}\int_0^{2\pi}\cdots\int_0^{2\pi} H(\theta_1,\cdots,\theta_k)$$

$$\cdot\,(\cos l_1\theta_1\cdots\cos l_k\theta_k - i^k\sin l_1\theta_1\cdots\sin l_k\theta_k)d\theta_1\cdots d\theta_k,$$

where

$$H(\theta_1,\cdots,\theta_k) = h^*(\theta_1,\cdots,\theta_k)c(k-1,\cdots,1,0). \tag{8.5.10}$$

Similarly

$$N^3(m)\mathrm{tr}(B_m B_m') = \frac{1}{C}\cdot\left(\frac{2^{k-1}}{(2k-2)!\cdots 4!2!}\right)^5\frac{1}{(2\pi)^k}$$

$$\cdot\int_0^{2\pi}\int_0^{2\pi}\left\{\left(D\left(\frac{\partial^2}{\partial\theta_1^2},\cdots,\frac{\partial^2}{\partial\theta_k^2}\right)\right)^5 H(\theta_1,\cdots,\theta_k)\right\}$$

$$\cdot\,(\cos l_1\theta_1\cdots\cos l_k\theta_k - i^k\sin l_1\theta_1\cdots\sin l_k\theta_k)d\theta_1\cdots d\theta_k$$

$$= \frac{1}{c}\left(\frac{2^{k-1}}{(2k-2)!\cdots 4!2!}\right)^5\left(\frac{1}{2}\right)^{k-1}\left(\sum_{1\leqslant i_1\leqslant k} c_{-l_1,\cdots,l_{i},\cdots,l_k}\right.$$

$$\left. + \sum_{1\leqslant i_1<i_2\leqslant k} c_{-l_1,\cdots,l_{i_1},\cdots,l_{i_2},\cdots,l_k} + \cdots\right),$$

where $c\alpha_1,\cdots,\alpha_k$ belongs to the coefficients of the multiple Fourier series of

$$\left(D\left(\frac{\partial^2}{\partial\theta_1^2},\cdots,\frac{\partial^2}{\partial\theta_k^2}\right)\right)^5 H(\theta_1,\cdots,\theta_k).$$

When k is even, the preceding expression ends at

$$c_{-l_1,l_2,\cdots,l_k} + c_{l_1,-l_2,\cdots,l_k} + \cdots + c_{l_1,\cdots,l_{k-1},-l_k}.$$

When k is odd, it ends at $c_{l_1,\cdots,l_k}$. So, by a similitude, we get

$$\left|\sum_{N\geqslant l_1>\cdots>l_k\geqslant 0} N^3(m)\mathrm{tr}(B_mB'_m)\right| \leqslant B_2 \sum_{N\geqslant l_1>\cdots>l_k\geqslant -N} |c_{l_1,\cdots,l_k}|, \quad (8.5.11)$$

where B_2 is an absolute constant. Similarly we may obtain the term $\mathrm{tr}(A_mA'_m)$.

Let $f(\theta_1,\cdots,\theta_k)$ be a continuous function defined in $0\leqslant\theta_1,\cdots,\theta_k\leqslant 2\pi$. The modulus of continuity, modulus of integral and Lipschitz condition can be well defined as in § 3.6 of Chapter 3. Thus we have (see Wang Shikun and Dong Daozheng [1])

Theorem 8.5.1 *Let $u(\Gamma)$ be integrable on $SO(n)$, $\Gamma\in SO(n)$. When $n=2k+1$, it follows that $u(\Gamma)\in c^{5k^2}$ and*

$$\left(\frac{\partial}{\partial\theta_1}\cdots\frac{\partial}{\partial\theta_k} D\left(\frac{\partial^2}{\partial\theta_1^2},\cdots,\frac{\partial^2}{\partial\theta_k^2}\right)\right)^5$$

$$\cdot H(\theta_1,\cdots,\theta_k)\in \mathrm{Lip}(2,\alpha), \quad \alpha>\frac{1}{2}, \quad (8.5.12)$$

where H is defined by (8.5.3);*when $n=2k$,it follows that $u(\Gamma)\in c^{5k(k-1)}$ and*

$$\left(D\left(\frac{\partial^2}{\partial\theta_1^2},\cdots,\frac{\partial^2}{\partial\theta_k^2}\right)\right)^5 H(\theta_1,\cdots,\theta_k)\in \mathrm{Lip}(2,\alpha), \quad \alpha>\frac{1}{2}, \quad (8.5.13)$$

where H is defined by (8.5.10). *Then the Fourier series* (6.1.14) *and* (6.1.12) *are absolutely convergent.*

Proof. For $n=2k+1$, we have by Schwarz' inequality that

$$\sum_{i,j}^{N(m)} |a_{i,j}^m|\cdot|\varphi_{i,j}^m(\Gamma)| \leqslant N^{\frac{1}{2}}(m)\mathrm{tr}(A_m\cdot A'_m).$$

This gives us

$$\left|\sum_{m_1\geqslant\cdots\geqslant m_k\geqslant 0} N^{\frac{1}{2}}(m)\operatorname{tr}(A_m A_m')^{1/2}\right|$$
$$\leqslant \left|\sum_{m_1\geqslant\cdots\geqslant m_k\geqslant 0} N^3(m)\operatorname{tr}(A_m A_m')\right|^{1/2}\left|\sum_{m_1\geqslant\cdots\geqslant m_k\geqslant 0}\frac{1}{N^2(m)}\right|^{1/2}.$$

The right side in (8.5.8) is convergent if (8.5.12) is satisfied (Musielak [1]). The proof is completed.

For $n=2k$ the theorem can be similarly deduced from (8.5.13) and (8.5.11).

§ 8.6 Some Remarks

Some results on $SO(n)$ can be stated by unified notations no matter how the order is.

Let

$$n=\begin{cases}2k, & \text{if } n \text{ is even,}\\ 2k+1, & \text{if } n \text{ is odd,}\end{cases}$$

and define

$$G_n(\lambda_1,\cdots,\lambda_k)=(\lambda_1\cdots\lambda_k)^{n-2k}D(\lambda_1^2,\cdots,\lambda_k^2),$$

where D is the Vandermonde determinant and $k=\left[\frac{n}{2}\right]$, $G_n(\lambda_1,\cdots,\lambda_k)$ is denoted by $G_n(\lambda)$ for short.

Consequently, the volume element on $SO(n)$ can be written as

$$\dot{\Gamma}_n=2^{nk-k^2-\frac{k}{2}}\left|G_n\left(2i\sin\frac{\theta_1}{2},\cdots,2i\sin\frac{\theta_k}{2}\right)\right|^2\dot{\Sigma}d\theta_1\cdots d\theta_k.$$

The dimension of the representation with signature $m=(m_1,\cdots,m_k)$ is connected with

$$\frac{2^{\left[\frac{n+1}{2}\right]-1}}{(n-2)!(n-4)!\cdots(n-2k)!}G_n\left(l+\frac{n-2k}{2}\right).$$

The Dirichlet kernel can be expressed in a unified way as

$$\mathscr{D}_N(\Gamma)=$$
$$\frac{2^{n-k-1}G_n\left(i\dfrac{\partial}{\partial\theta}\right)\left(\dfrac{\sin\left(N+\dfrac{n-2k+1}{2}\right)\theta_1}{\sin\dfrac{\theta_1}{2}}\cdots\dfrac{\sin\left(N+\dfrac{n-2k+1}{2}\right)\theta_k}{\sin\dfrac{\theta_k}{2}}\right)}{(n-2)!\cdots(n-2k)!\,G_n\left(-2i\sin\dfrac{\theta_1}{2},\cdots,-2i\sin\dfrac{\theta_k}{2}\right)}.$$

The convergence theorem can be stated as follows:

If $u(\Gamma) \in C^{[\frac{n}{2}][\frac{n-1}{2}]+p}$, $0 < p < 1$, then the partial sum $s_N(\Gamma)$ of its Fourier series converges to itself.

Finally, the absolute convergence theorem can be stated as follows:

If $u(\Gamma) \in C^{s[\frac{n}{2}][\frac{n-1}{2}]}$, $\Gamma \in SO(n)$, and

$$G_n^s\left(\frac{\partial}{\partial\theta_1}, \cdots, \frac{\partial}{\partial\theta_k}\right) H(\theta_1, \cdots, \theta_k) \in \mathrm{Lip}(2, \alpha), \quad \alpha > \frac{1}{2}$$

then the Fourier series of $u(\Gamma)$ is absolutely convergent.

The related proofs can also be treated by unified notations.

As stated in Chapter 3, by the help of some other absolute convergence theorems of the multiple Fourier series, such as the generalized forms of Zygmund's Theorem and Szasz' Theorem, the absolute convergence theorem of the Fourier series on rotation groups is obtained.

Combining Chapter 9 of this book and Chapter 9 of Chandrasekharan-Minakshisundaran [1], we can deduce some results on absolute convergence of the spherical summation for the Fourier series.

Chapter 9. Spherical Summation of Fourier Series on Rotation Groups

§ 9.1 Spherical Summation of Fourier Series

As in the case of unitary groups, we can take the spherical summation of the Fourier series on rotation groups into account and show the relevant theorems. Namely, we can prove that, if $u(\Gamma)$ is continuous over $SO(n)$, then its Fourier series is Abel, Gauss-Sommerfeld and Riesz-summable of order δ to itself provided

$$\delta > \frac{\dim SO(n) - 1}{2}.$$

Here all summations, of course, mean spherical ones, which are the topic of this chapter. The method used here is the same as that used in Chapter 5 and the materials are taken from Wang Shikun and Dong Daozhen [3].

As we know, the Fourier series of any integrable function $u(\Gamma)$ over $SO(n)$ is given by

$$\sum_{m_1 \geqslant \cdots \geqslant m_k \geqslant 0} \sum_{i,j}^{N(m)} a_{ij}^m \varphi_{ij}^m(\Gamma) \tag{9.1.1}$$

for $n = 2k + 1$, and

$$\sum_{m_1 \geqslant \cdots \geqslant m_k \geqslant 0} \sum_{i,j}^{N(m)} a_{ij}^m \phi_{ij}^m(\Gamma) + \sum_{m_1 \geqslant \cdots \geqslant m_k \geqslant 0} \sum_{i,j}^{N(m)} (a_{ij}^m \phi_{ij}^m(\Gamma) + b_{ij}^m \psi_{ij}^m(\Gamma)) \tag{9.1.2}$$

for $n = 2k$. As for the definitions of $\varphi_{ij}^m, \phi_{ij}^m, \psi_{ij}^m, a_{ij}^m$, and b_{ij}^m, the reader is referred to (6.1.10), (6.1.11), (6.1.13) etc. of § 6.1 in Chapter 6.

The spherical summation is to consider the Fourier series (9.1.1) and (9.1.2) as

$$\sum_{q=0}^{\infty} \left(\sum_{\substack{m_1 \geqslant \cdots \geqslant m_k \geqslant 0 \\ (l_1+\frac{1}{2})^2 + \cdots + (l_k+\frac{1}{2})^2 = \frac{q}{4}}} \sum_{i,j}^{N(m)} a_{ij}^m \varphi_{ij}^m(\Gamma) \right) \tag{9.1.3}$$

for $n = 2k + 1$, and

$$\sum_{q=0}^{\infty}\left(\sum_{\substack{m_1\geqslant\cdots\geqslant m_k>0\\ l_1^2+\cdots+l_k^2=q}}\sum_{i,j}^{N(m)} a_{ij}^m \phi_{ij}^m(\Gamma) + \sum_{\substack{m_1\geqslant\cdots\geqslant m_k>0\\ l_1^2+\cdots+l_k^2=q}}\sum_{i,j}^{N(m)} (a_{ij}^m \phi_{ij}^m(\Gamma) + b_{ij}^m \psi_{ij}^m(\Gamma))\right) \quad (9.1.4)$$

for $n = 2k$. Let $\varphi(t)$ be a given function defined on $[0, \infty)$ such that $\varphi(0) = 1$. Then the spherical summation of a kind corresponds to a kind of means which takes the form

$$\sum_{q=0}^{\infty}\Phi\left(\frac{\sqrt{q}}{2R}\right)\sum_{\substack{m_1\geqslant\cdots\geqslant m_k\geqslant 0\\ (l_1+\frac{1}{2})^2+\cdots+(l_k+\frac{1}{2})^2=\frac{q}{4}}}\sum_{i,j}^{N(m)} a_{ij}^m \varphi_{ij}^m(\Gamma), \quad (9.1.5)$$

or

$$\sum_{q=0}^{\infty}\Phi\left(\frac{\sqrt{q}}{R}\right)\left(\sum_{\substack{m_1\geqslant\cdots\geqslant m_k=0\\ l_1^2+\cdots+l_k^2=q}}\sum_{i,j}^{N(m)} a_{ij}^m \phi_{ij}^m(\Gamma)\right.$$

$$\left. + \sum_{\substack{m_1\geqslant\cdots\geqslant m_k>0\\ l_1^2+\cdots+l_k^2=q}}\sum_{i,j}^{N(m)} (a_{ij}^m \phi_{ij}^m(\Gamma) + b_{ij}^m \psi_{ij}^m(\Gamma))\right), \quad (9.1.6)$$

according as $n = 2k + 1$ or $n = 2k$, where

$$\Phi(t) = \varphi(t)/\varphi(\sqrt{q_0}/R) \quad (9.1.7)$$

and

$$q_0 = \begin{cases} \dfrac{(2k-1)k(2k+1)}{12}, & \text{for } n = 2k+1; \\ \dfrac{(2k-1)k(k-1)}{6}, & \text{for } n - 2k. \end{cases} \quad (9.1.8)$$

The Fourier series of $u(\Gamma)$ is said to be φ-spherical summable to its limit, if (9.1.5) and (9.1.6) converge as $R \to \infty$.

As in Chapter 5, here we only consider the following examples of $\varphi(t)$:

(1) $\varphi(t) = e^{-t}$——the Abel summation;

(2) $\varphi(t) = e^{-t^2}$——the Gauss-Sommerfeld sumation;

(3) $\varphi(t) = \begin{cases} (1-t^2)^\delta, & \text{if } 0 \leqslant t < 1 \\ 0, & \text{if } 1 \leqslant t \end{cases}$ ——the Riesz summation of order δ.

§ 9.2 Expression by Integral

Let (9.1.5) and (9.1.6) be denoted by $s_R^\phi(\Gamma)$ which can be written as

$$\frac{1}{C}\int_{SO(n)} u(K\Gamma)\sum_{m\geqslant 0}\phi\left(\frac{\sqrt{q}}{R}\right)N(m)\sigma_m(K)\dot{K}, \tag{9.2.1}$$

where m denotes the signature $m_1 \geqslant \cdots \geqslant m_k \geqslant 0$; $\sigma_m(\Gamma) = \sigma_m^\phi(\Gamma) + \sigma_m^\psi(\Gamma)$, when $n = 2k$ and the signature $m_1 \geqslant \cdots \geqslant m_k > 0$; and

$$q = \begin{cases} \left(l_1 + \frac{1}{2}\right)^2 + \cdots + \left(l_k + \frac{1}{2}\right)^2, & \text{if } n = 2k+1, \\ l_1^2 + \cdots + l_k^2, & \text{if } n = 2k. \end{cases}$$

Lemma 9.2.1 *Let* $\varphi(t)$ *be*

(1) *absolutely continuous over any finite interval*; (9.2.2)

and satisfy that

$$(2) \int_0^\infty |\varphi(t)| t^{\frac{k-1}{2}} dt < \infty. \tag{9.2.3}$$

Then, for any real numbers $p_1, \cdots, p_k$, *we have*

$$\sum_{q=0}^{\infty}\ \sum_{q=(l_1+p_1)^2+\cdots+(l_k+p_k)^2} \phi\left(\frac{\sqrt{q}}{R}\right) a_{l_1,\cdots,l_k} e^{i((l_1+p_1)\theta_1+\cdots+(l_k+p_k)\theta_k)}$$

$$= R\int_0^\infty g_\theta(t)H_\theta(tR)\,dt, \tag{9.2.4}$$

where ϕ *is given by* (9.1.7),

$$a_{l_1,\cdots,l_k} = \left(\frac{1}{2\pi}\right)^k \int_0^{2\pi}\cdots\int_0^{2\pi} g(\theta_1,\cdots,\theta_k)$$

$$\cdot\, e^{-i((\theta_1+p_1)\theta_1+\cdots+(l_k+p_k)\theta_k)}d\theta_1\cdots d\theta_k, \tag{9.2.5}$$

$$g_\theta(t) = \frac{1}{\omega_{k-1}}\int_\sigma g(\theta_1 + t\eta_1, \cdots, \theta_k + t\eta_k)d\sigma_\eta, \tag{9.2.6}$$

$$H_\phi(tR) = \frac{(tR)^{\frac{k}{2}}}{2^{\frac{k}{2}-1}\Gamma(k/2)}\int_0^\infty \phi(u)u^{\frac{k}{2}}J_{\frac{k-2}{2}}(utR)du. \tag{9.2.7}$$

$g(\theta_1,\cdots,\theta_k)$ *is a periodic integrable function defined over* $0 \leqslant \theta_1, \cdots, \theta_k \leqslant 2\pi$. $J_\mu(s)$ *represents a Bessel function of the first kind.* σ *indicates the sphere* $\eta_1^2 + \cdots + \eta_k^2 = 1$ *and its volume reads* $\omega_{k-1} = 2\pi^{\frac{k}{2}}/\Gamma(k/2.)$

When $p_1 = \cdots = p_k = 0$, the lemma just turns out to be Bochner's formula (see S. Bochner [1]) and the conclusion in general case can be deduced from the proof of Bochner's formula.

Any $\Gamma \in SO(n)$ can be expressed as

$$\Gamma = K(c_1 \dotplus c_2 \dotplus \cdots)K, c_i = \begin{pmatrix} \cos\theta_i, & \sin\theta_i \\ -\sin\theta_i, & \cos\theta_i \end{pmatrix},$$

$$\pi \geqslant \theta_1 \geqslant \cdots \geqslant \theta_k \geqslant 0,$$

where the direct sum ends at $c_k + 1$ or at c_k in accordance with $n = 2k+1$ or $n = 2k, K \in SO(n)$.

Write

$$G_n(\xi) = (\xi_1 \cdots \xi_n)^{n-2k} D(\xi_1^2 \cdots \xi_k^2), \xi = (\xi_1, \cdots, \xi_k). \tag{9.2.8}$$

Using the algebraic method given by Hua Luogeng (cf. § 1.9 in Chapter 1), we arrive at the following

Lemma 9.2.2 *The identity*

$$\sum_{m \geq 0} r^{l_1 + \cdots + l_k} N(m) \sigma_m(\Gamma) G_n\left(2i \sin\frac{\theta_1}{2}, \cdots, 2i \sin\frac{\theta_k}{2}\right)$$

$$= \frac{2^{n-k-1}}{(n-2)!(n-4)!\cdots(n-2k)!}$$

$$\cdot G_n\left(-i\frac{\partial}{\partial\theta_1}, \cdots -i\frac{\partial}{\partial\theta_k}\right) \cdot (p_r(\theta_1)p_r(\theta_2)\cdots p_r(\theta_k)) \tag{9.2.9}$$

holds, where

$$p_r(\theta) = \begin{cases} \dfrac{2(1-r)\cos\dfrac{\theta}{2}}{1 - 2r\cos\theta + r^2}, & \text{if } n = 2k+1; \\ \dfrac{1-r^2}{1 - 2r\cos\theta + r^2} + 1, & \text{if } n = 2k. \end{cases}$$

We can easily prove that

$$G_{2k+1}\left(2i \sin\frac{\theta_1}{2}, \cdots, 2i \sin\frac{\theta_k}{2}\right) = s\left(k - \frac{1}{2}, \cdots \frac{3}{2}, \frac{1}{2}\right);$$

$$G_{2k}\left(2i \sin\frac{\theta_1}{2}, \cdots, 2i \sin\frac{\theta_k}{2}\right) = c(k-1, \cdots, 1, 0).$$

By $g(\theta)$, we denote the right side of (9.2.9), where $\theta = (\theta_1, \cdots, \theta_k)$. Put

$$F(r, R, \Gamma) = \sum_{m \geq 0} \phi\left(\frac{\sqrt{q}}{R}\right) r^{l_1 + \cdots + l_k} N(m) \sigma_m(\Gamma).$$

Then we propose the following

Lemma 9.2.3 *For any $\varphi(t)$ that satisfies (9.2.2) and (9.2.3), the following equality holds*

$$F(r, R, \Gamma) = R\left[G_n\left(2i\sin\frac{\theta_1}{2}, \cdots, 2i\sin\frac{\theta_k}{2}\right)\right]^{-1}\int_0^\infty g_\theta(t)H_\phi(tR)dt. \qquad (9.2.10)$$

Proof. (1) When $n = 2k+1$, taking $p_1 = \cdots = p_k = 1/2$ in (9.2.4) and substituting $g(\theta)$ into (9.2.5), we obtain

$$a_{l_1,\cdots,l_k} = \sum_{l_1'>\cdots>l_k'\geqslant 0} r^{l_1'+\cdots+l_k'}N(m')\left(\frac{1}{2\pi}\right)^k\int_0^{2\pi}\cdots\int_0^{2\pi} s\left(l_1'+\frac{1}{2},\cdots,l_k'+\frac{1}{2}\right)$$

$$\cdot e^{-i((l_1+\frac{1}{2})\theta_1+\cdots+(l_k+\frac{1}{2})\theta_k)}d\theta_1\cdots d\theta_k.$$

By substitution, the left side of (9.∟.4) becomes

$$\sum_{\substack{l_1\cdots l_k\\ q=(l_1+\frac{1}{2})^2+\cdots+(l_k+\frac{1}{2})^2}} \phi(q/R) \sum_{l_1'>\cdots>l_k'\geqslant 0} r^{l_1'+\cdots+l_k'} N(m')$$

$$\cdot\left(\frac{1}{2\pi}\right)^k\int_0^{2\pi}\cdots\int_0^{2\pi} s\left(l_1'+\frac{1}{2},\cdots,l_k'+\frac{1}{2}\right)$$

$$\cdot e^{-i((l_1+\frac{1}{2})\theta_1+\cdots+(l_k+\frac{1}{2})\theta_k)}\cdot e^{i((l_1+\frac{1}{2})\varphi_1+\cdots+(l_k+\frac{1}{2})\varphi_k)}d\theta_1\cdots d\theta_k$$

$$= \sum_{l_1,\cdots,l_k>0}\phi(\sqrt{q/R})\sum_{l_1'>\cdots>l_k'\geqslant 0} r^{l_1'+\cdots+l_k'}N(m')$$

$$\cdot\left(\frac{1}{2\pi}\right)^k\int_0^{2\pi}\cdots\int_0^{2\pi} s\left(l_1'+\frac{1}{2},\cdots,l_k'+\frac{1}{2}\right)$$

$$\cdot 2^k\cos\left(l_1+\frac{1}{2}\right)(\theta_1-\varphi_1)\cdots\cos\left(l_k+\frac{1}{2}\right)(\theta_k-\varphi_k)d\theta_1\cdots d\theta_k$$

$$= \sum_{l_1\cdots l_k\geqslant 0}\phi(\sqrt{q/R})\sum_{l_1'>\cdots>l_k'\geqslant 0} r^{l_1'+\cdots+l_k'}N(m')(2i)^k$$

$$\cdot\left(\frac{1}{2\pi}\right)^k\int_0^{2\pi}\cdots\int_0^{2\pi}\sum_{(\mu_1\cdots\mu_k)}\delta_{1\cdots k}^{\mu_1\cdots\mu_k}\sin\left(l_{\mu_1}'+\frac{1}{2}\right)\theta_1\cdots$$

$$\cdot\sin\left(l_{\mu_k}'+\frac{1}{2}\right)\theta_k\cdot 2^k\cos\left(l_1+\frac{1}{2}\right)(\theta_1-\varphi_1)\cdots$$

$$\cdot\cos\left(l_k+\frac{1}{2}\right)(\theta_k-\varphi_k)d\theta_1\cdots d\theta_k$$

$$= \sum_{l_1>\cdots>l_k\geqslant 0}\phi(\sqrt{q/R})r^{l_1+\cdots+l_k}N(m)s\left(l_1+\frac{1}{2},\cdots,l_k+\frac{1}{2}\right).$$

From Lemma 9.2.1, (9.2.10) holds.

(2) When $n=2k$, the method is the same as above, so it is omitted.

By the definition of $g_\theta(t)$, (9.2.6) and (9.2.10) can also be expressed as

$$\frac{R}{G_n\left(2i\sin\frac{\theta_1}{2},\cdots,2i\sin\frac{\theta_k}{2}\right)\omega_{k-1}}\cdot\int_0^\infty\int_\sigma g(\theta+t\eta)H_\phi(tR)d\sigma_\eta\cdot dt \quad (9.2.11)$$

or

$$\frac{R}{G_n\left(2i\sin\frac{\theta_1}{2},\cdots,2i\sin\frac{\theta_k}{2}\right)\omega_{k-1}}\cdot\int_{-\infty}^\infty\cdots\int_{-\infty}^\infty g(\theta+\xi)\frac{H_\phi(|\xi|R)}{|\xi|^{k-1}}d\xi, \quad (9.2.12)$$

where $\theta=(\theta_1,\cdots,\theta_k)$, $\eta=(\eta_1,\cdots,\eta_k)$, $\xi=(\xi_1,\cdots,\xi_k)$ and $d\xi=d\xi_1\cdots d\xi_k$. By Lemma 9.2.2, (9.2.12) relates well to

$$\frac{2^{n-k-1}R}{(n-2)!(n-4)!\cdots(n-2k)!G_n\left(2i\sin\frac{\theta_1}{2},\cdots,2i\sin\frac{\theta_k}{2}\right)\omega_{k-1}}$$
$$\cdot\int_{-\infty}^\infty\cdots\int_{-\infty}^\infty\left[G_n\left(-i\frac{\partial}{\partial\theta_1},\cdots,-i\frac{\partial}{\partial\theta_k}\right)\cdot(p_r(\theta_1+\xi_1)\cdot p_r(\theta_2+\xi_2)\right.$$
$$\left.\cdots p_r(\theta_k+\xi_k))\right]\frac{H_\phi(|\xi|R)}{|\xi|^{k-1}}d\xi. \quad (9.2.13)$$

Obviously

$$G_n\left(i\frac{\partial}{\partial\theta}\right)(p_r(\theta_1+\xi_1)\cdots p_r(\theta_k+\xi_k))$$
$$=G_n\left(i\frac{\partial}{\partial\xi}\right)(p_r(\theta_1+\xi_1)\cdots p_r(\theta_k+\xi_k)),$$

for $0\leqslant\lambda_1,\cdots,\lambda_k\leqslant n-2$. We have

$$\left.\frac{\partial^{\lambda_1}}{\partial\xi_1^{\lambda_1}}\cdots\frac{\partial_k^{\lambda}}{\partial\xi_k^{\lambda_k}}\frac{H_\phi(|\xi|R)}{|\xi|^{k-1}}\right|_{|\xi|=\infty}=0. \quad (9.2.14)$$

Applying integration by parts to (9.2.13) leads to

$$\frac{2^{n-k-1}R}{(n-2)!(n-4)!\cdots(n-2k)!G_n\left(2i\sin\frac{\theta_1}{2},\cdots,2i\sin\frac{\theta_k}{2}\right)\omega_{k-1}}$$
$$\cdot\int_{-\infty}^\infty\cdots\int_{-\infty}^\infty p_r(\theta_1+\xi_1)\cdots p_r(\theta_k+\xi_k)$$
$$\cdot G_n\left(i\frac{\partial}{\partial\xi_1},\cdots,i\frac{\partial}{\partial\xi_k}\right)\frac{H_\phi(|\xi|R)}{|\xi|^{k-1}}d\xi. \quad (9.2.15)$$

Put

$$s_R^{\phi}(r\cdot\Gamma)=\frac{1}{C}\int_{SO(n)}u(k\Gamma)\sum_{m\geqslant 0}\phi\left(\frac{\sqrt{q}}{R}\right)r^{l_1+\cdots+l_k}$$
$$\cdot N(m)\sigma_m(K)\dot{K}.$$

If $u^*(k\Gamma)$ is defined as in § 8.3 of Chapter 8, then

$$s_R^{\phi}(r,\Gamma)=\frac{1}{C}\int_{so(n)}u^*(K\Gamma)F(r,R,K)\dot{K}.$$

Write

$$\psi_\Gamma(-\theta_1,\cdots,-\theta_k)=\frac{1}{V(\Sigma)}\int_\Sigma u^*(K\Gamma)\dot{\Sigma}. \tag{9.2.16}$$

Then ψ_Γ, obviously, is a symmetric function of $\theta_1,\cdots,\theta_k$. Thus

$$s_R^{\phi}(r\cdot\Gamma)=\frac{1}{2^{n-k-1}\pi^k}\int\cdots\int_{\pi\geqslant\theta_1>\cdots>\theta_k\geqslant 0}\psi_\Gamma(-\theta)F(r,R,K)$$
$$\cdot\left|G_n^2\left(2i\sin\frac{\theta_1}{2},\cdots,2i\sin\frac{\theta_k}{2}\right)\right|^2 d\theta_1\cdots d\theta_k. \tag{9.2.17}$$

We now proceed to prove

Theorem 9.2.1 *Let $u(\Gamma)$ be integrable on $SO(n)$ and $\varphi(t)$ satisfy (9.2.2), (9.2.3) and (9.2.14). Then the φ-spherical mean $s_R^{\phi}(\Gamma)$ of its Fourier series can be expressed as*

$$\frac{R}{\omega_{k-1}(n-2)!(n-4)!\cdots(n-2k)!}$$
$$\cdot\int\cdots\int_{\infty>\xi_1\geqslant\cdots\geqslant\xi_k>-\infty}\psi_\Gamma(\xi)G_n\left(2i\sin\frac{\xi}{2}\right)\left(G_n\left(i\frac{\partial}{\partial\xi}\right)\frac{H_\phi(|\xi|R)}{|\xi|^{k-1}}\right)d\xi, \tag{9.2.18}$$

where G_n is defined by (9.2.8), ψ_Γ is defined by (9.2.16) and $2i\sin\frac{\xi}{2}$ denotes

$$\left(2i\sin\frac{\xi_1}{2},\cdots,2i\sin\frac{\xi_k}{2}\right).$$

(9.2.18) can also be expressed as

$$\frac{R}{k!(n-2)!(n-4)!\cdots(n-2k)!}\int_0^\infty h_\Gamma(t)t^{k-1}dt, \tag{9.2.19}$$

where $$h_\Gamma(t)=\frac{1}{\omega_{k-1}}\int\psi_\Gamma(t\eta)G_n\left(2i\sin\frac{t\eta}{2}\right)$$
$$\cdot\left(G_n\left(i\frac{\partial}{\partial\xi}\right)\frac{H_\phi(|\xi|R)}{|\xi|^{k-1}}\right)\bigg|_{\xi=t\eta}d\eta. \tag{9.2.20}$$

Proof. Substituting (9.2.15) into (9.2.17), we obtain by the Fubini Theorem that $s_k^\phi(r,\Gamma)$ is equal to

$$\frac{(-1)^{kn-k(k+1)}R}{(n-2)!(n-4)!\cdots(n-2k)!\omega_{k-1}}$$

$$\cdot\int_{-\infty}^{\infty}\cdots\int_{-\infty}^{\infty}\left(G_n\left(i\frac{\partial}{\partial\xi}\right)\frac{H_\phi(|\xi|R)}{|\xi|^{k-1}}\right)d\xi\,\frac{1}{\pi^k}$$

$$\cdot\int\limits_{\pi\geqslant\theta_1>\cdots>\theta_k\geqslant 0}\cdots\int p_r(\theta_1+\xi_1)\cdots p_r(\theta_k+\xi_k)\psi_r(-\theta)$$

$$\cdot\, G_n\left(2i\sin\frac{\theta_1}{2},\cdots 2i\sin\frac{\theta_k}{2}\right)d\theta.$$

In the preceding integral, we make all possible permutations of θ and, correspondingly, permutations of ξ. Then it follows that $s_k^\phi(r,\Gamma)$ is equal to

$$\frac{(-1)^{kn-k(k+1)}R}{k!(n-2)!(n-4)!\cdots(n-2k)!\omega_{k-1}}$$

$$\cdot\int_{-\infty}^{\infty}\cdots\int_{-\infty}^{\infty}\left(G_n\left(i\frac{\partial}{\partial\xi}\right)\frac{H_\phi(|\xi|R)}{|\xi|^{k-1}}\right)d\xi$$

$$\cdot\frac{1}{\pi^k}\int_0^\pi\cdots\int_0^\pi p_r(\theta_1+\xi_1)\cdots p_r(\theta_k+\xi_k)\psi_r(-\theta)$$

$$\cdot\, G_n\left(2i\sin\frac{\theta_1}{2},\cdots,2i\sin\frac{\theta_k}{2}\right)d\theta.$$

Let $r\to 1$. By the help of the Lebesgue convergence theorem and the Abel theorem of several variables, we find that (9.2.18) holds true. However, for even n, the condition (9.2.14)′ should be used in the proof.

Taking substitution $\xi_1=t\eta_1,\cdots,\xi_k=t\eta_k$, where $\eta_1^2+\cdots+\eta_k^2=1$, and calculating Jacobian for this substitution, we get (9.2.19).

§ 9.3 Riesz Mean

The so-called Riesz summation of order δ is the limiting case of the mean (9.1.5) and (9.1.6) as $R\to\infty$, in which

$$\varphi(t)=\begin{cases}(1-t^2)^\delta, & \text{if } 0\leqslant t<1\\ 0, & \text{if } 1\leqslant t.\end{cases}$$

$\varphi(t)$, obviously, satisfies the condition (9.2.2). When $\delta > \frac{k-1}{2}$, it satisfies (9.2.3). Moreover,

$$H_\phi(|\xi| R) = \frac{2^{\delta-\frac{k}{2}+1}\Gamma(\delta+1)}{\Gamma\left(\frac{k}{2}\right)}\left(1-\frac{q_0}{R^2}\right)^{-\delta} R^{k-1} \cdot |\xi|^{k-1} V_{\delta+\frac{k}{2}}(|\xi| R). \tag{9.3.1}$$

It is easily shown that (9.2.14) is valid (cf. Chapter 5). Thus, by Theorem 9.2.1, the Riesz mean of order δ for any integrable function over $SO(n)$, when $\delta > 1/2((n-1)n/2-1)$, can be expressed as

$$\frac{2^{\delta-k}\Gamma(\delta+1)R^k\left(1-\frac{q_0}{R^2}\right)^{-\delta}}{(n-2)!(n-4)!\cdots(n-2k)!\pi^{\frac{k}{2}}}\int\cdots\int_{\infty>\xi_1\geq\cdots\xi_n\geq-\infty} \cdot \psi_T(\xi) G_n\left(2i\sin\frac{\xi}{2}\right)\left(G_n\left(i\frac{\partial}{\partial\xi}\right)V_{\delta+\frac{k}{2}}(|\xi| R)\right)d\xi \tag{9.3.2}$$

or as (9.2.19), where $h_T(t)$ refers to

$$\frac{2^{\delta-k}\Gamma(\delta+1)R^{k-1}\left(1-\frac{q_0}{R^2}\right)^{-\delta}}{\pi^{k/2}}\int_\sigma \psi_T(t\eta) \cdot G_n\left(2i\sin\frac{t\eta}{2}\right)\left(G_n\left(i\frac{\partial}{\partial\xi}\right)V_{\delta+\frac{k}{2}}(|\xi| R)\right)\Big|_{\xi=t\eta} d\sigma_\eta. \tag{9.3.3}$$

By the properties of Bessel functions, we have

$$\frac{\partial}{\partial\xi_{i_1}} V_{\delta+\frac{k}{2}}(|\xi| R) = -R^2\xi_{i_1}V_{\delta+\frac{k}{2}+1}(|\xi| R);$$

$$\frac{\partial^3}{\partial\xi_{i_2}^3} V_{\delta+\frac{k}{2}+1}(|\xi| R) = -R^6\xi_{i_2}V_{\delta+\frac{k}{2}+4}(|\xi| R) + 3R^4\xi_{i_2}V_{\delta+\frac{k}{2}+3};$$

....

$$\frac{\partial^{2k-1}}{\partial\xi_{ik}^{2k-1}} V_{\delta+\frac{k}{2}+(k-1)^2}(|\xi| R) = -R^{-2(2k-1)}\xi_{ik}^{2k-1}V_{\delta+\frac{k}{2}+k^2} \cdot (|\xi| R) + A^1_{2k-1}R^{4k-3}\xi_{ik}^{2k-3}V_{\delta+\frac{k}{2}+k^2-1}(|\xi| R) + \cdots + (-1)^k A^{k-1}_{2k-1}R^{2k}\xi_{ik}V_{\delta+\frac{k}{2}+k^2-k-1}(|\xi| R),$$

where $A^l_m = A^{l-1}_{m-1}(k-2l+1) + A_{m-1}$, $A^0_m = 1$ and for $m \leq l$, $A^l_m = 0$.

From properties of the determinant, it follows that

$$G_{2k+1}\left(i\frac{\partial}{\partial\xi}\right)V_{\delta+\frac{k}{2}}(|\xi|R) = (-i)^{k^2}R^{2k^2}G_{2k+1}(\xi)\cdot V_{\delta+\frac{k}{2}+k^2}(|\xi|R).$$

Similarly

$$G_{2k}\left(i\frac{\partial}{\partial\xi}\right)V_{\delta+\frac{k}{2}}(|\xi|R) = i^{k(k-1)}R^{2k(k-1)}G_n(\xi)\cdot V_{\delta+\frac{k}{2}+k(k-1)}(|\xi|R).$$

Thus

$$G_n\left(i\frac{\partial}{\partial\xi}\right)V_{\delta+\frac{k}{2}}(|\xi|R) = R^{2kn-2k(k+1)}G_n(-i\xi)\cdot V_{\delta+\frac{k}{2}+kn-k(k+1)}(|\xi|R). \tag{9.3.4}$$

With the understanding of (9.3.4),(9.3.3) can be expressed as

$$\frac{2^{\delta-\frac{k}{2}}\Gamma(\delta+1)R^{2kn-2k^2-k-1}\left(1-\frac{q_0}{R^2}\right)^{-\delta}}{\pi^{\frac{k}{2}}}\, t^{kn-k(k+1)}$$

$$\cdot V_{\delta+\frac{k}{2}+kn-k(k+1)}(tR)\cdot\int_\sigma \phi_\Gamma(t\eta)G_n\left(2\sin\frac{t\eta}{2}\right)G_n(\eta)d\sigma_\eta. \tag{9.3.5}$$

We now are in a position to prove

Theorem 9.3.1 *If $u(\Gamma)$ is continuous over $SO(n)$, then, when $\delta > 1/2\cdot(1/2n(n-1)-1)$, its Fourier series is Riesz summable of order δ to itself.*

To begin with, we give

Lemma 9.3.1 *Let $\varphi(t)$ satisfy (9.2.2), (9.2.3) and(9.2.14). Then, when $R^2\geqslant q_0$, we have*

$$\frac{R}{(n-2)!(n-4)!\cdots(n-2k)!\omega_{k-1}}$$

$$\cdot\int\cdots\int_{\infty>\xi_1\geqslant\cdots\geqslant\xi_k<-\infty} G_n\left(2i\sin\frac{\xi}{2}\right)\left(G_n\left(i\frac{\partial}{\partial\xi}\right)\frac{H_\phi(|\xi|k)}{|\xi|^{k+1}}\right)d\xi = 1, \tag{9.3.6}$$

where q_0 is defined by (9.1.8).

Proof. When $n=2k+1$, we have

$$G_{2k+1}\left(2i\sin\frac{\xi}{2}\right) = s\left(k-\frac{1}{2},\cdots,\frac{3}{2},\frac{1}{2}\right).$$

Since H_ϕ satisfies (9.2.14), the left side of (9.3.6), after being integrated by parts, k^2 times is equal to

$$\frac{(-1)^{k^2}R}{(2k-1)!\cdots 3!1!w_{k-1}}\int_{-\infty}^{\infty}\cdots\int_{-\infty}^{\infty}\left\{G_n\left(i\frac{\partial}{\partial\xi}\right)(2i)^k\right.$$

$$\left.\cdot \sin\left(k-\frac{1}{2}\right)\xi_1\cdots\sin\frac{1}{2}\xi_k\right\}\frac{H_\phi(|\xi|R)}{|\xi|^{k-1}}\,d\xi,$$

which can be rewritten as

$$\frac{(-1)^{\frac{k(k-1)}{2}}2^kR}{(2k-1)!\cdots 3!1!w_{k-1}}\int_0^\infty H_\phi(tR)dt$$

$$\cdot\int_\sigma\left\{G_n\left(\frac{\partial}{\partial\xi}\right)\left(\sin\left(k-\frac{1}{2}\right)\xi_1\cdots\sin\frac{1}{2}\xi_k\right)\right\}\Big|_{\xi=t\eta}d\sigma_\eta. \tag{9.3.7}$$

However, we have

$$G_n\left(\frac{\partial}{\partial\xi}\right)\left(\sin\left(k-\frac{1}{2}\right)\xi_1\cdots\sin\frac{1}{2}\xi_k\right)\Big|_{\xi=t\eta}$$

$$=(-1)^{\frac{k(k-1)}{2}}\frac{(2k-1)!\cdots 3!1!}{2^k}\cos\left(k-\frac{1}{2}\right)t\eta_1\cdots\cos\frac{1}{2}t\eta_k,$$

and

$$V_{\frac{k-2}{2}}(\sqrt{q_0}\,t)=\frac{1}{(2\pi)^{k/2}}\int_\sigma\cos\left(k-\frac{1}{2}\right)t\eta_1\cdots\cos\frac{1}{2}t\eta_k d\sigma_\eta$$

(cf. S. Bochner [1]). So(9.3.7) is equal to

$$R\Gamma\left(\frac{k}{2}\right)2^{\frac{k}{2}-1}\int_0^\infty V_{\frac{k-2}{2}}(\sqrt{q_0}\,t)H_\phi(tR)dt, \tag{9.3.8}$$

which comes up to 1 (cf. Bochner[1]).

The case $n=2k$ can be similarly treated.

We now return to Theorem 9.3.1. When $\delta>[n(n-1)-2]/4$, from Lemma 9.3.1 and (9.3.5), itfollows that the difference between $u(\Gamma)$ and the Riesz mean $s_R^\delta(\Gamma)$ of order δ of the Fourier series of $u(\Gamma)$, i. e. $s_R^\delta(\Gamma)-u(\Gamma)$, is equal to

$$\frac{R}{k!(n-2)!(n-4)!\cdots(n-2k)!}\int_0^\infty h_\Gamma(t)t^{k-1}dt, \tag{9.3.9}$$

where

$$h_\Gamma(t)=\frac{2^{\delta-\frac{k}{2}}\Gamma(\delta+1)R^{2kn-2k^2-k-1}}{\pi^{\frac{k}{2}}\left(1-\frac{q_0}{R^2}\right)^\delta}t^{kn-k(k+1)}V_{\delta+\frac{k}{2}+kn-k(k+1)}(tR)$$

$$\cdot\int_\sigma\phi_\Gamma(t\eta)G_n\left(2\sin\frac{t\eta}{2}\right)G_n(\eta)d\sigma_\eta \tag{9.3.10}$$

and

$$\phi_\Gamma(-\theta) = \frac{1}{V(\Sigma)} \int_\Sigma (u^*(w\Gamma) - u(\Gamma))\dot{\Sigma}.$$

Divide the integral in (9.3.9) into three parts. Then $s_R^\delta(\Gamma) - u(\Gamma)$ is equal to

$$\frac{R}{k!(n-2)!(n-4)! - (n-2k)!}\left(\int_0^{\frac{1}{R}} + \int_{\frac{1}{R}}^{\tau} + \int_\tau^\infty\right) = I_1 + I_2 + I_3,$$

where τ is a fixed constant greater than $1/R$.

First consider I_1. Making substitution $Rt = u$ leads to

$$I_1 = \frac{R}{k!(n-2)!(n-4)!\cdots(n-2k)!} \cdot \int_0^1 h_\Gamma\left(\frac{u}{R}\right)\frac{u^{k-1}}{R^{k-1}}\,du.$$

From a theorem given by Hua (Hua Luogeng [1]), we are apt to obtain

$$\left|G_n\left(2\sin\frac{u\eta}{2R}\right)\right| = O(R^{-kn+k(k+1)}). \tag{9.3.11}$$

Since $u(\Gamma)$ is continuous on $SO(n)$,

$$\phi_\Gamma\left(\frac{u\eta}{R}\right) = o(1). \tag{9.3.12}$$

Therefore

$$I_1 = o(1).$$

Next consider I_2, for which we have $\frac{1}{R} \leqslant t \leqslant \tau$.

Set

$$\lambda(t) = \int_0^t t^{k-1}dt\left|\int_\sigma t^{nk-k(k+1)}\phi_\Gamma(t\eta)G_n(\eta)G_n\left(2i\sin\frac{t\eta}{2}\right)d\sigma_\eta\right|. \tag{9.3.13}$$

Making substitution $t\eta = \eta'$, by (9.3.11) and (9.3.12) we obtain

$$\begin{aligned}\lambda(t) &= O\left(\int\cdots\int_{\eta_1^2+\cdots+\eta_k^2 \leqslant t^2}\left|\phi_\Gamma(t\eta)G_n(\eta)G_n\left(2i\sin\frac{t\eta}{2}\right)\right|d\eta_1\cdots d\eta_k\right)\\ &= o(t^{2nk-2k^2-k}).\end{aligned} \tag{9.3.14}$$

Letting $t \to 0$, we get

$$|I_2| = O\left(R^{2kn-2k^2-k}\int_{\frac{1}{R}}^{\tau} V_{\delta+\frac{k}{2}+kn-k(k+1)}(tR)d\lambda(t)\right).$$

By (9.3.14) and the properties of Bessel functions (cf. Chandrasekharan-Minakshisundaran [1]), we have

$$|I_2| = O\left(R^{2kn-2k^2-k}\int_{\frac{1}{R}}^{\tau}(Rt)^{-\delta-\frac{k}{2}-kn+k(k+1)-\frac{1}{2}}\,d\lambda(t)\right)$$

$$= O\left(R^{-\delta+kn-k^2-\frac{k}{2}-\frac{1}{2}} t^{-\delta-\frac{k}{2}-kn+k(k+1)-\frac{1}{2}} \lambda(t) \Big|_{\frac{1}{k}}^{r} \right)$$

$$+ o\left(R^{-\delta+kn-k^2-\frac{k}{2}-\frac{1}{2}} \int_{\frac{1}{R}}^{r} t^{2kn-2k^2-k} \cdot t^{-\delta-\frac{k}{2}-kn+k(k+1)-\frac{3}{2}} dt \right).$$

Thus we obtain

$$I_2 = o(1),$$

since $\delta > \frac{1}{2}\left(\frac{1}{2} n(n-1) - 1\right) = kn - k^2 - \frac{k}{2} - \frac{1}{2}$.

Finally we consider I_3. This can be done as we did for I_2, but here we have

$$\lambda(t) = O(t^{kn-k^2})$$

as $t \to \infty$. Thus

$$I_3 = O(R^{-\delta+\frac{1}{2}(\frac{1}{2}n(n-1)-1)}).$$

Theorem 9.3.1 has been proved. In estimating I_3, we have not touched upon the property of $u(\Gamma)$. Therefore, the Riesz summation of order δ of the Fourier series of integrable function $u(\Gamma)$ over $SO(n)$, for

$$\delta > [n(n-1) - 2]/4$$

is a local property.

From the proof of Theorem 9.3.1 it can be seen that the following result holds.

Corollary 9.3.1 *If* $u(\Gamma) \in \mathrm{Lip}\gamma$, *then*

$$s_R^\delta(\Gamma) - u(\Gamma) = O(R^{-\gamma})$$

for

$$\delta > [n(n-1) - 2]/4 + \gamma.$$

From this corollary a convergence theorem can be deduced.

Theorem 9.3.2 *If* $u(\Gamma) \in \mathrm{Lip}\gamma$ *and*

$$N(m)\mathrm{tr}(A_m A_m') = o((l_1^2 + \cdots + l_k^2)^{-\alpha}), \quad \textit{for } n = 2k+1;$$

$$N(m)(\mathrm{tr}(A_m A_m') + \mathrm{tr}(B_m B_m')) = o((l_1^2 + \cdots + l_k^2)^{-\alpha}), \quad \textit{for } n = 2k,$$

where

$$\alpha = k - \frac{4\gamma(1+\beta)}{n(n-1) - 2 - 4\beta + 4p + \sigma}, \quad \beta \geqslant 0, \sigma > 0,$$

then

$$s_R^\beta(\Gamma) - u(\Gamma) = o(1).$$

The proof is of the same as that of Theorem 5.8.2. From this we immediately arrive at the following theorem.

Theorem 9.3.3 *Suppose that* $u(\Gamma) \in \mathrm{Lip}\gamma$ *and*

$$N(m)\mathrm{tr}(A_m A_m') = o(1), \quad \text{for } n = 2k+1;$$

$$N(m)(\mathrm{tr}(A_m A_m') + \mathrm{tr}(B_m B_m')) = o(1), \quad \text{for } n = 2k.$$

Then $s_R(\Gamma)$ *is convergent to* $u(\Gamma)$ *whenever*

$$[n(n-1)-2]/4 \geqslant \delta \geqslant [n(n-1)-2]/4 - \gamma[n(n-1)-4k+2]/[4(\gamma+k)].$$

Similarly, taking $\beta = 0$ in Theorem 9.3.2, we obtain a Tauber-type theorem.

Theorem 9.3.4 *Suppose that* $u(\Gamma) \in \mathrm{Lip}\gamma$ *and*

$$N(m)\mathrm{tr}(A_m A_m') = o((l_1^2 + \cdots + l_k^2)^{-\alpha}), \quad \text{for } n = 2k+1;$$

$$N(m)(\mathrm{tr}(A_m A_m') + \mathrm{tr}(B_m B_m')) = o((l_1^2 + \cdots + l_k^2)^{-\alpha}), \quad \text{for } n = 2k;$$

Then the Fourier series of $u(\Gamma)$ *converges to* $u(\Gamma)$ *for*

$$\alpha = k - \frac{4\gamma}{n(n-1) - 2 + 4\gamma + \varepsilon}, \quad \varepsilon > 0.$$

§ 9.4 A General Convergence Theorem

From the proof of Theorem 9.3.1, the following theorem can be proved.

Theorem 9.4.1 *Let* $u(\Gamma)$ *be continuous on* $SO(n)$, $\varphi(t)$ *satisfy the conditions* (9.2.2), (9.2.3) *and* (9.2.14), *and*

(1) *for* $0 \leqslant t \leqslant \frac{1}{R}$, *we thve*

$$G_n\left(\frac{\partial}{\partial \xi}\right) \frac{H_\phi(|\xi| R)}{|\xi|^{k-1}}\bigg|_{\xi = t\eta} = O(R^{kn-k^2-1}); \tag{9.4.1}$$

(2) *for* $\frac{1}{R} \leqslant t < \infty$, *we have*

$$G_n\left(\frac{\partial}{\partial \xi}\right) \frac{H_\phi(|\xi| R)}{|\xi|^{k-1}}\bigg|_{\xi = t\eta} = O(R^{-p-1} t^{-p-kn+k^2}) \ (p > 0). \tag{9.4.2}$$

Let $s_R^\phi(\Gamma)$ *denote the spherical means* (9.1.5) *and* (9.1.6) *of the Fourier series of* $u(\Gamma)$ *in the sense of* φ. *Then* $s_R^\phi(\Gamma)$ *converges to* $u(\Gamma)$ *as* $R \to \infty$ *and the sperical summability in the sense of* φ *of the Fourier series of* $u(\Gamma)$ *is a local property.*

Taking

$$\varphi(t) = e^{-t}(\text{the Abel summation}),$$

and $\varphi(t) = e^{-t^2}$(the Gauss-Sommerfeld summation), we propose

Theorem 9.4.2 *If $u(\Gamma)$ is continuous on $SO(n)$, then its Fourier series is Abel and Gauss-Sommerfeld summable to itself.*

Proof. When $\varphi(t) = e^{-t}$ (cf. Chapter 5),

$$H_\phi(|\xi|R) = \frac{(n-1)!}{2^{k-2}\left(\Gamma\left(\frac{k}{2}\right)\right)^2} R^{k-1}|\xi|^{k-1} \frac{\exp\left(\frac{\sqrt{q_0}}{R}\right)}{(1+|\xi|^2R^2)^{\frac{k+1}{2}}},$$

and when $\varphi(t) = e^{-t^2}$,

$$H_\phi(|\xi|R) = \frac{R^{k-1}\exp\left(\frac{q_0}{R^2}\right)}{2^{k-1}\Gamma\left(\frac{k}{2}\right)}|\xi|^{k-1}\exp\left(\frac{-|\xi|^2R^2}{4}\right).$$

They, obviously, satisfy conditions (9.2.2), (9.2.3) and (9.2.14). As for conditions (9.4.1) and (9.4.2), by a similitude in deducing (9.3.4), a general result can be found. More precisely, if $\lambda(t)$ is a function differentiable of sufficiently high order, then

$$G_n\left(\frac{\partial}{\partial\xi}\right)\lambda\left(\frac{|\xi|^2}{2}\right) = G_n(\xi)\lambda^{(kn-k(k+1))}(t)\Big|_{t=\frac{|\xi|^2}{2}}.$$

Thus, when $\varphi(t) = e^{-t}$, we have

$$G_n\left(\frac{\partial}{\partial\xi}\right)\frac{H_\phi(|\xi|R)}{|\xi|^{k-1}}\Bigg|_{\xi=t\eta} = \frac{(k-1)!\exp\frac{\sqrt{q_0}}{R}}{2^{k-2}\left(\Gamma\left(\frac{k}{2}\right)\right)^2}R^{k-1}$$

$$\cdot R^{nk-k(k-1)}G_n\left(\frac{\partial}{\partial(R\xi)}\right)\frac{1}{(1+(\xi R)^2)^{\frac{k+1}{2}}}\Bigg|_{\xi=t\eta}$$

$$= \frac{(k-1)!\exp\frac{\sqrt{q_0}}{R}}{2^{k-2}\left(\Gamma\left(\frac{k}{2}\right)\right)^2}\cdot R^{nk-k^2-1}G_n(Rt\eta)(-1)^{nk-k(k+1)}$$

$$\cdot\frac{(k+1)(k+3)\cdots(k+2nk-2k(k-1)-1)}{(1+R^2t^2)^{\frac{k+1}{2}+nk-k(k+1)}},$$

and, when $\varphi(t) = e^{-t^2}$, we have

$$G_n\left(\frac{\partial}{\partial\xi}\right)\frac{H_\phi(|\xi|R)}{|\xi|^{k-1}}\Big|_{\xi=t\eta}$$

$$= \frac{\exp\frac{q_0}{R^2}}{2^{k-1}\Gamma\left(\frac{k}{2}\right)} R^{k-1}R^{nk-k(k+1)} \cdot G_n\left(\frac{\partial}{\partial(R\xi)}\right)e^{\frac{-|\xi|^2R^2}{4}}\Big|_{\xi=t\eta}$$

$$= \frac{(-1)^{kn-k(k+1)}\exp\frac{q_0}{R^2}}{2^{k-1}\Gamma\left(\frac{k}{2}\right)2^{kn-k(k+1)}} \cdot R^{kn-k^2-1}G(Rt\eta)\exp\left(\frac{-R^2t^2}{4}\right).$$

It is easily shown that both conditions (9.4.1) and (9.4.2) are satisfied here.

PART III

HARMONIC ANALYSIS ON UNITARY SYMPLECTIC GROUPS

Chapter 10. The Volume of Unitary Symplectic Group and Criteria of Convergence of Fourier Series

§ 10.1 The Volume of Unitary Symplectic Group

In the first two parts, the harmonic analysis on both unitary and rotation groups has been discussed respectively. In this part the harmonic analysis on unitary symplectic groups will be treated. In the present chapter, first of all, we give the volumes of unitary symplectic groups and their cosets, and then find out the Dirichlet kernel of the Fourier series on unitary symplectic groups and the criteria for convergence. In Chapter 11, we shall discuss Cesàro summation and Abel summation of Fourier series on unitary symplectic groups, where the Abel summation is deduced from the Cesàro summation. In Chapter 12, we will give some results on the spherical summation of Fourier series on unitary symplectic groups. As unitary groups and orthogonal groups can be regarded as the characteristic manifolds of the complex and the real classical domains of the first class, respectively, unitary symplectic groups also can be regarded as characteristic manifolds of classical domains of the first class on the quaternion field. In the last chapter, we shall discuss the harmonic analysis on classical domains of the first class on the quaternion field. Starting from this point of view, the same Poisson kernel and Abel summation can also be obtained.

The results of this part are made by my graduate students Chen Guangxiao and He Zuqi (see Chen Guangxiao [1] [2]; He Zuqi and Chen Guangxiao [1] [2] [3] [4]).

The unitary symplectic group $USp(2n)$ of order $2n$ refers to the group consisting of all complex square matrices U of order $2n$ satisfying

$$UJU' = J, \qquad U\bar{U}' = I,$$

where J is a matrix of order $2n$, i. e.

$$J = [J_0, J_0, \cdots, J_0], \qquad J_0 = \begin{pmatrix} 0 & 1 \\ -1 & 0 \end{pmatrix}.$$

In this section, we will find the volume of $USp(2n)$ to be

$$\omega_{2n} = \frac{2^{2n^2+\frac{n}{2}}\pi^{n^2+n}}{\Gamma(2n)\Gamma(2n-2)\cdots\Gamma(2)},$$

which is calculated out by Chen Guangxiao [1] on the basis of the method in Hua Luogeng [1].

Introduce the expression of $USp(2n)$ by auxiliary variables. If $U \in USp(2n)$ and $\det(I+U) \neq 0$, set

$$H = i(I-U)(I+U)^{-1}, \tag{10.1.1}$$

of which the inverse reads

$$U = (I+iH)(I-iH)^{-1}.$$

It is easy to prove that H is a Hermite matrix of order $2n$ and satisfies

$$HJ + JH' = 0. \tag{10.1.2}$$

As it is impossible to have $\det(I-iH) = 0$, the transformation (10.1.1) from $USp(2n)$ onto the set of Hermite matrices satisfying (10.1.2) is one-to-one except some submanifold of lower dimension of $USp(2n)$.

From Hua Luogeng [1], we know further that

$$\operatorname{tr}(\delta U\delta \bar{U}') = 4\operatorname{tr}((I+H^2)^{-1}dH(I+H^2)^{-1}dH),$$

where $\delta U = U^{-1}dU$, since U is an element of the unitary group. On the other hand, U also satisfies (10.1.2), as U belongs to the symplectic group. Those H satisfying the preceding conditions are called symplectic Hermite matrices as well.

Lemma 10.1.1 *we have*

$$\dot{U} = 2^{n(2n-1)}(\det(I+H^2))^{-(n+1/2)}\dot{H}. \tag{10.1.3}$$

Proof. Write dH as $(dH_{jk})_{1\leq j,k\leq n}$, where H_{jk} is a 2×2 square matrix. Then

$$d\bar{H}'_{kj} = dH_{jk} \text{ and } J_0 dH_{jk} + d\bar{H}_{jk}J_0 = 0.$$

Thus

$$dH_{jk} = \begin{pmatrix} d\alpha_{jk} + id\sigma_{jk}, & d\beta_{jk} + id\gamma_{jk} \\ d\beta_{jk} - id\gamma_{jk}, & -d\alpha_{jk} + id\sigma_{jk} \end{pmatrix}, \tag{10.1.4}$$

where α_{jk}, β_{jk} and γ_{jk} are symmetric in j and k, and σ_{jk} is anti-symmetric. Therefore,

$$\dot{H} = 2^{\frac{n(4n-1)}{2}} \prod_{1\leq j\leq n} (d\alpha_{jj}d\beta_{jj}d\gamma_{jj}) \prod_{1\leq j<k\leq n} (d\alpha_{jk}d\beta_{jk}d\gamma_{jk}d\sigma_{jk}). \tag{10.1.5}$$

If P is a fixed unitary symplectic square matrix and

$$dK = PdHP^{-1}, \tag{10.1.6}$$

then $\dot{K} = \dot{H}$. Let

$$dL = TdKT, \tag{10.1.7}$$

where

$$T = [a_1 I_0, \cdots, a_n I_0], \quad I_0 = \begin{pmatrix} 1 & 0 \\ 0 & 1 \end{pmatrix}, \quad a_j > 0, \tag{10.1.8}$$

dL can be written as $(dL_{jk})_{1\leqslant j,k\leqslant n}$, dL_{jk} still takes the form of (10.1.4), and so does dK. From (10.1.7) and (10.1.8), we admit

$$dL_{jk} = a_j a_k dK_{jk},$$

$$\dot{K} = 2^{n(n-1)} \prod_{1\leqslant j\leqslant n} \dot{K}_{jj} \prod_{1\leqslant j<k\leqslant n} \dot{K}_{jk}, \tag{10.1.9}$$

and L has the same expression. Thus

$$\dot{L} = (\det T)^{2n+1}\dot{K}. \tag{10.1.10}$$

For any symplectic Hermite square matrix H, there exists a unitary symplectic square matrix P such that

$$PHP^{-1} = [b_1, -b_1, \cdots, b_n, -b_n], \quad (b_j \geqslant 0, j = 1, \cdots, n).$$

Setting

$$T = [\sqrt{1+b_1^2}\, I_0, \cdots, \sqrt{1+b_n^2}\, I_0], \tag{10.1.11}$$

we have

$$P(I + H^2)^{-1}P^{-1} = (T^{-1})^2. \tag{10.1.12}$$

Thus (10.1.12) and (10.1.6) yield

$$\begin{aligned} \operatorname{tr}(\delta U \overline{\delta U}') &= 4\operatorname{tr}(P^{-1}(T^{-1})^2 PP^{-1}dKPP^{-1}(T^{-1})^2PP^{-1}dKP) \\ &= 4\operatorname{tr}((T^{-1})^2 dK (T^{-1})^2 dK) \\ &= 4\operatorname{tr}(T^{-1}dKT^{-1}T^{-1}dKT^{-1}). \end{aligned}$$

By (10.1.11) and (10.1.10), we know that

$$\dot{U} = 2^{n(2n+1)}(\det T)^{-(2n+1)}, \quad \dot{K} = 2^{n(2n+1)}\det{}^{-\frac{2n+1}{2}}(I + H^2)\dot{H},$$

which is just (10.1.3).

Assume that H is a symplectic Hermite square matrix of order $2n$, $s > n - 1/4$, and

$$\mathscr{H}_n(s) = \int_H \frac{\dot{H}}{\det(I + H^2)^s}.$$

Then we propose

Lemma 10.1.2 *When $s > n - 1/4$, we have*

$$\frac{\mathscr{H}_n(s)}{\mathscr{H}_{n-1}(s-1)} = 2^{4n-4s+\frac{3}{2}} \cdot \pi^{2n} \frac{\Gamma(4s-2n-1)}{\Gamma(2s)\Gamma(2s-1)}. \qquad (10.1.13)$$

Proof. Imitating Hua Luogeng [1], we write the symplectic Hermite square matrix H of order $2n$ to be

$$H = \begin{pmatrix} H_1 & \bar{v}' \\ v & \hat{h} \end{pmatrix},$$

where H_1 and $\hat{h}$ are symplectic Hermite matrices of order $2n-2$ and 2, respectively. If

$$J = \begin{pmatrix} J_1 & 0 \\ 0 & J_0 \end{pmatrix},$$

then, by (10.1.2), the rectangular matrix v satisfies

$$vJ_1 + J_0\bar{v} = 0. \qquad (10.1.14)$$

Let
$$v = (A_1, \cdots, A_{n-1}),$$

and then from (10.1.14), it can be easily shown that

$$A_j = \begin{pmatrix} \alpha_j + i\sigma_j, & \beta_j + i\gamma_j \\ \beta_j - i\gamma_j, & -\alpha_j + i\sigma_j \end{pmatrix}.$$

Thus

$$v\bar{v}' = I_0 \sum_{1 \leqslant j \leqslant n-1} (\alpha_j^2 + \beta_j^2 + \gamma_j^2 + \sigma_j^2).$$

It can be readily seen that

$$I + H^2 = \begin{pmatrix} I_1 + H_1^2 + \bar{v}'v, & \bar{p}' \\ p, & I + v\bar{v}' + \hat{h}^2 \end{pmatrix}, \quad p = vH_1 + \hat{h}v.$$

Therefore

$$\det(I + H^2)$$
$$= \det(I_1 + H_1^2 + \bar{v}'v)\det(I_0 + \hat{h}^2 + v\bar{v}' - p(I_1 + H_1^2 + \bar{v}'v)^{-1}\bar{p}').$$

There exists a unitary symplectic square matrix P_1 such that

$$H_1 = P_1\Lambda_1P_1^{-1}, \ \Lambda_1 = [\lambda_1, -\lambda_1, \cdots, \lambda_{n-1}, -\lambda_{n-1}].$$

Let

$$T_1 = P_1[\sqrt{1+\lambda_1^2}\, I_0, \cdots, \sqrt{1+\lambda_{n-1}^2}\, I_0]P_1^{-1},$$
$$v = uT_1.$$

Then it is easy to see that

$$\bar{T}_1' = T_1, \ T_1H_1 = H_1T_1, \ I_1 + H_1^2 = T_1^2. \qquad (10.1.15)$$

Assume that

$$\tau = P_1[\sqrt[4]{1+\lambda_1^2}, \sqrt[4]{1+\lambda_1^2}, \cdots, \sqrt[4]{1+\lambda_{n-1}^2}, \sqrt[4]{1+\lambda_{n-1}^2}]P_1^{-1}.$$

Then $T_1 = \tau^2$ and τ is a symplectic Hermite matrix. Thus τ satisfies

$$\tau J_1 = -J_1\bar{\tau},$$

which leads to $$T_1J_1 = \tau^2 J_1 = -\tau J_1\bar{\tau} = J_1\bar{\tau}^2 = J_1\bar{T}_1.$$

So, by (10.1.14), we arrive at

$$uJ_1 + J_0\bar{u} = 0. \tag{10.1.16}$$

In addition, we have

$$\dot{v} = |\det T_1|^2\dot{u} = \det(I_1 + H_1^2)\dot{u}, \tag{10.1.17}$$

and

$$I_1 + H_1^2 + \bar{v}'v = T_1(I_1 + \bar{u}'u)T_1. \tag{10.1.18}$$

Set

$$u = (B_1, \cdots, B_{n-1}).$$

Here, by (10.1.16),

$$B_j = \begin{pmatrix} l_j + ik_j, & f_j + ig_j \\ f_j - ig_j, & -l_j + ik_j \end{pmatrix},$$

we conclude

$$v\bar{v}' = \mu^2 I_0, \tag{10.1.19}$$

where

$$\mu^2 = \sum_{1\leqslant j\leqslant n-1} (l_j^2 + f_j^2 + g_j^2 + k_j^2).$$

By (10.1.18), (10.1.15) and

$$(I_1 + \bar{u}'u)^{-1} = I_1 - \bar{u}'(I_0 + u\bar{u}')^{-1}u = I_1 - \frac{1}{1+\mu^2}\bar{u}'u,$$

we obtain

$$\begin{aligned}
&-p(I_1 + H_1^2 + \bar{v}'v)^{-1}\bar{p}' \\
&= -(vH_1 + \hat{h}v)T_1^{-1}(I_1 + \bar{u}'u)^{-1}T_1^{-1}(\bar{H}_1'\bar{v}' + \bar{v}\hat{h}') \\
&= -(uT_1H_1T_1^{-1} + \hat{h}u)(I_1 + \bar{u}'u)^{-1}(\overline{T_1H_1T_1^{-1}}'\bar{u}' + \bar{u}'\hat{h}) \\
&= -(uH_1 + \hat{h}u)(I_1 + \bar{u}'u)^{-1}(H_1\bar{u}' + \bar{u}'\hat{h}) \\
&= -(uH_1 + \hat{h}u)\left(I_1 - \frac{1}{1+\mu^2}\bar{u}'u\right)(H_1\bar{u}' + \bar{u}'\hat{h}) \\
&= -uH_1^2\bar{u}' - \frac{\mu^2}{1+\mu^2}\hat{h}^2 + \frac{1}{1+\mu^2}\hat{h}_0^2 - \frac{1}{1+\mu^2}(\hat{h}\hat{h}_0 + \hat{h}_0\hat{h}),
\end{aligned}$$

where

$$\hat{h}_0 = uH_1\bar{u}',$$

which, by (10.1.16), is a symplectic Hermite square matrix.

By (10.1.18), we have

$$\det(I_1 + H_1^2 + \bar{v}'v) = \det T_1^2 \det(I_1 + \bar{u}'u) = \det(I + H_1^2)(1 + \mu^2)^2.$$

However, we have

$$\begin{aligned}
&\det(I_0 + \hat{h}^2 + v\bar{v}' - p(I_1 + H_1^2 + \bar{v}'v)^{-1}\bar{p}') \\
&= \det\left(I_0 + \frac{1}{1+\mu^2}(\hat{h} - \hat{h}_0)^2 + v\bar{v}' - u(H_1^2\bar{u}') \right) \\
&= \det\left(\frac{1}{1+\mu^2}(\hat{h} - \hat{h}_0)^2 + I_0 + u(T_1^2 - H_1^2)\bar{u}'\right) \\
&= \det\left(\frac{1}{1+\mu^2}(\hat{h} - \hat{h}_0)^2 + (1+\mu^2)I_0\right).
\end{aligned}$$

Therefore

$$\begin{aligned}
\det(I + H^2) &= \det(I_1 + H_1^2)\ (1+\mu^2)^2 \\
&\cdot \det\left(\frac{1}{1+\mu^2}(\hat{h} - \hat{h}_0)^2 + (1+\mu^2)I_0\right) \\
&= \det(I_1 + H_1^2)\det((\hat{h} - \hat{h}_0)^2 + (1+\mu^2)^2 I_0).
\end{aligned}$$

Making a change

$$\hat{h}_1 = \frac{\hat{h} - \hat{h}_0}{1+\mu^2},$$

we get

$$\dot{\hat{h}}_1 = (1+\mu^2)^{-3}\dot{\hat{h}}.$$

So

$$\begin{aligned}
&\det[(1+\mu^2)^2 I_0 + (\hat{h} - \hat{h}_0)^2]^{-s}\dot{\hat{h}} \\
&= (1+\mu^2)^{3-4s}\det(I_0 + \hat{h}_1^2)^{-s}\dot{\hat{h}}_1.
\end{aligned}$$

Thus, by (10.1.17), we obtain

$$\mathscr{H}_n(s) = 2^{2(n-1)}\int_{H_1} \det{}^{1-s}(I_1 + H_1^2)\dot{H}_1 \int_u \frac{\dot{u}}{(H\mu^2)^{4s-3}} \cdot \mathscr{H}_1(s). \quad (10.1.20)$$

We now turn to calculate $\mathscr{H}_1(s)$. Here

$$H = \begin{pmatrix} \alpha & \beta + i\gamma \\ \beta - i\gamma & -\alpha \end{pmatrix},$$

α, β and γ are real numbers.

Consequently

$$H^2 = (\alpha^2 + \beta^2 + \gamma^2) I_0$$

and

$$\mathscr{H}_1(s) = \int_H \frac{\dot{H}}{\det(I + H^2)^s}$$

$$= (\sqrt{2})^3 \iiint_{-\infty}^{\infty} \frac{d\alpha d\beta d\gamma}{(1 + \alpha^2 + \beta^2 + \gamma^2)^{2s}} = 4\pi \cdot 2\sqrt{2} \int_0^\infty \frac{R^2 dR}{(1 + R^2)^{2s}}$$

$$= 2\sqrt{2} \cdot 2\pi B\left(\frac{3}{2}, 2s - \frac{3}{2}\right) = (2\pi)^{3/2} \frac{\Gamma(2s - 3/2)}{\Gamma(2s)}.$$

Next we treat $\int_u \frac{\dot{u}}{(1 + \mu^2)^{4s-3}}$, which is equal to

$$\int_{-\infty}^{\infty} \cdots \int \frac{(\sqrt{2})^{4(n-1)} \prod_{j=1}^{n-1} (dl_j df_j dg_j dh_j)}{\left\{1 + \sum_{j=1}^{n-1} (e_j^2 + f_j^2 + g_j^2 + h_j^2)\right\}^{4s-3}}$$

$$= 2^{2(n+1)} \cdot \frac{2\pi^{2n-2}}{\Gamma(2n-2)} \int_0^\infty \frac{R^{4n-5} dR}{(1 + R^2)^{4s-3}}$$

$$= 2^{2(n-1)} \pi^{2n-2} \frac{\Gamma(4s - 2n - 1)}{\Gamma(4s - 3)}$$

if $s \geqslant n - 1/4$. Substituting it into (10.1.20), we find

$$\frac{\mathscr{H}_n(s)}{\mathscr{H}_{n-1}(s-1)} = 4^{2(n-1)} (2\pi)^{\frac{3}{2}} \pi^{2n-2} \frac{\Gamma(2s - 3/2)}{\Gamma(2s)} \cdot \frac{\Gamma(4s - 2n - 1)}{\Gamma(4s - 3)}. \tag{10.1.21}$$

Since

$$\frac{\Gamma(x)\Gamma\left(x + \frac{1}{2}\right)}{\Gamma(2x)\Gamma\left(\frac{1}{2}\right)} = 2^{1-2x} \qquad (x > 0),$$

(10.1.21) is barely (10.1.13).

We take $s = n + 1/2$, and (10.1.13) becomes

$$\frac{\mathscr{H}_n\left(n + \frac{1}{2}\right)}{\mathscr{H}_n\left(n - \frac{1}{2}\right)} = 2^{-\frac{1}{2}} \pi^{2n} \frac{1}{\Gamma(2n)}, \qquad (n > 1).$$

If $H_0(s) = 1$, then the above formula is also valid for $n = 1$. Thus

$$\mathscr{H}_n\left(n + \frac{1}{2}\right) = 2^{-n/2}\pi^{n(n+1)} \frac{1}{\Gamma(2n)\Gamma(2n-2)\cdots\Gamma(2)}.$$

Therefore, we obtain (Chen Guangxiao [1])

Theorem 10.1.1 *The volumes of $USp(2n)$ amounts to*

$$\omega_n = \frac{2^{\frac{n}{2}+2n^2}\pi^{n+n^2}}{\Gamma(2n)\Gamma(2n-2)\cdots\Gamma(2)}. \tag{10.1.22}$$

§ 10.2 The Volume of the Coset Space of Unitary Symplectic Group

Any unitary symplectic square matrix w can be expressed as

$$w = P\Lambda P^{-1},\ \Lambda = [e^{i\theta_1}, e^{-i\theta_1}, \cdots, e^{i\theta_n}, e^{-i\theta_n}],$$

with $\pi \geqslant \theta_1 \geqslant \cdots \geqslant \theta_n \geqslant 0$, where P stands for some unitary symplectic square matrix and Λ is uniquely determined, since $e^{\pm i\theta_j}$ $(1 \leqslant j \leqslant n)$ refer to the characteristic roots of w. Besides, if

$$w = P\Lambda P^{-1} = P_1\Lambda P_1^{-1},$$

and all characteristic roots of w are distinct, then

$$P_1^{-1}P = [e^{i\varphi_1}, e^{-i\varphi_1}, \cdots, e^{i\varphi_n}, e^{-i\varphi_n}].$$

All those unitary symplectic square matrices of the form

$$[e^{i\varphi_1}, e^{-i\varphi_1}, \cdots, e^{i\varphi_n}, e^{-i\varphi_n}]$$

constitute a subgroup of $USp(2n)$. Let $[USp(2n)]$ denote the set consisting of all cosets of $USp(2n)$ relative to the subgroup.

Then P_1 and P belong to the same coset. Thus, any unitary symplectic square matrix must be one to one corresponding to some Λ and some coset of $USp(2n)$, except some manifold of lower dimension consisting of those unitary symplectic square matrices having multiple characteristic roots.

Let $\delta P = P^{-1}dP$, then

$$\overline{\delta P}' = -\delta P,\ J\delta P + \delta P'J = 0. \tag{10.2.1}$$

Thus

$$\delta P = i\left(\begin{pmatrix} \delta\alpha_{jk} + i\delta\sigma_{jk}, & \delta\beta_{jk} + i\delta\gamma_{jk} \\ \delta\beta_{jk} - i\delta\gamma_{jk}, & -\delta\alpha_{jk} + i\delta\sigma_{jk} \end{pmatrix}\right)_{1\leqslant j,k\leqslant n}, \tag{10.2.2}$$

where $\delta\alpha_{jk}$, $\delta\beta_{jk}$ and $\delta\gamma_{jk}$ are symmetric in j and k, and $\delta\sigma_{jk}$ is anti-symmetric in j and k. Moreover

$$\operatorname{tr}(\delta w\overline{\delta w}') = \operatorname{tr}(\delta\Lambda\overline{\delta\Lambda}') + \operatorname{tr}([\delta P, \Lambda]\,\overline{[\delta P, \Lambda]}'), \qquad (10.2.3)$$

where

$$[\delta P, \Lambda] = \delta P \cdot \Lambda - a \cdot \delta P.$$

We now prove (10.2.3). Since

$$\delta w = P\Lambda^{-1}P^{-1}dP\Lambda P^{-1} - dP \cdot P^{-1} + P\Lambda^{-1}d\Lambda P^{-1},$$

we have

$$\begin{aligned}\operatorname{tr}(\delta w\bar{\delta} w') &= 2[-\operatorname{tr}(\delta P\delta P) + \operatorname{tr}(\Lambda^{-1}\delta P\Lambda\delta P)\\ &\quad - \operatorname{tr}(\Lambda^{-1}\delta P d\Lambda) + \operatorname{tr}(\delta P\delta\Lambda) - \operatorname{tr}(\delta\Lambda\delta\Lambda)\\ &= 2(-\operatorname{tr}(\delta P\delta P) + \operatorname{tr}(\Lambda^{-1}\delta P\Lambda\delta P))\end{aligned}$$

$$\operatorname{tr}(\delta\Lambda\delta\Lambda) = \operatorname{tr}(\delta\Lambda\overline{\delta\Lambda}') + \operatorname{tr}([\delta P, \Lambda]\,\overline{[\delta P, \Lambda]}').$$

This leads to

$$\begin{aligned}\operatorname{tr}(\delta\bar{w}\delta\bar{w}') &= 2\sum_{j=1}^{n} d\theta_j^2 + 4\sum_{j<k} |e^{i\theta_j} - e^{i\theta_k}|^2(d\alpha_{jk}^2 + d\sigma_{jk}^2)\\ &\quad + 4\sum_{j<k} |e^{i\theta_j} - e^{i\theta_k}|^2(d\beta_{jk}^2 + d\gamma_{jk}^2)\\ &\quad + 2\sum_{j=1}^{n} |e^{i\theta_j} - e^{i\theta_j}|^2(d\beta_{jj}^2 + d\gamma_{jj}^2),\end{aligned}$$

where $d\alpha_{jj}$ does not appear. Let $\dot{V}$ denote the integral element spanned by the differential vectors $d\alpha_{jk}$, $d\beta_{jk}$, $d\gamma_{jk}$ and $d\sigma_{jk}$, and $[\dot{P}] = 2^{n(2n-\frac{1}{2})}\dot{V}$ denote the integral element over $[USp(2n)]$. Then

$$\begin{aligned}\dot{W} &= [\dot{P}]\prod_{j<k} |(e^{i\theta_j} - e^{i\theta_k})(e^{i\theta_j} - e^{-i\theta_k})|^2\\ &\quad \cdot \prod_{j=1}^{n} |e^{i\theta_j} - e^{-i\theta_j}|^2 d\theta_1 d\theta_2 \cdot \cdots \cdot d\theta_n.\end{aligned} \qquad (10.2.4)$$

We now come to calculate the volume ω_n' of $[USp(2n)]$. Integrating (10.2.4) over $USp(2n)$ gives

$$\omega_n = \omega_n' 2^{\frac{n}{2}} \int\cdots\int_{\pi\geq\theta_1\geq\cdots\theta_n\geq 0} J(\theta_1, \cdots, \theta_n)d\theta_1\cdots d\theta_n, \qquad (10.2.5)$$

where

$$\begin{aligned}J(\theta_1, \cdots, \theta_n) &= \prod_{j=1}^{n} |e^{i\theta_j} - e^{-i\theta_j}|^2\\ &\quad \cdot \prod_{j<k} |(e^{i\theta_j} - e^{i\theta_k})(e^{i\theta_j} - e^{-i\theta_k})|^2.\end{aligned} \qquad (10.2.6)$$

This is equal to

$$(-1)^n \prod_{j=1}^{n} (e^{i\theta_j} - e^{-i\theta_j}) D(e^{i\theta_1}, e^{-i\theta_1}, \cdots, e^{i\theta_n}, e^{-i\theta_n}),$$

where D is a Vandermonde determinant multiplied by $(-1)^{\frac{2n(n-1)}{2}} = (-1)^n$. It is easy to see that

$$J = s \cdot \bar{s},$$

where $s = s(1, 2, \cdots, n)$ is defined in (6.1.6).

Since $J(\theta_1, \cdots, \theta_n)$ is a symmetric function in $\theta_1, \cdots, \theta_n$, we have

$$\int \cdots \int_{\pi \geq \theta_1 \geq \cdots \geq \theta_n \geq 0} J(\theta_1, \cdots, \theta_n) d\theta_1 \cdots d\theta_n$$
$$= \frac{1}{n!} \cdot \int_0^\pi \cdots \int_0^\pi J(\theta_1, \theta_2, \cdots, \theta_n) d\theta_1 \cdots d\theta_n.$$

In addition, J means even function in each θ_j, so we have

$$\int \cdots \int_{\pi \geq \theta_1 \geq \cdots \geq \theta_n \geq 0} J(\theta_1, \cdots, \theta_n) d\theta_1 \cdots d\theta_n$$
$$= \frac{1}{2^n n!} \int_{-\pi}^{\pi} \cdots \int_{-\pi}^{\pi} J(\theta_1, \cdots, \theta_n) d\theta_1 \cdots d\theta_n$$
$$= \frac{1}{2^n n!} \int_{-\pi}^{\pi} \cdots \int_{-\pi}^{\pi} \prod_{j=1}^{n} (e^{i\theta_j} - e^{-i\theta_j}) \sum_{k_1, L_1, \cdots, k_n, L_n} \delta_{k_1, L_1, \cdots, k_n, L_n},$$
$$\cdot e^{i(k_1 - L_1)\theta_1} \cdots e^{i(k_n - L_n)\theta_n} d\theta_1 \cdots d\theta_n$$

where $(k_1, L_1, \cdots, k_n, L_n)$ signifies a permutation of $(1, 2, \cdots, 2n)$ and

$$\int_{-\pi}^{\pi} (e^{i\theta} - e^{-i\theta}) e^{i\theta(k-L)} d\theta = \begin{cases} 0, & \text{if } k - L \neq \pm 1, \\ 2\pi, & \text{if } k - L = -1, \\ -2\pi, & \text{if } k - L = 1. \end{cases}$$

Thus we obtain

$$\int \cdots \int_{\pi \geq \theta_1 \geq \cdots \geq \theta_n \geq 0} J d\theta_1 \cdots d\theta_n = (2\pi)^n,$$

which leads to

$$w_n' = \frac{w_n}{(2\pi)^n}.$$

At last we formulate (Chen Guangxiao [1])

Theorem 10.2.1 *The volume w_n' of the coset space of $USp(2n)$ is equal to*

$$\frac{2^{n^2 - \frac{n}{2}} (2\pi)^{n^2}}{\Gamma(2n)\Gamma(2n-2) \cdots \Gamma(2)}. \tag{10.2.7}$$

§ 10.3 Fourier Series on Unitary Symplectic Groups

Let $w \in USp(2n)$ and $A_f(w)$ be the irreducible representation with signature $f=(f_1,\cdots,f_n)$, $f_1 \geqslant \cdots f_n \geqslant 0$. Then $A_f(w)$ is still a unitary symplectic square matrix and its order, denoted by $N(f)$, is equal to

$$\frac{\prod_{1\leqslant j\leqslant n} l_j \prod_{j<k}(l_j^2-l_k^2)}{(2n-1)!(2n-3)!\cdots 3!1!} = \frac{l_1\cdots l_n D(l_1^2,\cdots,l_n^2)}{n!D(n^2,\cdots,1^2)}, \quad (10.3.1)$$

where

$$l_1=f_1+n,\cdots,l_n=f_n+1$$

(see Weyl [1]).

Let $\chi_f(\omega)$ be the character of the representation $A_f(w)$. If the characteristic roots of w are $e^{\pm i\theta_1},\cdots,e^{\pm i\theta_n}$, then

$$x_f(w)=\frac{S_\theta(l_1,\cdots,l_n)}{S_\theta(n,\cdots,1)}=\frac{S_\theta(l)}{S_\theta(n)}, \quad (10.3.2)$$

where $s_\theta(l_1,\cdots,l_n)$ is defined in (6.1.6).

Write

$$\Phi_f(U)=\sqrt{\frac{N(f)}{C}}A_f(U)=(\phi_{ij}^f(U))_{1\leqslant i,j\leqslant N(f)},$$

where C is the volume of $USp(2n)$, and $\{\phi_{ij}^f(U)\}$ forms an orthogonal system over $USp(2n)$. If $u(U)$ is integrable over $USp(2n)$, then $u(U)$ takes

$$\sum_{f_1\geqslant\cdots\geqslant f_n\geqslant 0} \mathrm{tr}(C_f\Phi_f'(U)) \quad (10.3.3)$$

as its Fourier series, where

$$C_f=\int_{USp(2n)} u(W)\Phi_f(\bar{w})\dot{W}. \quad (10.3.4)$$

§ 10.4 Dirichlet Kernels and Convergence Criterion of Fourier Series

Let $u(U)$ be an integrable function over $USp(2n)$. As in Chapter 3, we may consider the partial sum

$$s_N(U) = \sum_{N \geqslant l_1 > \cdots > l_n > 0} \operatorname{tr}(C_{f_1 \cdots f_n} \Phi'_{f_1 \cdots f_n}(U)) \tag{10.4.1}$$

of the Fourier series (10.3.3) of $u(U)$, where $l = (l_1, \cdots, l_n)$, $l_1 = f_1 + n - 1, \cdots, l_n = f_n + 1$.

Similarly, $s_N(U)$ can be put into

$$\frac{1}{c} \int_{USp(2n)} u(VU) \mathscr{D}_N(\bar{V}) \dot{V}, \tag{10.4.2}$$

where

$$\mathscr{D}_N(\bar{V}) = \sum_{N \geqslant l_1 > \cdots > l_n > 0} N(f) \chi_f(\bar{V}) \tag{10.4.3}$$

is the Dirichlet kernel. We can prove the following

Theorem 10.4.1 *Let $u(U)$ be integrable over $USp(2n)$. Then the partial sum (10.4.1) of its Fourier series (10.3.3) can be expressed by (10.4.2) and its Dirichlet kernel (10.4.3) is equal to*

$$\frac{(-1)^n \det(d_N^{(2i-1)}(\theta_j))_{1 \leqslant i,j \leqslant n}}{\prod_{k=1}^{n} (2n - 2k + 1)! \det(\sin(n - i + 1)\theta_j)_{1 \leqslant i,j \leqslant n}} \tag{10.4.4}$$

where $e^{i\theta_1}, e^{-i\theta_1}, \cdots, e^{i\theta_n}, e^{-i\theta_n}$ mean the characteristic roots of $\bar{V}$ and

$$d_N(\theta) = \frac{\sin\left(N + \frac{1}{2}\right)\theta}{2 \sin \frac{1}{2} \theta}$$

indicates the Dirichlet kernel in one variable (He Zuqi and Chen Guangxiao [1]).

Proof. We have $$\chi_f(\bar{V}) = s(l)/s(n),$$

where $l = (l_1, \cdots, l_n)$, $n = (n, n-1, \cdots, 1)$ and the definition of $s(t)$ is given in(6.1.6). Again

$$N(f) = \lim_{\substack{\theta_1 \to 0 \\ \cdots \\ \theta_n \to 0}} \chi_f,$$

which is equal to

$$\begin{vmatrix} l_1^{2n-1}, & l_2^{2n-1}, & \cdots, & l_n^{2n-1} \\ \cdots & \cdots & \cdots & \cdots \\ l_1^3, & l_2^3, & \cdots, & l_n^3 \\ l_1, & l_2, & \cdots, & l_n \end{vmatrix} \Big/ \prod_{k=1}^{n} (2n - 2k + 1)!.$$

Consequently

$$\mathscr{D}_N(\bar{V}) = \sum_{N \geqslant l_1 > l_2 > \cdots > l_n > 0} \begin{vmatrix} \sin l_1\theta_1, & \cdots, & \sin l_1\theta_n \\ \sin l_2\theta_1, & \cdots, & \sin l_2\theta_n \\ \cdots & \cdots & \cdots \\ \sin l_n\theta_1, & \cdots, & \sin l_n\theta_n \end{vmatrix} \cdot \begin{vmatrix} l_1^{2n-1}, & l_2^{2n-1}, & \cdots, & l_n^{2n-1} \\ \cdots & \cdots & \cdots & \cdots \\ l_1^3, & l_2^3, & \cdots, & l_n^3 \\ l_1, & l_2, & \cdots & l_n \end{vmatrix}$$

$$\cdot \frac{1}{\prod_{k=1}^{n} (2n-2k+1)!\det(\sin(n-i+1)\theta_i)_{1 \leqslant i,j \leqslant n}}.$$

From Lemma 8.1.1, we arrive at

$$\mathscr{D}_N(\bar{V}) = \frac{(-1)^{\frac{1}{2}n(n-1)}\det\left(\sum_{l=1}^{N} l^{2i-1}\sin l\theta_j\right)_{1 \leqslant i,j \leqslant n}}{(2n-1)!(2n-3)!\cdots 3!1!\det(\sin(n-i+1)\theta)_{1 \leqslant i,j \leqslant n}}.$$

By

$$\sum_{l=1}^{N} l^{2i-1}\sin l\theta_j = (-1)^i d_N^{(2i-1)}(\theta_j),$$

we are led to (10.4.4)

Starting from this point, we come to deduce a convergence criterion of the Fourier series.

If $\bar{V} \in USp(2n)$, it can be written as

$$V_1 \begin{pmatrix} e^{i\theta_1} & & & & \\ & e^{-i\theta_1} & & & \\ & & \ddots & & \\ & & & e^{i\theta_n} & \\ & & & & e^{-i\theta_n} \end{pmatrix} V_1^{-1}, \quad V_1 \in USp(2n).$$

Take $\Gamma \in USp(2n)$ and $\Gamma_1 \in USp(2n)$ such that

$$\Gamma \begin{pmatrix} e^{i\theta_1} & & & & \\ & e^{-i\theta_1} & & & \\ & & \ddots & & \\ & & & e^{i\theta_n} & \\ & & & & e^{-i\theta_n} \end{pmatrix} \Gamma^{-1}$$

$$= \begin{pmatrix} e^{i\theta_{\nu_1}} & & & & \\ & e^{-i\theta_{\nu_1}} & & & \\ & & \ddots & & \\ & & & e^{i\theta_{\nu_n}} & \\ & & & & e^{-i\theta_{\nu_n}} \end{pmatrix},$$

and

$$\Gamma_1\begin{pmatrix} e^{i\theta_{\nu_1}} & & & & & & & \\ & e^{-i\theta_{\nu_1}} & & & & & & \\ & & \ddots & & & & & \\ & & & e^{i\theta_{\nu_k}} & & & & \\ & & & & e^{-i\theta_{\nu_k}} & & & \\ & & & & & \ddots & & \\ & & & & & & e^{i\theta_{\nu_n}} & \\ & & & & & & & e^{-i\theta_{\nu_n}} \end{pmatrix}\Gamma_1^{-1}$$

$$=\begin{pmatrix} e^{i\theta_{\nu_1}} & & & & & & & \\ & e^{-i\theta_{\nu_1}} & & & & & & \\ & & \ddots & & & & & \\ & & & e^{-i\theta_{\nu_k}} & & & & \\ & & & & e^{i\theta_{\nu_k}} & & & \\ & & & & & \ddots & & \\ & & & & & & e^{i\theta_{\nu_n}} & \\ & & & & & & & e^{-i\theta_{\nu_n}} \end{pmatrix},$$

respectively, where $(\nu_1, \cdots, \nu_n)$ is a permutation of $(1, 2, \cdots, n)$. Write

$$\overline{V}^* = V_1\Gamma_1\Gamma\begin{pmatrix} e^{i\theta_1} & & & & \\ & e^{-i\theta_1} & & & \\ & & \ddots & & \\ & & & e^{i\theta_n} & \\ & & & & e^{-i\theta_n} \end{pmatrix}\Gamma^{-1}\Gamma_1^{-1}V_1^{-1}.$$

Thus

$$\frac{1}{C}\int_{USp(2n)} u(V^*U)\mathscr{D}_N(V)\dot{V} = \frac{1}{C}\int_{USP(2n)} u(VU)\mathscr{D}_N(V)\dot{V}.$$

Making all possible V^*, adding all $u(V^*U)$ together and dividing the sum by $2^n n!$ (the result thus obtained is denoted by $u^*(VU)$), we observe

$$s_N(U) = \frac{1}{C}\int_{USp(2n)} u^*(VU)\mathscr{D}_N(\overline{V})\dot{V}.$$

Setting

$$g(\theta_1, \cdots, \theta_n) = \frac{1}{C}\int_{USp(2n)} u^*(VU)[\dot{V}],$$

which is an even symmetric and periodic function in $\theta_1, \cdots, \theta_n$, we have

$$s_N(U) = 2^{n^2+n} \int\cdots\int_{\pi\geqslant\varphi_1\geqslant\cdots\varphi_n\geqslant-\pi} g(\theta_1, \cdots, \theta_n)\mathscr{D}_N(\overline{V})$$

$$\cdot \prod_{j=1}^{n} \sin^2\theta_j \prod_{1\leqslant i,j\leqslant n} (\cos\theta_i - \cos\theta_j)^2 d\theta_1\cdots d\theta_n.$$

As

$$\frac{1}{C}\int_{USp(2n)} \mathscr{D}_N(V)\dot{V} = 1,$$

$s_N(U) - u(U)$ is equal to

$$\frac{(-1)^n 2^{n^2+n}}{\prod_{k=1}^{n}(2n-2k+1)!}\int_{-\pi}^{\pi}\cdots\int (g(\theta_1,\cdots,\theta_n) - u(U))$$

$$\cdot \frac{\det(d_N^{(2i-1)}(\theta_j))_{1\leqslant i,j\leqslant n}}{\det(\sin(n-i+1)\theta_j)_{1\leqslant i,j\leqslant n}}$$

$$\cdot \prod_{j=1}^{n}\sin^2\theta_j \prod_{1\leqslant i,j\leqslant n}(\cos\theta_i - \cos\theta_j)^2 d\theta_1\cdots d\theta_n.$$

On the other hand

$$\prod_{j=1}^{n}\sin^2\theta_j \prod_{1\leqslant i<j\leqslant n}(\cos\theta_i - \cos\theta_j)^2 = 2^{-n(n-1)}$$

$$\cdot \begin{vmatrix} \sin n\theta_1, & \sin n\theta_2, & \cdots, & \sin n\theta_n \\ \sin(n-1)\theta_1, & \sin(n-1)\theta_2, & \cdots, & \sin(n-1)\theta_n \\ \cdots & \cdots & \cdots & \cdots \\ \sin\theta_1, & \sin\theta_2, & \cdots, & \sin\theta_n \end{vmatrix}^2.$$

Therefore $s_N(U) - u(U)$ is equal to

$$\frac{(-1)^n 2^{2n}}{\prod_{k=1}^{n}(2n-2k+1)!}\int_{-\pi}^{\pi}\cdots\int (g(\theta_1,\cdots,\theta_n) - u(U))$$

$$\cdot \det(d_N^{(2i-1)}(\theta_j))_{1\leqslant i,j\leqslant n} \det(\sin(n-i+1)\theta_j)_{1\leqslant i,j\leqslant n} \cdot d\theta_1 d\theta_2\cdots d\theta_n.$$

From the properties of determinants, we can easily show that the preceding expression exactly turns out to be

$$\frac{(-1)^n 2^{2n}}{\prod_{k=1}^{n}(2n-2k+1)!}\int_{-\pi}^{\pi}\cdots\int (g(\theta_1,\cdots,\theta_n) - u(U))$$

$$\cdot \det(\sin(n-i+1)\theta_j)_{1\leqslant i,j\leqslant n} d_N'(\theta_1) d_N^{(3)}(\theta_2)\cdots d_N^{(2n-1)}(\theta_n)\, d\theta_1\cdots d\theta_n.$$

If $u(U)\in C^{n^2}$, applying integration by parts to the above expression, by the periodicity of $g(\theta_1,\cdots,\theta_m)$, we obtain that $s_N(U) - u(U)$ is equal to

$$\frac{(-1)^n 2^{2n}}{\prod_{k=1}^{n}(2n-2k+1)!}\int_{-\pi}^{\pi}\cdots\int \frac{\partial}{\partial\theta_1}\frac{\partial^3}{\partial\theta_2^3}\cdots\frac{\partial^{2n-1}}{\partial\theta_n^{2n-1}}$$

$$\cdot\, [(g(\theta_1, \cdots, \theta_n) - u(U))\det(\sin(n-i+1)\theta_j)_{1\le i,j\le n}]$$
$$\cdot\, d_N(\theta_1)d_N(\theta_2)\cdots d_N(\theta_N)d\theta_1\cdots d\theta_n.$$

As in Chapter 3, we can prove the following

Theorem 10.4.2 *If $u(U)$ is integrable over $USp(2n)$ and $u(U) \in C^{n^2+p}$ $(0 < p \leqslant 1)$, then the partial sum (10.4.1) of its Fourier serie converges to $u(U)$ and*

$$|s_N(U) - u(U)| \leqslant A \cdot \max\left(\left(\frac{\ln^{n^2} N}{N}\right)^{\frac{1}{n+1}}, \frac{\ln^{n-1} N}{N^p}\right)$$

(He Zuqi & Chen Guangxiao [1]).

Proof. The proof is all alike to that of Theorem 3.4.1. Let

$$I_{\mu_1,\mu_2,\cdots,\mu_n} = \int_{-\pi}^{\pi}\int \frac{\partial^{\mu_1}}{\partial\theta_1^{\mu_1}}\cdots\frac{\partial^{\mu_n}}{\partial\theta_n^{\mu_n}}(g(\theta_1,\cdots,\theta_n) - u(U))$$
$$\cdot\,\frac{\partial^{v_1}}{\partial\theta_1^{v_1}}\cdots\frac{\partial^{v_n}}{\partial\theta_n^{v_n}}\begin{vmatrix}\sin n\theta_1, & \cdots, & \sin n\theta_n\\ \cdots & \cdots & \cdots\\ \sin\theta_1, & \cdots, & \sin\theta_n\end{vmatrix}$$
$$\cdot\, d_N(\theta_1)\cdots d_N(\theta_n)d\theta_1\cdots d\theta_n,$$

where

$$\mu_k + U_k = 2k-1,\ \mu_k \geqslant 0,\ v_k \geqslant 0,\ k = 1, 2, \cdots, n.$$

First consider $I_{0,\cdots,0}$ and decompose the integral domain into

$$\pi \geqslant \theta_{i_1} \geqslant \theta_{i_2} \geqslant \cdots \geqslant \theta_{i_n} \geqslant -\pi,$$

where $(i_1, \cdots, i_n)$ is a permutation of $(1, \cdots, n)$. We single one of them out, say

$$\pi \geqslant \theta_1 \geqslant \theta_2 \geqslant \cdots \geqslant \theta_n \geqslant -\pi.$$

The other sub-domains can be treated in the same way. Then decompose the above sub-domain into

$$R_1:\ \delta \geqslant \theta_1 \geqslant \theta_2 \geqslant \cdots \geqslant \theta_n \geqslant -\pi,$$
$$R_2:\ \pi \geqslant \theta_1 \geqslant \delta \geqslant \cdots \geqslant \theta_n \geqslant -\pi$$
$$\cdots\cdots\cdots\cdots$$
$$R_n:\ \pi \geqslant \theta_1 \geqslant \theta_2 \geqslant \cdots \geqslant \delta \geqslant \theta_n \geqslant -\pi,$$
$$R_{n+1}:\ \pi \geqslant \theta_1 \geqslant \theta_2 \geqslant \cdots \geqslant \theta_n \geqslant \delta,$$

where $0 < \delta < 1$.

Let $I_j = \int_{R_j} \cdots \int (g(\theta_1, \cdots, \theta_n) - u(U)) \frac{\partial}{\partial \theta_1} \frac{\partial^3}{\partial \theta_2^3} \cdots \frac{\partial^{2n-1}}{\partial \theta_n^{2n-1}}$

$$\cdot \begin{vmatrix} \sin n\theta_1, & \cdots, & \sin n\theta_n \\ \cdots & \cdots & \cdots \\ \sin \theta_1, & \cdots, & \sin \theta_n \end{vmatrix} d_N(\theta_1) \cdots d_N(\theta_n) d\theta_1 \cdots d\theta_n.$$

Then

$$I_{0,0,\cdots,0} = I_1 + I_2 + \cdots + I_{n+1}.$$

Since

$$\int_\delta^\pi f(\theta) d_N(\theta) d\theta = O\left(\frac{1}{\delta N}\right)$$

and

$$\int_{-\pi}^{\pi} |d_N(\theta)| d\theta = O(\ln N),$$

we have

$$I_2 = O\left(\frac{\ln^{n-1} N}{\delta N}\right), \quad \text{if } f(\theta) \in C^1.$$

By similar argument, it follows

$$I_p = O\left(\frac{\ln^{n-p+1} N}{\delta^{p-1} N}\right), \quad p = 3, \cdots, n+1.$$

Again decompose R_1 into

$$Q_1:\ \delta \geqslant \theta_1 \geqslant \theta_2 \cdots \geqslant \cdots \geqslant \theta_n \geqslant -\delta,$$
$$Q_2:\ \delta \geqslant \theta_1 \geqslant \theta_2 \cdots \geqslant \cdots \geqslant \theta_{n-1} \geqslant -\delta \geqslant \theta_n \geqslant -\pi,$$
$$\cdots\cdots\cdots\cdots\cdots\cdots\cdots\cdots\cdots\cdots\cdots\cdots$$
$$Q_{n+1}:\ \delta \geqslant \theta_1 \geqslant -\delta \geqslant \theta_2 \geqslant \cdots \geqslant \theta_n \geqslant -\pi.$$

Thus I_1 is divided into J_p, $p = 1, 2, \cdots, n+1$, which is equal to

$$\int_{\theta_p} \cdots \int (g(\theta_1, \cdots, \theta_n) - u(U)) \frac{\partial}{\partial \theta_1} \frac{\partial^3}{\partial \theta_2^3} \cdots \frac{\partial^{2n-1}}{\partial \vartheta_n^{2n-1}}$$

$$\cdot \begin{vmatrix} \sin n\theta_1, & \cdots, & \sin n\theta_n \\ \cdots & \cdots & \cdots \\ \sin \theta_1, & \cdots, & \sin \theta_n \end{vmatrix} d_N(\theta_1) \cdots d_N(\theta_n) d\theta_1 \cdots d\theta_n,$$

$$p = 1, 2, \cdots, n+1.$$

In the same way we can prove

$$J_p = O\left(\frac{\ln^{n-p+1} N}{\delta^{p-1} N}\right), \quad p = 2, \cdots, n+1.$$

On account of

$$\delta \geqslant \theta_1 \geqslant \theta_2 \geqslant \cdots \geqslant \theta_n \geqslant -\delta,$$

we have

$$g(\theta_1, \cdots \theta_n) - u(U) = \omega(\delta),$$

where ω is the modulus of continuity of $u(U)$. In virtue of $u(U) \in C^{n^2}$, we get

$$J_1 = O(\delta \ln^n N).$$

To sum up, we obtain

$$I_{0,\cdots,0} = O(\delta \ln^n N) + O\left(\frac{\ln^{n-1} N}{\delta N}\right) + \cdots$$

$$+ O\left(\frac{\ln^{n-p+1} N}{\delta^{p-1} N}\right) + \cdots + O\left(\frac{1}{\delta^n N}\right).$$

Taking $\delta = (N \ln^n N)^{\frac{-1}{n+1}}$, we have

$$I_{0,\cdots,0} = O\left(\left(\frac{\ln^{n^2} N}{N}\right)^{\frac{1}{n+1}}\right).$$

Next we consider $I_{\mu_1,\mu_2,\cdots,\mu_n}$, where all of $\mu_1, \cdots, \mu_n$ are not zero.

If some of $\mu_1, \mu_2, \cdots, \mu_n$ are odd, then among $\nu_1, \nu_2, \cdots, \nu_n$ there exist even numbers, say $\nu_j (1 \leqslant j \leqslant n)$. Thus we have

$$I_{\mu_1,\mu_2,\cdots,\mu_n} = \int_{-\pi}^{\pi} \cdots \int \frac{\partial^{\mu_1}}{\partial \theta_1^{\mu_1}} \cdots \frac{\partial^{\mu_n}}{\partial \theta_n^{\mu_n}} (g(\theta_1, \cdots, \theta_n) - u(U))$$

$$\cdot \sin \theta_j F(\theta_1, \cdots, \theta_n) d_N(\theta_1) \cdots d_N(\theta_n) d\theta_1 \cdots d\theta_n,$$

where $F(\theta_1, \cdots, \theta_n)$ is a differentiable function. As a consequence, we obtain

$$I_{\mu_1 \cdots \mu_n} = O\left(\frac{\ln^{n-1} N}{N}\right).$$

If all of $\mu_1, \cdots, \mu_n$ are even, then all of $\nu_1, \nu_2, \cdots, \nu_n$ are odd. In this case, there are at least two, say ν_i and ν_j $(i \neq j)$ which are equal. We can easily prove that

$$\frac{\partial^{\nu_1}}{\partial \theta_1^{\nu_1}} \cdots \frac{\partial^{\nu_n}}{\partial \theta_n^{\nu_n}} \begin{vmatrix} \sin n\theta_1, & \cdots, & \sin n\theta_n \\ \cdots & \cdots & \cdots \\ \sin \theta_1, & \cdots, & \sin \theta_n \end{vmatrix}$$

$$= \sin \frac{\theta_i - \theta_j}{2} \cdot G(\theta_1, \cdots, \theta_n),$$

where $G(\theta_1, \cdots, \theta_n)$ is a differentiatiable function. From this it is apt

to show that
$$I_{\mu_1,\cdots,\mu_n} = O\left(\frac{\ln^{n-1}N}{N^p}\right).$$

The proof of Theorem 10.4.2 is thus finished.

§ 10.5 Absolute Convergence of Fourier Series

As in Chapter 3, we may discuss the absolute convergence of the Fourier series (10.3.3) of the integrable function $u(U)$ over $US_p(2n)$. If C_f signifies its Fourier coefficient (10.3.4), then

$$C_f\bar{C}_f' = \iint u(V)u(U)\overline{\Phi_f(V)}\Phi_f'(U)\dot{V}\dot{U}$$
$$= \frac{N(f)}{C}\iint u(V)u(WV)A_f(W')\dot{V}\dot{W}.$$

Consequently,
$$\mathrm{tr}(C_f\bar{C}_f') = \frac{N(f)}{C}\iint u(V)u(WV)\chi_f(W)\dot{W}\dot{V}.$$

Letting
$$g(W) = \int_{USp(2n)} u(V)u(WV)\dot{V},$$

we have
$$\frac{\mathrm{tr}(C_f\bar{C}_f')}{N(f)} = \frac{1}{C}\int_{USp(2n)} g(W)\chi_f(W)\dot{W}.$$

Let $e^{i\theta_1}, e^{-i\theta_1}, \cdots, e^{i\theta_n}, e^{-i\theta_n}$ be the characteristic roots of W and
$$h(\theta_1,\cdots,\theta_n) = \frac{2^{3n}\pi^n}{C}\int_{[USp(2n)]} g(W)[\dot{W}].$$

Then
$$\frac{\mathrm{tr}(C_f\bar{C}_f')}{N(f)} = \frac{1}{(2\pi)^n}\int\cdots\int_{\pi>\theta_1>\cdots\theta_n>-\pi} h(\theta_1,\cdots,\theta_n)$$
$$\cdot\begin{vmatrix} \sin l_1\theta_1, & \cdots, & \sin l_1\theta_n \\ \cdots & \cdots & \cdots \\ \sin l_n\theta_1, & \cdots, & \sin l_n\theta_n \end{vmatrix}\begin{vmatrix} \sin n\theta_1, & \cdots, & \sin n\theta_n \\ \cdots & \cdots & \cdots \\ \sin \theta_1, & \cdots, & \sin \theta_n \end{vmatrix} d\theta_1\cdots d\theta_n.$$

Just as in § 10.4, replacing h by h^*, h^* being an even symmetric function, we get

$$\frac{\mathrm{tr}(C_f\bar{C}_f')}{N(f)} = \frac{(2i)^{-2n}}{(2\pi)^n}\int\cdots\int_{\pi>\theta_1>\cdots>\theta_n>-\pi} h^*(\theta_1,\cdots,\theta_n)s_\theta(l)s_\theta(n)\cdot d\theta_1\cdots d\theta_n$$
$$= \frac{(2i)^{-2n}}{(2\pi)^n}\int_{-\pi}^{\pi}\cdots\int h^*(\theta_1,\cdots,\theta_n)\cdot \sin l_1\theta_1\cdots \sin l_n\theta_n s_\theta(n)d\theta_1\cdots d\theta_n.$$

Set

$$H(\theta_1,\cdots,\theta_n)=h^*(\theta_1,\cdots,\theta_n)s_\theta(n), \tag{10.5.1}$$

which is an odd function. If $(i_1,\cdots,i_n)$ denote a permutation of $(1,2,\cdots,n)$, then

$$\frac{1}{(2\pi)^n}\int_{-\pi}^{\pi}\int H(\theta_1,\cdots,\theta_n)\sin l_{i_1}\theta_1\cdots\sin l_{i_n}\theta_n\cdot d\theta_1\cdots d\theta_n$$

$$=\delta^{i_1\cdots i_n}_{1\cdots n}\frac{\operatorname{tr}(C_f\bar{C}_f')}{N(f)}. \tag{10.5.2}$$

The multiple Fourier series of $H(\theta_1,\cdots,\theta_n)$ gives

$$\sum_{l_1,l_2,\cdots,l_n} d_{l_1,l_2,\cdots l_n}\sin l_1\theta_1\cdots\sin l_n\theta_n, \tag{10.5.3}$$

where

$$d_{l_1\cdots l_n}=\begin{cases}\dfrac{\operatorname{tr}(C_f\bar{C}_f')}{N(f)}, & \text{if } l_1>l_2>\cdots\cdots>l_n>0;\\ 0, & \text{if at least two of } l_1,\cdots,l_n \text{ are equal};\\ \delta^{i_1\cdots i_n}_{1\cdots n}d_{l_1\cdots l_n}, & \text{if } (i_1,\cdots,i_n) \text{ is a permutation of } (1,2,\cdots\cdots,n)\\ & \text{and } l_1>l_2>\cdots>l_n>0.\end{cases}$$

Thus (10.5.3) can be rewritten as

$$\sum_{l_1>l_2>\cdots>l_n>0}\frac{\operatorname{tr}(C_f\bar{C}_f')}{N(f)}(2i)^n s_\theta(l). \tag{10.5.4}$$

Let

$$P\left(\frac{\partial}{\partial\theta_1},\cdots,\frac{\partial}{\partial\theta_n}\right)=(-1)^{\frac{n(n-1)}{2}}\frac{\partial}{\partial\theta_1}\cdots\frac{\partial}{\partial\theta_n}\cdot D\left(\frac{\partial}{\partial\theta_1},\cdots,\frac{\partial}{\partial\theta_n}\right).$$

If $l_1>l_2>\cdots>l_n>0$, we have

$$\frac{1}{(2\pi)^n}\int_{-\pi}^{\pi}\cdots\int_{-\pi}^{\pi}\left(P\left(\frac{\partial}{\partial\theta_1},\cdots,\frac{\partial}{\partial\theta_n}\right)H(\theta_1,\cdots,\theta_n)\right)\cos l_1\theta_1\cdots\cos l_n\theta_n d\theta_1\cdots d\theta_n$$

$$=\sum_{(i_1,\cdots,i_n)}\delta^{i_1\cdots i_n}_{1\cdots n}\frac{1}{(2\pi)^n}\int_{-\pi}^{\pi}\cdots\int_{-\pi}^{\pi}\left(\frac{\partial}{\partial\theta_{i_1}}\frac{\partial^3}{\partial\theta_{i_2}^3}\cdots\frac{\partial^{2n-1}}{\partial\theta_{i_n}^{2n-1}}H(\theta_1,\theta_2,\cdots,\theta_n)\right)$$

$$\cdot\cos l_1\theta_1\cdots\cos l_n\theta_n d\theta_1\cdots\cdots d\theta_n$$

$$= \frac{1}{(2\pi)^n}\int_{-\pi}^{\pi}\cdots\int_{-\pi}^{\pi} H(\theta_1,\cdots,\theta_n)\sin l_1\theta_1\cdots\sin l_n\theta_n d\theta_1\cdots d\theta_n$$

$$\cdot(-1)^{\frac{n(n+1)}{2}}\sum_{(i_1,\cdots,i_n)}\delta^{i_1,\cdots,i_n}_{1,\cdots,n} l_{i_1} l^3_{i_2}\cdots l^{2n-1}_{i_n}$$

$$= (-1)^n(2n-1)!(2n-3)!\cdots 3!1!\operatorname{tr}(C_f\bar{C}_f').$$

That is to say that the multiple Fourier series of $P\left(\frac{\partial}{\partial\theta_1},\cdots,\frac{\partial}{\partial\theta_n}\right)$ $\cdot H(\theta_1,\cdots,\theta_n)$ is written as

$$2^{-n}(-1)^n(2n-1)!(2n-3)!\cdots 3!1!$$
$$\cdot\sum_{l>\cdots>l_n>0}\operatorname{tr}(C_f\bar{C}_f')C_\theta(l_1,\cdots,l_n),$$

where $C_\theta(l)$ is defined by (6.1.1). Similarly, the multiple Fourier series of $\left(P\left(\frac{\partial}{\partial\theta_1},\cdots,\frac{\partial}{\partial\theta_n}\right)\right)^3 H(\theta_1,\cdots,\theta_n)$ is

$$2^{-n}(-1)^{3n}((2n-1)!(2n-3)!\cdots 3!1!)^3$$
$$\cdot\sum_{l_1>\cdots>l_n>0}(N(f))^2\operatorname{tr}(C_f\bar{C}_f')C_\theta(l_1,\cdots,l_n),$$

and the one of $\left(P\left(\frac{\partial}{\partial\theta_1},\cdots,\frac{\partial}{\partial\theta_n}\right)\right)^5 H(\theta_1,\cdots,\theta_n)$ is

$$2^{-n}((2n-1)!(2n-3)!\cdots 3!1!)$$
$$\cdot\sum_{l_1>\cdots>l_n>0}(N(f))^4\operatorname{tr}(C_f\bar{C}_f')C_\theta(l_1,\cdots,l_n).$$

As in Chapter 3, by these results, we can prove

Theorem 10.5.1 *If* $u(U)\in C^{n^2}$ *and*

$$P\left(\frac{\partial}{\partial\theta_1},\cdots,\frac{\partial}{\partial\theta_n}\right)H(\theta_1,\cdots,\theta_n)\in \operatorname{Lip}(2,\alpha),$$

with

$$\alpha>\frac{1}{r}-\frac{1}{2},$$

then the series

$$\sum_{f_1>f_2\cdots>f_n>0}|\operatorname{tr}(C_f\bar{C}_f')|^r$$

is convergent for $1>r>\frac{2}{3}$ *(He Zuqi & Chen Guangxiao [1], the same below).*

Theorem 10.5.2 *If $u(U)\in C^{3n^2}$ and*

$$\left(P\left(\frac{\partial}{\partial\theta_1},\cdots,\frac{\partial}{\partial\theta_n}\right)\right)^3 H(\theta_1,\cdots,\theta_n)\in \mathrm{Lip}\,(2,\alpha),$$

with

$$\alpha>\frac{1}{r}-\frac{1}{2},$$

then the series

$$\sum_{f_1\geqslant f_2\geqslant\cdots\geqslant f_n\geqslant 0}\left(\sum_{j,k=1}^{N(f)}|C_{jk}^f||\varphi_{jk}^f|\right)^{2r}$$

is convergent for $1>r>2/3$.

Theorem 10.5.3 *If $u(U)\in C^{5n^2}$ and*

$$\left(P\left(\frac{\partial}{\partial\theta_1},\cdots,\frac{\partial}{\partial\theta_n}\right)\right)^5 H(\theta_1,\cdots,\theta_n)\in \mathrm{Lip}(2,\alpha),$$

with $\alpha>1/2$, then the Fourier series (10.3.3) of $u(U)$ is absolutely convergent.

Chapter 11. Cesàro and Abel Summation of Fourier Series on Unitary Symplectic Groups

§ 11.1 Definition of Cesàro Sum

Let $u(U)$ be integrable on $USp(2n)$. As in Chapter 2, we may consider the Cesàro summation of the Fourier series (10.3.3) of $u(U)$. Here the Cesàro (c, α)-sum which is defined by

$$\sum_{nN \geq f_1 \geq f_2 \geq \cdots \geq f_n \geq 0} B^{\alpha}_{f_1 \cdots f_n} \operatorname{tr}(C_{f_1 \cdots f_n} \Phi'_{f_1 \cdots f_n}(U)), \tag{11.1.1}$$

where

$$B^{\alpha}_{f_1 \cdots f_n} = \frac{2^{-2n}}{CN(f)} \int_{USp(2n)} \chi_{f_1 \cdots f_n}(\bar{V}) K^{\alpha}_N(V) \dot{V}. \tag{11.1.2}$$

And $K^{\alpha}_N(V)$ represents the Cesàro (c, α)-kernel which is equal to

$$\frac{1}{B^{\alpha}_N (2A^{\lambda}_N)^{n(2n+1)}} \left\{ \frac{\det\left[\sum_{k=0}^{N} A^{\alpha-1}_{N-k} V^k (I - \bar{V}'^{2k+1})\right]}{\det(I - \bar{V}')} \right\}^{n+\frac{1}{2}}, \tag{11.1.3}$$

where

$$B^{\alpha}_N = \frac{1}{C} \int_{USp(2n)} \frac{1}{(2A^{\alpha}_N)^{n(2n+1)}} \cdot \left\{ \frac{\det\left[\sum_{k=0}^{N} A^{\alpha-1}_{N-k} V^k (I - \bar{V}'^{2k+1})\right]}{\det(I - \bar{V}')} \right\}^{n+\frac{1}{2}} \dot{V}, \tag{11.1.4}$$

$$A^{\alpha}_N = (\alpha + N)(\alpha + N - 1) \cdots (\alpha + 1)/N!, \quad \alpha > -1,$$

and C refers to the volume of $USp(2n)$.

First we prove

Theorem 11.1.1 *The Cesàro (c, α)-sum (11.1.1) of the Fourier series (10.3.3) of an integrable function $u(U)$ on $USp(2n)$ can be expressed* *as*

$$\frac{1}{C}\int_{USp(2n)} u(VU)K_N^{\alpha}(V)\dot{V}, \tag{11.1.5}$$

where $K_N^{\alpha}(V)$ *is defined by* (11.1.3).

Proof. Let $e^{i\theta_1}, e^{-i\theta_1}, \cdots, e^{i\theta_n}\ e^{-i\theta_n}$ be the characteristic roots of $\bar{V}$. Then

$$\sum_{nN\geq f_1\geq\cdots\geq f_n\geq 0} B^{\alpha}_{f_1\cdots f_n}N(f)\begin{vmatrix}\sin l_1\theta_1, & \cdots, & \sin l_1\theta_n\\ \cdots & \cdots & \cdots\\ \sin l_n\theta_1, & \cdots, & \sin l_n\theta_n\end{vmatrix}$$

$$= K_N^{\alpha}(V)\begin{vmatrix}\sin n\theta_1, & \cdots, & \sin n\theta_n\\ \cdots & \cdots & \cdots\\ \sin\theta_1, & \cdots, & \sin\theta_n\end{vmatrix}. \tag{11.1.6}$$

In fact, the left side of the above expression is equal to

$$\sum_{nN\geq f_1\geq\cdots\geq f_n\geq 0}\ \sum_{(i_1,\cdots,i_n)} \delta^{i_1 i_2,\cdots,i_n}_{1,2,\cdots,n} B^{\alpha}_{f_1\cdots f_n}N(f)\sin l_1\theta_{i_1}\cdots\sin l_n\theta_{i_n},$$

where

$$B^{\alpha}_{f_1\cdots f_n}N(f) = \frac{2^{-2n}}{C}\int_{USp(2n)} \chi_{f_1\cdots f_n}(\bar{V})K_N^{\alpha}(V)\dot{V}$$

$$= \frac{1}{(2\pi)^n}\int\cdots\int_{\pi\geq\theta_1\geq\cdots\geq\theta_n\geq-\pi}\begin{vmatrix}\sin l_1\theta_1, & \cdots, & \sin l_1\theta_n\\ \cdots & \cdots & \cdots\\ \cdots & \cdots & \cdots\\ \sin l_n\theta_1, & \cdots, & \sin l_n\theta_n\end{vmatrix}$$

$$\cdot\begin{vmatrix}\sin n\theta_1, & \cdots, & \sin n\theta_n\\ \cdots & \cdots & \cdots\\ \cdots & \cdots & \cdots\\ \sin\theta_1, & \cdots, & \sin\theta_n\end{vmatrix} K_N^{\alpha}(V)d\theta_1\cdots d\theta_n$$

$$= \frac{1}{(2\pi)^n}\int_{-\pi}^{\pi}\cdots\int_{-\pi}^{\pi} \sin l_1\theta_1\cdots\sin l_n\theta_n$$

$$\cdot\begin{vmatrix}\sin n\theta_1, & \cdots, & \sin n\theta_n\\ \cdots & \cdots & \cdots\\ \cdots & \cdots & \cdots\\ \sin\theta_1, & \cdots, & \sin\theta_n\end{vmatrix} K_N^{\alpha}(V)d\theta_1\cdots d\theta_n.$$

From this, we could easily get (11.1.6), and so (11.1.5) follows.

The Cesàro (c, α)-sum thus defined is just the Abel sum as α tends to infinity and the (c, α)-kernel becomes to the Poisson kernel. This would be examined in § 11.6—11.7.

§ 11.2 Semi-Positivity of Cesàro Kernel

We will prove the following result (He Zuqi & Chen Guangxiao [2]).

Theorem 11.2.1 *Let $u(U)$ be continuous on $USp(2n)$, and $U \in USp(2n)$. Then, when*

$$\alpha > \frac{2n-2}{2n+1},$$

its Fourier series (10.3.3) *is (c, α)-summable to itself.*

To prove the above-mentioned Riesz-type theorem, we first show that the Cesàro kernel, when $\alpha > (2n-2)/(2n+1)$, is "semi-positive". In this way we are led to prove the following.

Theorem 11.2.2 *When $\alpha > \dfrac{2n-2}{2n+1}$,*

$$\frac{1}{C}\int_{USp(2n)} |K_N^\alpha(V)|\dot{V} \leqslant M, \tag{11.2.1}$$

Where M is a constant dependent only on n and α and independent of N.

Proof. We resort to induction method on n. When $n=1$, the conclusion (11.2.1) holds true evidently. As $|B_N^\alpha|$ does not tend to zero as $N \to \infty$, it suffices to prove that

$$I = \frac{2^{n^2+n}}{(2\pi)^n} \int\cdots\int_{\pi \geqslant \theta_1 \geqslant \cdots \geqslant \theta_n \geqslant -\pi} |\sigma_N^\alpha(\theta_1)\cdots\sigma_N^\alpha(\theta_n)|^{2n+1}$$

$$\cdot \prod_{j=1}^{n} \sin^2\theta_j \prod_{1 \leqslant i < j \leqslant n} (\cos\theta_i - \cos\theta_j)^2 d\theta_1 \cdots d\theta_n \tag{11.2.2}$$

is bounded, where

$$\sigma_N^\alpha(\theta) = \frac{1}{2A_N^\alpha \sin\frac{1}{2}\theta} \sum_{k=0}^{N} A_{N-k}^{\alpha-1} \sin\left(k+\frac{1}{2}\right)\theta.$$

As we did in § 2.3 of Chapter 2, taking $\delta \geqslant 1/N$ and decomposing the integral domain into $R_1 \cdots R_n$ and R_{n+1} (see § 2.3), letting I_j denote the part of the integral in (11.2.2) corresponding to sub-domain R_j, we find

$$I = I_1 + \cdots + I_{n+1}.$$

As stated above, we first consider I_2. Decompose R_2 into R_{2n}, $R_{\overline{2n-1}}$, $\cdots$ R_{22} and R_{21}, and let I_{ij} denote the part of the integral in (11.2.2) corresponding to the sub-domain R_{ij}. Using the method in § 2.3, we can show

$$I_{2n} \leqslant A \sum_{s_1=0}^{2n-2} \sum_{s_2=0}^{2} \cdots \sum_{s_n=0}^{2} a_{s_1 \cdots s_n} \int \cdots \int_{\delta > \theta_2 > \cdots > \theta_n > -\delta} |1 - \cos\theta_2|^{s_2} \cdots$$

$$\cdot |1 - \cos\theta_n|^{s_n} |\sigma_N^\alpha(\theta_2) \cdots \sigma_N^\alpha(\theta_n)|^{2n+1}$$

$$\cdot \prod_{j=2}^{n} \sin^2\theta_j \prod_{2 \leqslant i < j \leqslant n} (\cos\theta_i - \cos\theta_j)^2 d\theta_2 \cdots d\theta_n$$

$$\cdot \int_\delta^\pi |1 - \cos\theta_1|^{s_1} |\sigma_N^\alpha(\theta_1)|^{2n+1} d\theta_1,$$

where A is a constant related only to n and $a_{s_1 \cdots s_n}$ is a constant related only to $s_1, \cdots, s_n$. The integral on the right hand of the above expression is related to $s_1, \cdots, s_n$ and is denoted by $I_{s_1, \cdots, s_n}$. As in § 2.3 we can show that, when $\alpha > \dfrac{2n-1}{2n+1}$, $I_{s_1 \cdots s_n} = O((N\delta)^{2n-2-(2n+1)\alpha})$ is valid (here the assumption for induction is used). Thus

$$I_{2n} = O((N\delta)^{2n-2-(2n+1)\alpha}).$$

Similarly

$$I_{\overline{2n-j}} = O((N\delta)^{(j+1)(2n-2(j+1)-(2n+1)\alpha)}), \; j = 0, 1, 2, \cdots, n-1.$$

Thus, it is inferred that

$$I_2 = O((N\delta)^{(2n-2-(2n+1)\alpha)}) + O((N\delta)^{2(2n-4-(2n+1)\alpha)}) + \cdots + O((N\delta)^{-n(2n+1)\alpha}).$$

The same method gives us

$$I_3 = O((N\delta)^{2(2n-4-(2n+1)\alpha)}) + O((N\delta)^{3(2n-6-(2n+1)\alpha)}) + \cdots + O((N\delta)^{-n(2n+1)\alpha}).$$

$$\cdots\cdots\cdots\cdots\cdots\cdots\cdots\cdots\cdots\cdots\cdots\cdots\cdots$$

$$I_{n+1} = O((N\delta)^{-n(2n+1)\alpha}).$$

Next we take I_1 into account. Decompose R_1 into $s_1, s_2, \cdots, s_{n+1}$ (for definition, see § 2.3) and let J_k denote the part of the integral in (11.2.2) corresponding to s_k. Like I_{ij}, it can be shown that

$$J_2 = O((N\delta)^{(2n-2-(2n+1)\alpha)}) + O((N\delta)^{2(2n-4-(2n+1)\alpha)}) + \cdots + O((N\delta)^{-n(2n+1)\alpha}),$$

$$J_3 = O((N\delta)^{2(2n-4-(2n+1)\alpha)}) + \cdots + O((N\delta)^{-n(2n+1)\alpha}),$$

$$\cdots\cdots\cdots\cdots\cdots\cdots\cdots\cdots\cdots$$

$$J_{n+1} = O((N\delta)^{-n(2n+1)\alpha}).$$

Finally examining J_1, We have

$$J_1 = \frac{2^{n^2+n}}{(2\pi)^n} \int\cdots\int_{\delta\geqslant\theta_1\geqslant\cdots\geqslant\theta_n\geqslant-\delta} |\sigma_N^\alpha(\theta_1)\cdots\sigma_N^\alpha(\theta_n)|^{2n+1}$$

$$\cdot \prod_{j=1}^{n} \sin^2\theta_j \prod_{1\leqslant i<j\leqslant n} (\cos\theta_i - \cos\theta_j)^2 d\theta_1\cdots d\theta_n$$

$$\leqslant A \int\cdots\int_{\delta\geqslant\theta_2\geqslant\cdots\geqslant\theta_n\geqslant-\delta} |\sigma_N^\alpha(\theta_2)\cdots\sigma_N^\alpha(\theta_n)|^{2n+1}$$

$$\cdot \prod_{j=2}^{n} \sin^2\theta_j \prod_{2\leqslant i<j\leqslant n} (\cos\theta_i - \cos\theta_j)^2 d\theta_2\cdots d\theta_n$$

$$\cdot \int_{\theta_2}^{\delta} |\sigma_N^\alpha(\theta_1)|^{2n+1}\sin^2\theta_1 \prod_{j=2}^{n} (\cos\theta_1 - \cos\theta_j)^2 d\theta_1$$

$$\leqslant A'N^{4n-1}\delta^{4n-1} = O((N\delta)^{4n-1}).$$

To sum up, we conclude

$$I = O((N\delta)^{4n-1}) + O((N\delta)^{2n-2-(2n+1)\alpha}) + O((N\delta)^{2(2n-4-(2n+1)\alpha)}) + \cdots + O((N\delta)^{-n(2n+1)\alpha}).$$

Taking $\delta = 1/N$, we obtain $I = O(1)$, which proves (11.2.1).

§11.3 Proof of Riesz-Type Theorem

By the definition we note

$$\frac{1}{C}\int_{USp(2n)} K_N^\alpha(V)\dot{V} = 1.$$

(11.1.1) is denoted by $\Sigma_N^\alpha(U)$ and can be rewritten as

$$\frac{1}{C}\int_{USp(2n)} u(VU)K_N^\alpha(V)\dot{V}.$$

Therefore

$$\Sigma_N^\alpha(U) - u(U) = \frac{1}{C}\int_{USp(2n)} (u(VU) - u(U))K_N^\alpha(V)\dot{V}. \quad (11.3.1)$$

We attempt to show that the right side of the preceding expression tends to zero as $N \to \infty$. The right side of (11.3.1) gives

$$\frac{1}{\beta_N^\alpha}\int\cdots\int_{\pi\geqslant\theta_1\geqslant\cdots\geqslant\theta_n\geqslant-\pi} g(\theta_1,\cdots,\theta_n)|\sigma_N^\alpha(\theta_1)\cdots\sigma_N^\alpha(\theta_n)|^{2n+1}$$

$$\cdot\prod_{j=1}^{n}\sin^2\theta_j\prod_{1\leqslant i<j\leqslant n}(\cos\theta_i-\cos\theta_j)^2 d\theta_1\cdots d\theta_n, \qquad (11.3.2)$$

where

$$g(\theta_1,\cdots,\theta_n)=\frac{2^{n^2+n}}{C}\int_{[USp(2n)]}(u(VU)-u(U))[\dot{V}].$$

Let us decompose the integral domain of (11.3.2) into $R_1, R_2,\cdots R_{n+1}$. By I_j' we denote the part of the integral in (11.3.2) corresponding to the subdomain R_j. Since $u(U)$ is continuous on $USp(2n)$, we confirm

$$|g(\theta_1,\cdots,\theta_n)|\leqslant L,\ \pi\geqslant\theta_1\geqslant\cdots\geqslant-\pi,$$

where L is an absolute constant. When N tends to infinity, $|B_N|$ does not tend to zero. Taking $\delta\geqslant 1/N$, we claim

$$\begin{aligned}
I_2' &= O((N\delta)^{2n-2-(2n+1)\alpha})+O((N\delta)^{2(2n-4-(2n+1)\alpha)})\\
&\quad+\cdots+O((N\delta)^{-n(2n+1)\alpha}),\\
I_3' &= O((N\delta)^{2(2n-4-(2n+1)\alpha)})+\cdots+O((N\delta)^{-n(2n+1)\alpha}),\\
&\cdots\cdots\cdots\cdots\cdots\cdots\cdots\cdots\cdots\cdots\cdots\cdots\\
I_{n+1}' &= O((N\delta)^{-n(2n+1)\alpha}).
\end{aligned}$$

As for I_1', we follow the way of treating I_1. Decompose R_1 into $s_1,\cdots, s_{n+1}$ and let J_k' denote the integral over s_k of (11.3.2). It can be shown that

$$\begin{aligned}
|J_2'|+\cdots+|J_{n+1}'| &= O((N\delta)^{2n-2-(2n+1)\alpha})\\
&\quad+O((N\delta)^{2(2n-4-(2n+1)\alpha)})\\
&\quad+\cdots+O((N\delta)^{-n(2n+1)\alpha}).
\end{aligned}$$

Finally, we come to J_1'. We have

$$J_1'=\frac{1}{\beta_N^\alpha}\int\cdots\int_{\delta\geqslant\theta_1\geqslant\cdots\geqslant\theta_n\geqslant-\delta} g(\theta_1,\cdots,\theta_n)|\sigma_N^\alpha(\theta_1)\cdots\sigma_N^\alpha(\theta_n)|^{2n+1}$$

$$\cdot\prod_{j=1}^{n}\sin^2\theta_j\prod_{1\leqslant i<j\leqslant n}(\cos\theta_i-\cos\theta_j)^2 d\theta_1\cdots d\theta_n.$$

Since $u(U)$ is continuous on $USp(2n)$, for any given $\eta>0$, we can choose $\delta>0$ sufficiently small such that

$$|g(\theta_1,\cdots,\theta_n)|<\eta$$

for $\delta\geqslant\theta_1\geqslant\cdots\geqslant\theta_n\geqslant-\delta$. By Theorem 11.2.2, we have

$$\frac{1}{C}\int_{USp(2n)}|K_N^\alpha(V)|\dot{V}<M,$$

when $\alpha > \frac{2n-2}{2n+1}$. Thus, for any given $\varepsilon > 0$, we can choose $\delta > 0$ such that

$$|J_1'| < \varepsilon/2.$$

As $(2n+1)\alpha - 2n - 2 > 0$, we can choose sufficiently large N satisfying

$$|I_2'| + \cdots + |I_{n+1}'| + |J_2'| + \cdots + |J_{n+1}'| < \varepsilon/2.$$

Therefore, for any $\varepsilon > 0$, we can choose N_0 such that

$$|I| < \varepsilon,$$

whenever $N \geqslant N_0$. This completes the Proof of Theorem 11.2.1.

§ 11.4 Fejér Summation

When $\alpha = 1$, the Cesàro summation is reduced to the important Fejér summation, and the Fejér kernel comes to be

$$\frac{1}{B_N(N+1)^{n(2n+1)}}\left(\frac{\det(I - V^{N+1})}{\det(I - V)}\right)^{2n+1}, \tag{11.4.1}$$

where

$$B_N = B_N^1 = \frac{1}{C(N+1)^{n(2n+1)}} \cdot \int_{USp(2n)} \left(\frac{\det(I - V^{N+1})}{\det(I - V)}\right)^{2n+1} \dot{V}. \tag{11.4.2}$$

The coefficient of the Fejér summation is given by

$$B_{f_1\cdots f_n} = B_{f_1\cdots f_n}^1 = \frac{2^{-2n}}{B_N N(f)(N+1)^{n(2n+1)}C} \cdot \int_{USp(2n)} \chi_{f_1\cdots f_n}(\bar{V})\left(\frac{\det(I - V^{N+1})}{\det(I - V)}\right)^{2n+1} \dot{V}. \tag{11.4.3}$$

The Fejér sum amounts to

$$\sum_{nN \geqslant f_1 \geqslant \cdots \geqslant f_n \geqslant 0} B_{f_1\cdots f_n} \operatorname{tr}(C_{f_1\cdots f_n} \Phi_{f_1\cdots f_n}'(U))$$
$$= \frac{1}{B_N(N+1)^{2(2n+1)}} \int_{USp(2n)} u(VU)\left(\frac{\det(I - V^{N+1})}{\det(I - V)}\right)^{2n+1} \dot{V}.$$

Besides, the Riesz-type Theorem 11.2.1 becomes the Fejér-type theorem: If $u(U)$ is continuous on $USp(2n)$, then the Fourier series (10.3.3) of $u(U)$ is Fejér summable to itself.

As in Chapter 2, we can give the explicit expressions for $B_{f_1\cdots f_n}$ and B_N.

From (11.4.3), it can be readily shown that

$$B_{l_1\cdots l_n}=\frac{2^{\frac{n(n-1)}{2}}}{B_N N(f)(N+1)^{n(2n+1)}}\frac{1}{(2\pi)^n}\int_{-\pi}^{\pi}\cdots\int_{-\pi}^{\pi}\sin l_1\theta_1\cdots\sin l_n\theta_n$$

$$\cdot\sin\theta_1\cdots\sin\theta_n\prod_{1\leqslant i<j\leqslant n}(\cos\theta_i-\cos\theta_j)\left(\frac{1-\cos(N+1)\theta_1}{1-\cos\theta_1}\right)^{2n+1}$$

$$\cdots\left(\frac{1-\cos(N+1)\theta_n}{1-\cos\theta_n}\right)^{2n+1}d\theta_1\cdots d\theta_n$$

$$=\frac{2^{\frac{n(n-1)}{2}}}{B_N N(f)(N+1)^{n(2n+1)}}\frac{1}{(2\pi)^n}$$

$$\cdot\int_{-\pi}^{\pi}\cdots\int\sin l_1\theta_1\cdots\sin l_2\theta_n\sin\theta_1\cdots\sin\theta_n$$

$$\cdot\left(\frac{1-\cos(N+1)\theta_1}{1-\cos\theta_1}\right)^{2n+1}\cdots\left(\frac{1-\cos(N+1)\theta_n}{1-\cos\theta_n}\right)^{2n+1}$$

$$\cdot\begin{vmatrix}1, & \cdots, & 1\\ 1-\cos\theta_1, & \cdots, & 1-\cos\theta_n\\ \cdots & \cdots & \cdots\\ (1-\cos\theta_1)^{n-1}, & \cdots, & (1-\cos\theta_n)^{n-1}\end{vmatrix}d\theta_1\cdots d\theta_n.$$

Thus $B_{l_1\cdots l_n}$ can be written as

$$\frac{2^{\frac{n(n-1)}{2}}}{B_N N(f)(N+1)^{n(2n+1)}}\begin{vmatrix}a^0_{l_1}, & a^0_{l_2}, & \cdots, & a^0_{l_n}\\ a^1_{l_1} & a^1_{l_2} & & a^1_{l_n}\\ \cdots & \cdots & \cdots & \cdots\\ a^{n-1}_{l_1} & a^{n-1}_{l_2} & & a^{n-1}_{l_n}\end{vmatrix}, \tag{11.4.4}$$

where

$$a^q_p=\frac{1}{2\pi}\int_{-\pi}^{\pi}(1-\cos\theta)^p\sin q\theta\sin\theta\cdot\left(\frac{1-\cos(N+1)\theta}{1-\cos\theta}\right)^{2n+1}d\theta.$$

When $p=0,1,2,\cdots,n-1$ and $q=l_1,l_2,\cdots,l_n$, we attempt to prove that

$$a^q_p=(-1)^p2^{-(p+1)}\sum_{\substack{s=0\\ k\geqslant 0\\ k=N(2n+1)-(N+1)s-q+p}}^{4n+2}(-1)^s$$

$$\cdot\frac{(4n-2p-1+k)!(4n+2)!(4n-2p+2k)}{k!s!(4n-2p)!(4n+2-s)!}. \tag{11.4.5}$$

In fact, a^q_p can be expressed as

$$(-1)^p2^{-(p+1)}\frac{1}{2\pi}\int_{-\pi}^{\pi}(1-e^{i\theta})^{2p-4n-2}(1-e^{i(N+1)\theta})^{4n+2}$$

$$\cdot e^{i(2n+1-p-(N+1)(2n+1))\theta}(e^{i(q-1)\theta}-e^{i(q+1)\theta})de^{i\theta}$$

$$= (-1)^p 2^{-(p+1)} \lim_{r\to 1} \frac{1}{2\pi} \int_{|z|=r} (1-z)^{2p-4n-1}(1-z^{N+1})^{4n+2}$$

$$\cdot z^{q-p-N(2n+1)-2}(1+z)dz,$$

where $z = re^{i\theta}$. Since

$$(1-z)^{-(4n+1-2p)}(1-z^{N+1})^{4n+2} z^{q-p-N(2n+1)-2}(1+z)$$

$$= \sum_{k=0}^{\infty} \sum_{s=0}^{4n+2} (-1)^s \frac{(4n-2p+k)!(4n+2)!}{k!s!(4n-2p)!(4n+2-s)!}$$

$$\cdot z^{k-N(2n+1)+(N+1)s+q-p-2} + \sum_{k=0}^{\infty} \sum_{s=0}^{4n+2} (-1)^s \cdot$$

$$\cdot \frac{(4n-2p+k)!(4n+2)!}{k!s!(4n-2p)!(4n+2-s)!} z^{k-N(2n+1)+(N+1)s+q-p-1},$$

substituting this into the right-hand side of the above equality and rearranging it, we obtain (11.4.5) immediately.

Substituting (11.4.5) into (11.4.4), we obtain that $B_{f_1\cdots f_n}$ is equal to

$$\frac{(-1)^{\frac{n(n-1)}{2}} 2^{-n}((4n+2)!)^n}{B_N N(f)(N+1)^{n(2n+1)}(4n)!(4n-2)!\cdots(2n+2)!}$$

$$\cdot \det\left(\sum_{\substack{s_j=0 \\ k_j=-(i-1)}}^{4n+2} (-1)^{s_j} \frac{(4n-i+k_j)!(4n+2k_j)}{(k_j+i-1)!s_j!(4n+2-s_j)!} \right)$$

where

$$1 \leqslant i \leqslant n,$$

$$k_1 = N(2n+1) - l_1 - (N+1)s_1, \cdots,$$

$$k_n = N(2n+1) - l_n - (N+1)s_n;$$

$$l_1 = f_1 + n,\ l_2 = f_2 + n - 1, \cdots, l_n = f_n + 1.$$

Let us agree on that $\dfrac{1}{(-m)!} = 0$ for $m > 0$. Then $B_{f_1\cdots f_n}$ is equal to

$$\frac{(-1)^{\frac{n(n-1)}{2}} 2^{-n}((2n+1)!)^n}{B_N N(f)(N+1)^{n(2n+1)}(4n)!(4n-2)!\cdots(2n+2)!}$$

$$\cdot \sum_{\substack{s_1=0 \\ k_1 \geqslant -(n+1)}}^{4n+2} \cdots \sum_{\substack{s_n=0 \\ k_n \geqslant -(n+1)}}^{4n+2} (-1)^{s_1+\cdots+s_n} C_{s_1}^{4n+2} \cdots C_{s_n}^{4n+2}$$

$$\cdot C_{k_1+n-1}^{3n+k_1} \cdots C_{k_n+n-1}^{3n+k_n} (4n+2k_1)\cdots(4n+2k_n)\det(d_{ij})_{1\leqslant i,j\leqslant n},$$

where

$$d_{ij} = \begin{cases} (4n-i+k_j)(4n-i-1+k_j)\cdots(3n+k_j+1) \\ \quad\cdot(k_j+n-1)(k_j+n-2)\cdots(k_j+i), \text{ for } i=1,2,\cdots,n-2; \\ (3n+k_j+1)(k_j+n-1), \text{ for } i=n-1; \\ 1, \text{ for } i=n. \end{cases}$$

This is also equal to

$$\frac{((2n+1)!)^n}{(4n)!(4n-2)!\cdots(2n-2)!B_N N(f)(N+1)^{n(2n+1)}}$$

$$\cdot \sum_{\substack{s_1=0 \\ k_1\geqslant -(n-1)}}^{4n+2} \cdots \sum_{\substack{s_n=0 \\ k_n\geqslant -(n-1)}}^{4n+2} (-1)^{s_1+\cdots+s_n} C_{s_1}^{4n+2}\cdots C_{s_n}^{4n+2}$$

$$\cdot C_{k_1+n-1}^{3n+k_1}\cdots C_{k_n+n-1}^{3n+k_n} \begin{vmatrix} 2n+k_1, & \cdots, & 2n+k_n \\ (2n+k_1)^3, & \cdots, & (2n+k_n)^3 \\ \cdots & \cdots & \cdots \\ (2n+k_1)^{2n-1}, & \cdots, & (2n+k_n)^{2n-1} \end{vmatrix}.$$

Thus, at last, we obtain (He Zuqi & Chen Guangxiao [2])

$$B_{f_1\cdots f_n} = \frac{(-1)^{\frac{n(n-1)}{2}}((2n+1)!)^n(2n-1)!(2n-3)!\cdots 3!1!}{B_N N(f)(N+1)^{n(2n+1)}(4n)!(4n-2)!\cdots(2n+2)!}$$

$$\cdot \sum_{\substack{s_1=0 \\ k_1\geqslant -(n-1)}}^{4n+4} \cdots \sum_{\substack{s_n=0 \\ k_n\geqslant -(n-1)}}^{4n+2} (-1)^{s_1+\cdots+s_n} C_{s_1}^{4n+2}\cdots C_{s_n}^{4n+2}$$

$$\cdot C_{k_1+n-1}^{3n+k_1}\cdots C_{k_n+n-1}^{3n+k_n} N(n+N(2n+1)-(N+1)s_1-f_1,\cdots$$

$$\cdots\cdots, 2n-1+N(2n+1)-(N+1)s_n-f_n),$$

where

$$k_j = N(2n+1)-(N+1)s_j-(f_j+n-j+1),\ j=1,2,\cdots,n.$$

Since $B_{0,\cdots 0}=1$, we have

$$B_N = \frac{(-1)^{\frac{n(n-1)}{2}}((2n+1)!)^n(2n-1)!(2n-3)!\cdots 3!1!}{(N+1)^{n(2n+1)}(4n)!(4n-2)!\cdots(2n+2)!}$$

$$\cdot \sum_{\substack{s_1=0 \\ k_1\geqslant -(n-1)}}^{4n+2} \cdots \sum_{\substack{s_n=0 \\ k_n\geqslant -(n-1)}}^{4n+2} (-1)^{s_1+\cdots+s_n} C_{s1}^{4n+2}\cdots C_{s_n}^{4n+2}$$

$$\cdot C_{k_1+n-1}^{3n+k_1}\cdots C_{k_n+n-1}^{3n+k_n} N(n+N(2n+1)-(N+1)s_1,\cdots,$$

$$2n-1+N(2n+1)-(N+1)s_n).$$

§ 11.5 Approximation by Cesàro Means

Let U and V be two points in $USp(2n)$,

$$U=(u_{ij})_{1\leqslant i,j\leqslant 2n},\ V=(v_{ij})_{1\leqslant i,j\leqslant 2n},$$

and then the square of the Euclidean distance between U and V is given by

$$d^2(U, V) = \sum_{i,j=1}^{2n} |u_{ij} - v_{ij}|^2 = \mathrm{tr}((U - V)(\overline{U - V})')$$
$$= \mathrm{tr}(2I - U\bar{V}' - V\bar{U}').$$

If the characteristic roots of $U\bar{V}'$ are $e^{i\theta_1}, e^{-i\theta_1}, \cdots, e^{i\theta_n}, e^{-i\theta_n}$, then

$$d^2(U, V) = 4 \sum_{j=1}^{n} (1 - \cos\theta_j).$$

Let $u(U)$ be a continuous function on $USp(2n)$. Then

$$\omega(u, \delta) = \max_{d(U,V)\leqslant\delta} |u(U) - u(V)|$$

is called the modulus of continuity of $u(U)$.

If $u(U)$ is continuous on $USp(2n)$ and $\omega(u, \delta) = 0(\delta^p)$, then $u(U)$ is said to be "satisfying the Lipschitz condition" and is denoted by "$u(U) \in \mathrm{Lip}\ p$".

The method used in Chapter 4 works in the following theorem (cf. He Zuqi & Chen Guangxiao [2])

Theorem 11.5.1 *If $u(U)$ is continuous on $USp(2n)$ and $u(U) \in \mathrm{Lip}\, p$ $(0 < p < 1)$, then the Nth term $\Sigma_N^a(U)$ of the (c, α) mean of its Fourier series* (10.3.3) *satisfies*

$$(1)\ \ u(U) - \sum_N^a (U) = O(N^{-p}),\ \textit{if}\ (2n + 1)\alpha - 2n + 2 > p;$$

$$(2)\ \ |u(U) - \sum_N^a (U)| = O(N^{-p}\ln N),\ \textit{if}\ (2n + 1)\alpha - 2n + 2 = p,$$

$$(3)\ \ u(U) - \sum_N^a (U) = O(N^{2n-2-(2n+1)\alpha}),\ \textit{if}\ (2n + 1)\alpha - 2n + 2 < p.$$

§ 11.6 Poisson Kernel and Abel Summation

The Cesàro (c,α) kernel (11.1.3) defined in § 11.3, when α tends to infinity, becomes the Poisson kernel

$$P_r(U) = \frac{(1 - r^2)^{n(2n+1)}}{\det^{2n+1}(I - rU)}, \tag{11.6.1}$$

where $0 \leqslant r < 1$ and $U \in USp(2n)$.

Let the characteristic roots of U be $e^{\pm i\theta_1}, \cdots e^{\pm i\theta_n}$, and then

$$\det(I - rU) = \prod_{i=1}^{n} (1 + r^2 - 2r\cos\theta_i). \tag{11.6.2}$$

Thus $P_r(U)$ is positive definite and

$$P_r(U) = P_r(\bar{U}) = P_r(U^{-1}). \tag{11.6.3}$$

In this section, we will prove the following theorem (see Chen Guangxiao and He Zuqi[3]).

Theorem 11.6.1 *If $u(U)$ is continuous on $USp(2n)$, then*

$$\frac{1}{C}\int_{USp(2n)} u(wU)P_r(\bar{w})\dot{w} \tag{11.6.4}$$

tends to $u(U)$ as $r \to 1$, where $U \in USp(2n)$ and C is the volume of $USp(2n)$.

To this end, we first prove the following two lemmas.

Lemma 11.6.1 *If $U \in USp(2n)$ and H is defined by (10.1.1), which, obviously, satisfies (10.1.2), then*

$$P_r(\bar{U})\dot{U} = 2^{n(2n+1)}\det{}^{-(n+1/2)}(I + \tilde{H}^2)\dot{\tilde{H}}, \tag{11.6.5}$$

where $\tilde{H} = \dfrac{1+r}{1-r}H.$

Proof. What we need to prove is

$$\frac{(1-r^2)^{n(2n+1)}\dot{U}}{(\det(I - rU)\det(I - r\bar{U}'))^{n+1/2}} = 2^{n(2n+1)}\det(I + \tilde{H}^2)^{-(n+1/2)}\dot{\tilde{H}}.$$

As

$$U = (I + iH)(I - iH)^{-1}, \tag{11.6.6}$$

we have

$$I - rU = [(1-r)I - i(1+r)H](I - iH)^{-1},$$
$$I - r\bar{U}' = [(1-r)I + i(1+r)H](I + iH)^{-1}.$$

Thus

$$\det(I - rU)\det(I - r\bar{U}') = \det((1-r^2)I + (1+r)^2H^2)\det(I + H^2)^{-1}.$$

From

$$\dot{\tilde{H}} = \left(\frac{1+r}{1-r}\right)^{n(2n+1)}\dot{H},$$

and Lemma 10.1.1 of Chapter 10, (11.6.5) follows.

By this lemma, we immediately obtain that

$$\frac{1}{C}\int_{USp(2n)} P_r(\bar{w})\dot{w} = 1 \text{ for } 0 \leqslant r < 1. \tag{11.6.7}$$

Let $U \in USp(2n)$. Then U can be expressed by

$$U = V[e^{i\theta_1}, e^{-i\theta_1}, \cdots, e^{i\theta_n}, e^{-i\theta_n}]V^{-1},$$

where $V \in USp(2n)$. H in (11.6.6) becomes

$$V\left[\operatorname{tg}\frac{\theta_1}{2}, -\operatorname{tg}\frac{\theta_1}{2}, \cdots, \operatorname{tg}\frac{\theta_n}{2}, -\operatorname{tg}\frac{\theta_n}{2}\right]V^{-1}.$$

If $\varepsilon > 0$, we have

$$\operatorname{tr}(H\bar{H}') = 2\sum_{k=1}^{n} \operatorname{tg}^2\frac{\theta_k}{2} < \varepsilon^2, \tag{11.6.8}$$

which is denoted by $H < \varepsilon$, or otherwise we adopt the notation $H \geqslant \varepsilon$. All those $U \in USp(2n)$, determined by all H satisfying both (11.6.6) and (11.6.8), form a neighborhood of the unit matrix in $USp(2n)$ denoted by S_ε. We now show

Lemma 11.6.2 *When* $r \to 1$, $\int_{USp(2n)\setminus S_\varepsilon} P_r(\bar{U})\dot{U}$ *tends to zero.* (11.6.9)

Proof. By (11.6.6), the integral (11.6.9) changes into

$$\int_{\tilde{H} > \frac{1+r}{1-r}\varepsilon} 2^{2n^2+n}\det{}^{-(n+\frac{1}{2})}(I + \tilde{H}^2)\dot{\tilde{H}}.$$

As $\varepsilon(1+r)/(1-r) \to \infty (r \to 1)$, our lemma is proved.

The proof of Theorem 11.6.1.

By (11.6.7) we have

$$\frac{1}{C}\int_{USp(2n)} u(\dot{w}U)P_r(\bar{w})\dot{w} - u(U)$$

$$= \frac{1}{C}\int_{USp(2n)} (u(wU) - u(U))P_r(\bar{w})\dot{w}. \tag{11.6.10}$$

Since $u(U)$ is continuous, for given $\delta > 0$, there exists $\varepsilon > 0$ such that

$$|u(wU) - u(U)| < \delta/2 \text{ for } w \in S_\varepsilon. \tag{11.6.11}$$

Dividing the integral on the right-hand side of (11.6.10) into two parts

$$\frac{1}{C}\left(\int_{USp(2n)\setminus S_\varepsilon} + \int_{S_\varepsilon}\right) = I_1 + I_2,$$

by (11.6.11), we admit

$$|I_2| < \frac{\delta}{2}\cdot\frac{1}{C}\int_{S_\varepsilon} P_r(\bar{w})\dot{w} < \delta/2.$$

On the other hand, when $U \in USp(2n)\backslash s_\varepsilon$, we can choose r approaching 1 sufficiently such that

$$\frac{1}{C}\int_{USp(2n)\backslash s_\varepsilon} P_r(\bar{w})\dot{w} < \frac{\delta}{4M},$$

where M is the upper bound of $|u(U)|$ over $USp(2n)$. Consequently

$$|I_1| < 2M\frac{\delta}{4M} = \delta/2.$$

Therefore, given $\delta > 0$, when r approaches 1 sufficiently, we have

$$\left|\frac{1}{C}\right|\int_{USp(2n)} u(wU)P_r(\bar{w})\dot{w} - u(U)| < \delta.$$

The proof of Lemma 11.6.2 is thus finished.

In the next section we will show that the Poisson kernel of $USp(2n)$ can be expanded to a uniformly convergent series in $\chi_f(\bar{w})$, i. e.

$$P_r(\bar{U}) = \sum_{f_1 \geqslant f_2 \geqslant \cdots \geqslant f_n \geqslant 0} \rho^f(r)N(f)\chi_f(\bar{U}), \tag{11.6.12}$$

where $0 \leqslant r \leqslant r_0 < 1$, and

$$\rho^f(r) = \frac{1}{N(f)}\frac{1}{C}\int_{USp(2n)} P_r(\bar{w})\chi_f(w)\dot{w}. \tag{11.6.13}$$

We call

$$\sum_{f_1 \geqslant f_2 \geqslant \cdots \geqslant f_n \geqslant 0} \rho^f(r)\mathrm{tr}(C_f\Phi_f'(U)) \tag{11.6.14}$$

the Abel sum of the Fourier series (10.3.3) of $u(U)$ and call $\rho^f(r)$ its coefficients. From Theorem 11.6.1, it is easy to see that

$$\rho^f(r) \to 1$$

as $r \to 1$. The Fourier series (10.3.3) of $u(w)$ is said to be Abel summable to A, if (11.6.14) converges to A as $r \to 1$.

Obviously, if (11.6.12) is proved, then, by Theorem 11.6.1, any continuous function on $USp(2n)$ is Abel summable to itself.

§ 11.7 Expansion for Poisson Kernel

In this section, following the generating function method in Zhong Jiaqing [2], we will find out the explict expression for Abel coefficient and show that any continuous function on $USp(2n)$ is Abel summable to itself.

Lemma 11.7.1 *If $U \in USp(2n)$, $t_1, \cdots, t_n$ denote n indenpendent real variables,* $\max_{1\leqslant j\leqslant n} |t_j| \leqslant r < 1$ *and*

$$g_j = \det(I - t_jU) = \prod_{k=1}^{n} (1 + t_j^2 - 2t_j\cos\theta_k), \tag{11.7.1}$$

then

$$\frac{\prod_{j<k}^{n} (1 - t_jt_k)}{g_1\cdots g_n} = \frac{\sum_{f_1\geqslant\cdots\geqslant f_n\geqslant 0} \chi_f(U)Q^f(t_1, \cdots, t_n)}{\prod_{j<k}^{n} (t_j - t_k)}, \tag{11.7.2}$$

where $Q^f(t_1, \cdots, t_n)$ is equal to

$$\begin{vmatrix} t_1^{l_1-1}, & \cdots, & t_n^{l_1-1} \\ \cdots & \cdots & \cdots \\ t_1^{l_n-1}, & \cdots, & t_n^{l_n-1} \end{vmatrix}. \tag{11.7.3}$$

Proof. Let $e^{\pm i\theta_1}, \cdots, e^{\pm i\theta_n}$ be the characteristic roots of U and $N > n$, then

$$\sum_{N\geqslant l_1>\cdots>l_n>0} \begin{vmatrix} \sin l_1\theta_1, & \cdots, & \sin l_n\theta_1 \\ \cdots & \cdots & \cdots \\ \sin l_1\theta_n, & \cdots, & \sin l_n\theta_n \end{vmatrix} \begin{vmatrix} t_1^{l_1-1}, & \cdots, & t_n^{l_1-1} \\ \cdots & \cdots & \cdots \\ t_1^{l_n-1}, & \cdots, & t_n^{l_n-1} \end{vmatrix}$$

$$= \begin{vmatrix} \sum_{m=1}^{N} t_1^{m-1}\sin m\theta_1, & \cdots, & \sum_{m=1}^{N} t_n^{m-1}\sin m\theta_1 \\ \cdots & \cdots & \cdots \\ \sum_{m=1}^{N} t_1^{m-1}\sin m\theta_n, & \cdots, & \sum_{m=1}^{N} t_n^{m-1}\sin m\theta_n \end{vmatrix}.$$

By (11.7.1), letting $N \to \infty$, we get

$$\sum_{f_1\geqslant f_2\geqslant\cdots\geqslant f_n\geqslant 0} \chi_f(U)Q_f(t) = \frac{1}{\sin\theta_1\cdots\sin\theta_n \prod_{j<k}^{n} (2\cos\theta_j - 2\cos\theta_k)}$$

$$\cdot \begin{vmatrix} \dfrac{\sin\theta_1}{1 + t_1^2 - 2t_1\cos\theta_1}, & \cdots, & \dfrac{\sin\theta_1}{1 + t_n^2 - 2t_n\cos\theta_1} \\ \cdots & \cdots & \cdots \\ \dfrac{\sin\theta_n}{1 + t_1^2 - 2t_1\cos\theta_n}, & \cdots, & \dfrac{\sin\theta_n}{1 + t_n^2 - 2t_n\cos\theta_n} \end{vmatrix}$$

$$= \frac{1}{(1 + t_1^2)\cdots(1 + t_n^2) \prod_{j<k}^{n} (2\cos\theta_j - 2\cos\theta_k)}$$

$$\cdot \begin{vmatrix} \dfrac{1}{1-\dfrac{2t_1}{1+t_1^2}\cos\theta_1}, & \cdots, & \dfrac{1}{1-\dfrac{2t_n}{1+t_n^2}\cos\theta_1} \\ \cdots & \cdots & \cdots \\ \dfrac{1}{1-\dfrac{2t_1}{1+t_1^2}\cos\theta_n}, & \cdots, & \dfrac{1}{1-\dfrac{2t_n}{1+t_n^2}\cos\theta_n} \end{vmatrix}. \tag{11.7.4}$$

By Theorem 1.1.4 of Hua Luogeng [1], we have

$$\begin{vmatrix} \dfrac{1}{1-x_1y_1}, & \cdots, & \dfrac{1}{1-x_ny_1} \\ \cdots & \cdots & \cdots \\ \dfrac{1}{1-x_1y_n}, & \cdots, & \dfrac{1}{1-x_ny_n} \end{vmatrix}$$

$$= \frac{D(x_1,\cdots,x_n)D(y_1,\cdots,y_n)}{\prod_{j=1}^{n}\prod_{k=1}^{n}(1-x_jy_k)}.$$

We note that (11.7.4) is equal to

$$\frac{\prod_{j<k}^{n}\left(\dfrac{t_j}{1+t_j^2}-\dfrac{t_k}{1+t_k^2}\right)}{(1+t_1^2)\cdots(1+t_n^2)\prod_{j=1}^{n}\prod_{k=1}^{n}\left(1-\dfrac{2t_j}{1+t_j^2}\cos\theta_k\right)}$$

$$= \frac{\prod_{j<k}^{n}((t_j-t_k)(1-t_jt_k))}{\prod_{j=1}^{n}\prod_{k=1}^{n}(1+t_j^2-2t_j\cos\theta_k)}$$

$$= \frac{\prod_{j<k}^{n}((t_j-t_k)(1-t_jt_k))}{g_1\cdots g_n}.$$

Thus (11.7.2) follows.

Lemma 11.7.2 *Let $s \geqslant n$, $t_1,\cdots,t_s$ be s indepentent real variables, $U \in USp(2n)$,*

$$g_j = g(t_j, U) = \det(I - t_jU),$$

and

$$F_s(t_1,\cdots,t_s) = \prod_{j<k}^{s}((t_j-t_k)(1-t_jt_k)),$$

and then

$$\frac{F_s(t_1, \cdots, t_s)}{g_1 \cdots g_s} = \sum_{f_1 \geqslant \cdots \geqslant f_n \geqslant 0} \chi_f(U) Q_s^f(t_1, \cdots, t_s), \tag{11.7.5}$$

where

$$Q_s^f(t_1, \cdots, t_s)$$

$$= \begin{vmatrix} t_1^{l_1+s-n-1}, & \cdots, & t_1^{l_2+s-n-1}, & t_1^{s-n-1}, & t_1^{s-n-2}T_1, & \cdots, & T_1^{s-n-1} \\ \cdots & \cdots & \cdots & \cdots & \cdots & \cdots & \cdots \\ t_s^{l_1+s-n-1}, & \cdots, & t_s^{l_n+s-n-1}, & t_s^{s-n-1} & t_s^{s-n-2}T_s, & \cdots, & T_s^{s-n-1} \end{vmatrix} \tag{11.7.6}$$

and $T_j = 1 + t_j^2$, $j = 1, 2, \cdots, s$.

Proof. We resort to induction method on s. When $s = n$, this is just Lemma 11.7.1. Assume that (11.7.5) holds for $s (\geqslant n)$. We should prove that (11.7.5) still holds for $s + 1$, namely

$$\frac{F_{s+1}(t_1, \cdots, t_{s+1})}{g_1 \cdots g_{s+1}} = \sum_{f_1 \geqslant \cdots \geqslant f_n \geqslant 0} \chi(U) Q_{s+1}^f(t_1, \cdots, t_{s+1}), \tag{11.7.7}$$

where

$$Q_{s+1}^f(t_1, \cdots, t_{s+1})$$

$$= \begin{vmatrix} t^{l_1+s-n}, & \cdots, & t_1^{l_n+s-n}, & t_1^{s-n}, & t_1^{s-n-1}T_1, & \cdots, & T_1^{s-n} \\ \cdots & \cdots & \cdots & \cdots & \cdots & \cdots & \cdots \\ t_{s+1}^{l_1+s-n}, & \cdots, & t_{s+1}^{l_n+s-n}, & t_{s+1}^{s-n}, & t_{s+1}^{s-n-1}T_{s+1}, & \cdots, & T_{s+1}^{s-n} \end{vmatrix}. \tag{11.7.8}$$

Expanding the determinant on the right-hand side of (11.7.8) in the last column, we get

$$Q_{s+1}^f(t_1, \cdots, t_{s+1}) = \sum_{j=1}^{s+1} (-1)^{s+1-j} T_j^{s-n} t_1 \cdots \hat{t}_j \cdots t_{s+1}$$

$$\cdot Q_s^f(t_1, \cdots, \hat{t}_j, \cdots, t_{s+1}),$$

where $\hat{t}_j$ means that t_j does not appear. Thus the right-hand side of (11.7.7) is equal to

$$\sum_{j=1}^{s+1} (-1)^{s+1-j} T_j^{s-n} t_1 \cdots \hat{t}_j \cdots t_{s+1} \sum_{f_1 \geqslant \cdots \geqslant f_n \geqslant 0} \chi_f(U) Q_s^f(t_1, \cdots, \hat{t}_j, \cdots, t_{s+1}).$$

The assumption for induction shows that the right-hand side of (11.7.7) is equal to

$$\sum_{j=1}^{s+1} (-1)^{s+1-j} T_j^{s-n} t_1 \cdots \hat{t}_j \cdots t_{s+1} \frac{g_j}{g_1 \cdots g_{s+1}} \cdot F_s(t_1, \cdots, \hat{t}_j, \cdots, t_{s+1}).$$

By the identity (Zhong Jiaqing [2])

$$F_s(t_1,\cdots,t_s)=\begin{vmatrix} t_1^{s-1}, & t_1^{s-2}T_1, & \cdots, & T_1^{s-1} \\ \cdots & \cdots & \cdots & \cdots \\ t_s^{s-1}, & t_s^{s-1}T_s, & \cdots, & T_s^{s-1}\end{vmatrix}, \tag{11.7.9}$$

the right-hand side of (11.7.7) is equal to

$$\frac{1}{g_1g_2\cdots g_{s+1}}\begin{vmatrix} t_1^s, & t_1^{s-1}T_1, & \cdots, & t_1T_1^{s-1}, & T_1^{s-n}g_1 \\ \cdots & \cdots & \cdots & \cdots & \cdots \\ t_{s+1}^s, & t_{s+1}^{s-1}T_{s+1}, & \cdots, & t_{s+1}T_{s+1}^{s-1}, & T_{s+1}^{s-n}g_{s+1}\end{vmatrix}. \tag{11.7.10}$$

From (11.7.9) it follows that the left-hand side of (11.7.7) is equal to

$$\frac{1}{g_1g_2\cdots g_{s+1}}\begin{vmatrix} t_1^s & t_1^{s-1}T_1, & \cdots, & t_1T_1^{s-1}, & T_1^s \\ \cdots & \cdots & \cdots & \cdots & \cdots \\ t_{s+1}^s, & t_{s+1}^{s-1}T_{s+1}, & \cdots, & t_{s+1}T_{s+1}^{s-1} & T_{s+1}^s\end{vmatrix}. \tag{11.7.11}$$

If we can show that (11.7.10) and (11.7.11) are identical, then Lemma 11.7.2 is the immediate consequence. As

$$g_j=\prod_{k=1}^{n}(T_j-2t_j\cos\theta_k)=T_j^k-\cdots,$$

whose missing term is a linear combination of $t_j^mT_j^{n-m}(m=1,2,\cdots,n)$, we have

$$T_j^s=T_j^{s-n}T_j^n=T_j^{s-n}(g_j+\cdots)=T_j^{s-n}g_j+\cdots,$$

where the missing term is a linear combination of $t_j^mT_j^{s-m}$ $(m=1,2,\cdots,n)$. So (11.7.10) and (11.7.11) are identical. The desired result follows.

Setting $s=2n+1$, we have

$$\frac{\prod\limits_{j<k}^{2n+1}(1-t_jt_k)}{g_1\cdots g_{2n+1}}=\sum_{f_1\geqslant\cdots\geqslant f_n\geqslant 0}\chi_f(U)\frac{Q_{2n+1}^f(t_1,\cdots,t_{2n+1})}{D(t_1,\cdots,t_{2n+1})}, \tag{11.7.12}$$

where $Q_{2n+1}^f(t_1,\cdots,t_{2n+1})$

$$=\begin{vmatrix} t_1^{f_1+2n}, & \cdots, & t_1^{f_n+n+1}, & t_1^n, & t_1^{n-1}T_1, & \cdots, & t_1T_1^{n-1}, & T_1^n \\ \cdots & \cdots & \cdots & \cdots & \cdots & \cdots & \cdots & \cdots \\ t_{2n+1}^{f_1+2n}, & \cdots, & t_{2n+1}^{f_n+n+1}t_{2n+1}^n, & t_{2n+1}^{n-1}T_{2n+1}, & \cdots, & t_{2n+1}T_{2n+1}^{n-1}, & T_{2n+1}^n & \end{vmatrix}. \tag{11.7.13}$$

Replacing $T_j^m(1\leqslant m\leqslant n)$ in each row of (11.7.13) by $1+t_j^{2m}$, we can see that the value of the determinant remains unchanged. Letting $t_j\to r$ $(1\leqslant j\leqslant 2n+1)$ in (11.7.12), by (11.7.1) we obtain that the left-hand side of (11.7.12) is equal to

$$\frac{(1-r^2)^{n(2n+1)}}{\det^{2n+1}(I-rU)},$$

which is just the Poisson kernel (11.6.1) of $USp(2n)$. The right-hand side of (11.7.12) becomes

$$\sum_{f_1\geqslant\cdots\geqslant f_n\geqslant 0} \chi_f(U) \lim_{\substack{t_j\to r\\ 1\leqslant j\leqslant 2n+1}} \frac{Q^f_{2n+1}(t_1,\cdots,t_{2n+1})}{D(t_1,\cdots,t_{2n+1})}.$$

This proves (11.6.12) and, besides

$$N(f)\rho^f(r) = \lim_{\substack{t_j\to r\\ 1\leqslant j\leqslant 2n+1}} \frac{Q^f_{2n+1}(t_1,\cdots,t_{2n+1})}{D(t_1,\cdots t_{2n+1})}. \tag{11.7.14}$$

Taking $t_j = r\tau_j$, we get (see Cheng Guangxiao & He Zhuqi [3])

Theorem 11.7.1 *The Poisson kernel $P_r(U)$ of $USp(2n)$ has expansion (11.6.12) and the coefficient $\rho^f(r)$ of the Abel summation of (11.6.13) is equal to*

$$\frac{r^{f_1+\cdots+f_n}}{N(f)} \lim_{\substack{\tau_j\to 1\\ 1\leqslant j\leqslant 2n+1}} \frac{\det(q_{ij})_{1\leqslant i,j\leqslant 2n+1}}{D(\tau_1,\cdots,\tau_{2n+1})},$$

where $\qquad q_{ij} = \tau_i^{f_j+2n+1-j}, \ j = 1,2,\cdots,n+1$

and $\qquad q_{ij} = \tau_i^{2n+1-j}(1+\tau_i^{2(j-1)-2n}), \ j = n+2,\cdots,2n+1. \qquad (11.7.15)$

The determinant in (11.7.15) can be written as

$$\sum_{m_1,\cdots,m_n=0}^{1} r^{2(m_1+2m_2+\cdots+nm_n)} \begin{vmatrix} \tau_1^{f_1+2n}, & \cdots, & \tau_1^{f_n+n+1}, & \tau_1^n, & \tau^{n-1+2m_1}, & \cdots, & \tau_1^{0+2nm_n} \\ \cdots & \cdots & \cdots & \cdots & \cdots & \cdots & \cdots \\ \tau_{2n+1}^{f_1+2n}, & \cdots, & \tau_{2n+1}^{f_n+n+1}, & \tau_{2n+1}^n, & \tau_{2n+1}^{n-1+2m_1}, & \cdots, & \tau_{2n+1}^{0+2nm_n} \end{vmatrix}.$$

By $M(f_1,\cdots,f_n,P_0,P_1,\cdots,P_n)$, we denote

$$\lim_{\substack{\tau_j\to 1\\ 1\leqslant j\leqslant 2n+1}} \frac{\begin{vmatrix} \tau_1^{f_1+2n+1}, & \cdots, & \tau_2^{f_n+n+2}, & \tau_1^{\rho_0+n+1}, & \cdots, & \tau^{\rho_n+1} \\ \cdots & \cdots & \cdots & \cdots & \cdots & \cdots \\ \tau_{2n+1}^{f_1+2n+1}, & \cdots, & \tau_{2n+1}^{f_n+n+2}, & \tau_{2n+1}^{\rho_0+n+1}, & \cdots, & \tau_{2n+1}^{\rho_0+1} \end{vmatrix}}{\tau_1\cdots\tau_{2n+1}D(\tau_1^2,\tau_2^2,\cdots,\tau_{2n}^2)}.$$

Thus we further obtain

Theorem 11.7.2 *Any continuous function on $USp(2n)$ is Abel summable to itself and the coefficient $\rho^f(r)$ of Abel summation in (11.6.13) is equal to*

$$\frac{2^{n(2n+1)}}{N(f)} \sum_{(m_1,\cdots,m_n)\in E} r^{f_1+\cdots+f_n+2(m_1+2m_2+\cdots+nm_n)}$$

$$\cdot M(f_1, \cdots, f_n, 0, 2m_1, 4m_2, \cdots, 2nm_n),$$

where the set E of indices satisfies the following conditions:

1° $0 \leqslant m_j \leqslant 1, \ j = 1, \cdots, n;$

2° *for any pair* $(j, k), \ 1 \leqslant j, \ k \leqslant n, \ f_k + 2n + 1 - k \neq n - j + 2jm_j.$

By Hua's identity (see Theorem 1.2.4, p. 14 of Hua [1])

$$\lim_{\substack{x_1 \to x \\ \cdots\cdots \\ x_n \to x}} \frac{\begin{vmatrix} f_1(x_1), & \cdots, & f_n(x_1) \\ f_1(x_n), & \cdots, & f_n(x_n) \end{vmatrix}}{D(x_1, \cdots, x_n)} = \frac{(-1)^{\frac{n(n+1)}{2}}}{1!2!\cdots(n-1)!} \begin{vmatrix} f_1(x), & \cdots, & f_n(x) \\ f_1'(x), & \cdots, & f_n'(x) \\ \cdots & \cdots & \cdots \\ f_1^{(n-1)}(x), & \cdots, & f_n^{(n-1)}(x) \end{vmatrix},$$

we obtain

Theorem 11.7.3 $\rho^j(r)$ *is equal to*

$$\frac{(-1)^n}{N(f)!2!\cdots(2n)!} \begin{vmatrix} \xi_1(r), & \cdots, & \xi_{2n+1}(r) \\ \xi_1'(r), & \cdots, & \xi_{2n+1}'(r) \\ \xi_1^{(2n)}(r), & \cdots, & \xi_{2n+1}^{(2n)}(r) \end{vmatrix},$$

where

$$\xi_1(r) = r^{f_1+2n}, \cdots, \xi_n(r) = r^{f_n+n+1}, \ \xi_{n+1}(r) = r^n;$$
$$\xi_{n+2}(r) = r^{n-1} + r^{n+1}, \ \xi_{n+3}(r) = r^{n-2} + r^{n+2}, \cdots,$$
$$\xi_{2n+1}(r) = 1 + r^{2n}.$$

Chapter 12. Spherical Summation of Fourier Series on Unitary Symplectic Groups

§ 12.1 Expression of Spherical Summation by Integral

As in Chapter 5, we may consider not only the cubical summation of the Fourier series on unitary symplectic groups, but also the spherical summation of the Fourier series on the groups. In the latter case, (10.3.3) is regarded as

$$\sum_{m=0}^{\infty} \sum_{\substack{f_1 \geqslant \cdots \geqslant f_n \geqslant 0 \\ l_1^2 + \cdots + l_n^2 = m}} \operatorname{tr}(C_f \Phi_f'(U)). \tag{12.1.1}$$

Let $\varphi(t)$ be a given function defined on $[0, \infty]$ with $\varphi(0) = 1$. The so-called spherical summation in a sense of φ is the limiting case of

$$\sum_{m=0}^{\infty} \phi\left(\frac{\sqrt{m}}{R}\right) \sum_{\substack{f_1 \geqslant \cdots \geqslant f_n \geqslant 0 \\ l_1^2 + \cdots + l_n^2 = m}} \operatorname{tr}(C_f \Phi_f'(U)) \quad as \ R \to \infty, \tag{12.1.2}$$

where

$$\phi(t) = \phi_R(t) = \frac{\varphi(t)}{\left(\varphi \dfrac{\sqrt{m_0}}{R}\right)}, \quad m_0 = \frac{n(n+1)(2n+1)}{6}.$$

Write

$$s_R^{\varphi}(r, U) = \sum_{\substack{f_1 \geqslant \cdots \geqslant f_n \geqslant 0 \\ m = l_1^2 + \cdots + l_n^2}} \phi\left(\frac{\sqrt{m}}{R}\right) r^{l_1 + \cdots + l_n} \operatorname{tr}(C_f \Phi_f'(U)). \tag{12.1.3}$$

By (10.3.4), $s_R^{\varphi}(r, U)$ can be written as

$$\frac{1}{\omega_n} \int_{USp(2n)} u(wU) F_R(r, \bar{w}) \dot{w}, \tag{12.1.4}$$

where

$$F_R(r, \bar{w}) = \sum_{\substack{f_1 \geqslant \cdots \geqslant f_n \geqslant 0 \\ m = l_1^2 + \cdots + l_n^2}} \phi\left(\frac{\sqrt{m}}{R}\right) r^{l_1 + \cdots + l_n} N(f) \chi_f(\bar{w}), \tag{12.1.5}$$

and ω_n is the volume of $USp(2n)$.

Let $e^{\pm i\theta_j}$ $(1 \leqslant j \leqslant n)$ be the characteristic roots of $\bar{w}$. We attempt to prove

$$\sum_{l_1 > \cdots > l_n \geqslant 1} r^{l_1 + \cdots + l_n} N(f) \chi_f(\bar{w}) = \frac{1}{i^n (2n-1)! \cdots 3!1!} \cdot \det(P_r^{(2j-1)}(\theta_k))_{1 \leqslant j,k \leqslant n}, \tag{12.1.6}$$

where

$$P_r(\theta) = \frac{1 - r^2}{1 + r^2 - 2r\cos\theta}.$$

In fact, as

$$\frac{1}{2} P_r^{(2j-1)}(\theta) = (-1)^{j+1} \sum_{l=1}^{\infty} r^l l^{2j-1} \sin l\theta,$$

making Laplace's expansion for $\det(P_r^{(2j-1)}(\theta_k))_{1 \leqslant j,k \leqslant n}$, we get

$$(-1)^{\frac{n(n+1)}{2}} \sum_{l_1 > \cdots > l_n \geqslant 1} 2^n \begin{vmatrix} l_1 r^{l_1}, & \cdots, & l_n r^{l_n} \\ l_1^3 r^{l_1}, & \cdots, & l_n^3 r^{l_n} \\ \cdots & \cdots & \cdots \\ l_1^{2n-1} r^{l_1}, & \cdots, & l_n^{2n-1} r^{l_n} \end{vmatrix}$$

$$\begin{vmatrix} \sin l_1\theta_1, & \cdots, & \sin l_1\theta_n \\ \sin l_2\theta_1, & \cdots, & \sin l_2\theta_n \\ \cdots & \cdots & \cdots \\ \sin l_n\theta_1, & \cdots, & \sin l_n\theta_n \end{vmatrix} = (-1)^{\frac{n(n+1)}{2}}$$

$$\cdot \sum_{l_1 > \cdots > l_n \geqslant 1} \frac{2^n r^{l_1 + \cdots + l_n} \prod_{j=1}^{n} l_j \prod_{j>k} (l_j^2 - l_k^2)}{(2i)^n} s_\theta(l_1, \cdots, l_n)$$

$$= i^n \sum_{l_1 > \cdots > l_n \geqslant 1} r^{l_1 + \cdots + l_n} N(f)(2n-1)! \cdots 3!1! s_\theta(l_1, \cdots, l_n).$$

By dividing this expression by $i^n(2n-1)!\cdots 3!1! s(n, \cdots, 2, 1)$, (12.1.6) follows. Multiplying the right-hand side of (12.1.6) by $s_\theta(n, \cdots, 2, 1)$ and denoting the result by $g(\theta_1, \cdots, \theta_n)$, we obtain its Fourier series

$$\sum_{\infty > \nu_1, \cdots, \nu_n > -\infty} a_{\nu_1, \cdots, \nu_n} e^{i(\nu_1\theta_1 + \cdots + \nu_n\theta_n)}. \tag{12.1.7}$$

By Bochner [1], if $\varphi(t)$ is absolutely continuous on every finite interval and satisfies

$$\int_0^\infty |\varphi(t)| t^{\frac{n-1}{2}} dt < \infty, \tag{12.1.8}$$

then we have

$$\sum_{m=0}^\infty \sum_{m=\nu_1^2+\cdots+\nu_n^2} \phi\left(\frac{\sqrt{m}}{R}\right) a_{\nu_1,\cdots,\nu_n} e^{i(\nu_1\theta_1+\cdots+\nu_n\theta_n)} \tag{12.1.9}$$

$$= R \int_0^\infty g_\theta(t) H_\phi(tR)\, dt,$$

where

$$g_\theta(t) = \frac{\Gamma\left(\frac{n}{2}\right)}{2\pi^{\frac{n}{2}}} \int_\sigma g(\theta_1 + t\eta_1, \cdots, \theta_n + t\eta_n) d\sigma_\eta, \tag{12.1.10}$$

$$H_\varphi(tR) = \frac{1}{2^{\frac{n}{2}-1}\Gamma\left(\frac{n}{2}\right)} \int_0^\infty \phi(u)(utR)^{\frac{n}{2}} J_{\frac{n-2}{2}}(utR) du. \tag{12.1.11}$$

$J_\mu(s)$ is the Bessel function of the first kind of order μ. σ denotes the sphere $\eta_1^2 + \cdots + \eta_n^2 = 1$ and $d\sigma_\eta$ denotes the volume element of the sphere.

However, the left side of (12.1.9) is

$$\sum_{m=0}^\infty \sum_{m=l_1^2+\cdots+l_n^2} \phi\left(\frac{\sqrt{m}}{R}\right) r^{l_1+\cdots+l_n} N(f) s_\theta(l_1, \cdots, l_n).$$

Thus

$$F_R(r, \bar{w}) = \frac{R}{s_\theta(n, \cdots, 2, 1)} \int_0^\infty g_0(t) H_\phi(tR) dt$$

$$= \frac{R\Gamma\left(\frac{n}{2}\right)}{2\pi^{\frac{n}{2}} s_\theta(n, \cdots, 2, 1)} \int_{-\infty}^\infty \cdots \int g(\theta + \xi)$$

$$\cdot \frac{H_\phi(|\xi|R)}{|\xi|^{n-1}} d\xi_1 \cdots d\xi_n, \tag{12.1.12}$$

where $\xi = (\xi_1, \cdots, \xi_n)$ and $|\xi| = (\xi_1^2 + \cdots + \xi_n^2)^{1/2}$.

Substitute the expression of $g(\theta_1, \cdots, \theta_n)$ into (12.1.12) and expand the determinant. Assume that

$$\left(\frac{\partial}{\partial \xi_1}\right)^{\alpha_1} \cdots \left(\frac{\partial}{\partial \xi_n}\right)^{d_n} \frac{H_\phi(|\xi|R)}{|\xi|^{n-1}}\Bigg|_{|\xi|=\infty} = 0, \tag{12.1.13}$$

for $1 \leqslant \alpha_1, \cdots, \alpha_n \leqslant 2n-1$. Integrating by parts, we obtain

$$F_R(r, \bar{w}) = \frac{(-i)^n}{D(n, \cdots, 2, 1)s(n, \cdots, 2, 1)} \frac{\Gamma\left(\frac{n}{2}\right)}{2\pi^{\frac{n}{2}}}$$

$$\cdot \int_{-\infty}^{\infty} \cdots \int P_r(\theta_1 + \xi_1) \cdots P_r(\theta_n + \xi_n) Q\left(\frac{\partial}{\partial \xi_1}, \cdots, \frac{\partial}{\partial \xi_n}\right)$$

$$\cdot \frac{H_\phi(|\xi| R)}{|\xi|^{n-1}} d\xi_1 \cdots d\xi_n, \tag{12.1.14}$$

where

$$Q\left(\frac{\partial}{\partial \xi_1}, \cdots, \frac{\partial}{\partial \xi_n}\right) = (-1)^{n^2} \begin{vmatrix} \frac{\partial}{\partial \xi_1}, & \cdots, & \frac{\partial}{\partial \xi_n} \\ \left(\frac{\partial}{\partial \xi_1}\right)^3, & \cdots, & \left(\frac{\partial}{\partial \xi_n}\right)^3 \\ \cdots & \cdots & \cdots \\ \left(\frac{\partial}{\partial \xi_1}\right)^{2n-1}, & \cdots, & \left(\frac{\partial}{\partial \xi_n}\right)^{2n-1} \end{vmatrix}$$

Lemma 12.1.1 *Let $f(\xi)$ be a real function having continuous derivatives up to order n^2* ($\xi > 0$). Then

$$Q\left(\frac{\partial}{\partial \xi_1}, \cdots, \frac{\partial}{\partial \xi_n}\right) f\left(\frac{\xi_1^2 + \cdots + \xi_n^2}{2}\right)$$

$$= f^{(n^2)}\left(\frac{\xi_1^2 + \cdots + \xi_n^2}{2}\right) Q(\xi_1, \cdots, \xi_n). \tag{12.1.15}$$

Proof. We have

$$\left(\frac{\partial}{\partial \xi_1}\right)^1 \left(\frac{\partial}{\partial \xi_2}\right)^3 \cdots \left(\frac{\partial}{\partial \xi_n}\right)^{2n-1} f\left(\frac{\xi_1^2 + \cdots + \xi_n^2}{2}\right)$$

$$= f^{(n^2)}\left(\frac{|\xi|^2}{2}\right) \xi_1 \xi_2^3 \cdots \xi_n^{2n-1} + \cdots. \tag{12.1.16}$$

Here the terms in the ellipsis dots take the form

$$c_{i_1 \cdots i_n} f^{(2(i_1 + \cdots + i_n) - n)}\left(\frac{|\xi|^2}{2}\right) \xi_1^{2i_1 - 1} \cdots \xi_n^{2i_n - 1},$$

where the constant $c_{i_1 \cdots i_n}$ is related only to $i_1, \cdots, i_n$; $i_1 = 1, i_2 \leqslant 2, \cdots, i_n \leqslant n$ and in at least one of these inequalities the sign of inequality holds. Thus among $i_1, \cdots, i_n$ there are at least two identical indices. Therefore, by substituting (12.1.16) into the left-hand side of (12.1.15), those terms including in ellipsis dots form vanishing determinants, and the listed terms just form the right-hand side of (12.1.15). Thus the conclusion is proved.

By Lemma 12.1.1 we get

$$F_R(n, \bar{w}) = \frac{(-i)^n \Gamma\left(\frac{n}{2}\right)}{D(n, \cdots, 1) s(n, \cdots, 1)\, 2\pi^{\frac{n}{2}}}$$

$$\cdot \int_{-\infty}^{\infty} \cdots \int P_r(\theta_1 + \xi_1) \cdots P_r(\theta_n + \xi_n) \left[\left(\frac{d}{\xi d\xi} \right)^{n^2} \frac{H_\phi(\xi R)}{\xi^{n-1}} \right]_{\xi = |\xi|}$$

$$\cdot Q(\xi_1, \cdots, \xi_n) d\xi_1 \cdots d\xi_n. \tag{12.1.17}$$

Decompose $\bar{w}$ into

$$\bar{w} = P\Lambda P^{-1}, \quad P \in US_p(2n),$$

$$\Lambda = [e^{i\theta_1}, e^{-i\theta_1}, \cdots, e^{i\theta_n}, e^{-i\theta_n}].$$

Put the element $[e^{i\theta_j}, e^{-i\theta_j}]$ on the diagonal line of Λ into one group. Make permutations bet ween those groups and exchange $e^{i\theta_j}$ with $e^{-i\theta_j}$ in some $[e^{i\theta_j}, e^{-i\theta_j}]$. All those permutations σ can be obtained by applying similar transformations to the square matrices of $USp(2n)$. By Λ_σ we denote the diagonal matrix thus obtained and set $\bar{w} = P\Lambda_\sigma P^{-1}$. Make all possible$\bar{w}_\sigma$, and the total number of $\bar{w}_\sigma$ is $2^n \cdot n!$. Adding all $u(w_\sigma U)$ together, then dividing it by $2^n \cdot n!$ and denoting the result by $u^*(wU)$, thus we obtain that

$$\phi_U(\theta_1, \cdots, \theta_n) = \frac{1}{\omega_n'} \int_{[USp(2n)]} u^*(wU)[\dot{w}] \tag{12.1.18}$$

is an even symmetric function in $\theta_1, \cdots, \theta_n$. By (12.1.4), we have

$$s_R^\varphi(r, U) = \frac{1}{(2\pi)^n} \int \cdots \int_{\pi \geqslant \theta_1 \geqslant \cdots \geqslant \theta_n \geqslant 0} \phi_U(\theta) F_R(r, \bar{w})$$

$$\cdot |s_\theta(n, \cdots, 1)|^2 d\theta_1 \cdots d\theta_n.$$

Substituting (12.1.17) into the foregoing expression, by the Fubini Theorem, we get

$$s_R^\varphi(r, U) = \frac{(-i)^n \Gamma\left(\frac{n}{2}\right) R}{D(n, \cdots, 2, 1)\pi^{\frac{n}{2}} \cdot 2} \frac{1}{(2\pi)^n}$$

$$\cdots \int_{-\infty}^{\infty} \cdots \iint_{\pi \geqslant \theta_1 > \cdots > \theta_n \geqslant 0} \cdots \int \phi_U(\theta_1, \cdots, \theta_n) \overline{s_\theta(n, \cdots, 2, 1)}$$

$$\cdot P_r(\theta_1 + \xi_1) \cdots P_r(\theta_n + \xi_n) Q(\xi_1, \cdots, \xi_n) \left(\frac{d}{\xi d\xi} \right)^{n^2}$$

$$\cdot \frac{H_\phi(\xi R)}{\xi^{n-1}} \bigg|_{\xi = |\xi|} d\xi_1 \cdots d\xi_n d\theta_1 \cdots d\theta_n. \tag{12.1.19}$$

The right-hand side is equal to

$$\frac{(-i)^n\Gamma\left(\frac{n}{2}\right)R}{D(n,\cdots,2,1)\pi^{\frac{n}{2}}\cdot 2(2\pi)^n}\int_{-\infty}^{\infty}\cdots\int\frac{1}{2^n\cdot n!}\int_{-\pi}^{\pi}\cdots\int\cdot\phi_U(\theta_1,\cdots,\theta_n)$$

$$\overline{s_\theta(n,\cdots,2,1)}\prod_{j=1}^{n}P_r(\theta_j+\xi_j)$$

$$\cdot Q(\xi_1,\cdots,\xi_n)\left(\frac{d}{\xi d\xi}\right)^{n^2}\frac{H_\phi(\xi R)}{\xi^{n-1}}\bigg|_{\xi=|\xi|}d\xi_1\cdots d\xi_n d\theta_1\cdots d\theta_n.$$

Ler $r\to 1$. From Lebesgue's convergence Theorem and Zygmund's Abel summation Theorem on multiple Fourier series (cf. Zygmund [1]), it follows that

$$s_R^\varphi(U)=\frac{(-i)^n\Gamma\left(\frac{n}{2}\right)R}{D(n,\cdots,2,1)2\cdot\pi^{\frac{n}{2}}2^n\cdot n!}\int_{-\infty}^{\infty}\cdots\int\phi_U(-\xi)s_\xi(n,\cdots,2,1)$$

$$\cdot Q(\xi_1,\cdots,\xi_n)\left(\frac{d}{\xi d\xi}\right)^{n^2}\frac{H_\varphi(\xi R)}{\xi^{n-1}}\bigg|_{\xi=|\xi|}d\xi_1\cdots d\xi_n$$

$$=\frac{(-i)^n\Gamma\left(\frac{n}{2}\right)R}{D(n,\cdots,2,1)2\pi^{\frac{n}{2}}}\int\cdots\int_{\infty<\xi_1>\cdots>\xi_n>0}\phi_U(-\xi)s_\xi(n,\cdots,2,1)$$

$$\cdot Q(\xi_1,\cdots,\xi_n)\left(\frac{d}{\xi d\xi}\right)^{n^2}\frac{H_\varphi(\xi R)}{\xi^{n-1}}\bigg|_{\xi=|\xi|}d\xi_1\cdots d\xi_n. \quad (12.1.20)$$

Thus we obtain (Chen Guangxiao & He Zuqi [4])

Theorem 12.1.1 *If $u(U)$ is integrable on $USp(2n)$, then the spherical mean (12.1.2) in a sense of φ of its Fourier series (12.1.1) can be expressed by the right-hand side of (12.1.20), where $\phi_U(\xi)$ is given by (12.1.18), $\varphi(t)$ must satisfy both of (12.1.8) and (12.1.13) and is absolutely continuous on every finite interval. Besides, $H_\varphi(\xi R)/\xi^{n-1}$ is continuously differentiable up to order n^2 with respect to ξ.*

By Chandrasekharan-Minakshisundaran[1], $s_R^\varphi(U)$ also can be expressed as

$$\frac{(-i)^n R}{D(n,\cdots,2,1)2^n\cdot n!}\int_0^\infty h_U(t)t^{n-1}dt, \quad (12.1.21)$$

where

$$h_U(t)=\frac{\Gamma\left(\frac{n}{2}\right)}{2\pi^{\frac{n}{2}}}\int_\sigma s_{t\eta}(n,\cdots,2,1)\phi_U(t\eta)Q(t\eta)$$

$$\cdot \left(\frac{d}{tdt}\right)^{n^2} \frac{H_\varphi(tR)}{t^{n-1}} d\sigma_\eta, \tag{12.1.22}$$

with $\eta = (\eta_1, \cdots, \eta_n)$.

As $Q(t\eta) = t^{n^2}Q(\eta)$, supposing $Rt = u$, we deduce that

$$s_R^\varphi(U) = \frac{(-i)^n\Gamma(n/2)R^{n^2}}{D(n,\cdots,2,1)2^n n!2\pi^{\frac{n}{2}}}\int_0^\infty\left(\int_{\sigma} s_{\frac{u\eta}{R}}(n,\cdots,2,1)\right)\phi_U\left(\frac{u}{R}\eta\right)$$

$$\cdot\, Q(\eta)d\sigma_\eta)u^{n^2+n-1}\left(\frac{d}{\mu du}\right)^{n^2}\left(\frac{H_\varphi(\mu)}{u^{n-1}}\right)du. \tag{12.1.23}$$

Taking $u(U) \equiv 1$ in particular, we can easily prove that

$$s_R^\varphi(U) = 1, \text{ for } R \geqslant m_0.$$

Thus, when $R \geqslant M_0, u(U) - s_R^\varphi(U)$ can be expressed as

$$\frac{(-i)^n\Gamma\left(\frac{n}{2}\right)R}{D(n,\cdots,2,1)2\cdot\pi^{\frac{n}{2}}}\int\cdots\int_{\infty>\xi_1\geqslant\cdots\geqslant\xi_n\geqslant 0}\phi_U^*(-\xi)s_\xi(n,\cdots,2,1)$$

$$\cdot\, Q(\xi_1,\cdots,\xi_n)\left(\frac{d}{\xi d\xi}\right)^{n^2}\frac{H_\phi(\xi R)}{\xi^{n-1}}\bigg|_{\xi=|\xi|}d\xi_1\cdots d\xi_n, \tag{12.1.24}$$

where

$$\phi_U^*(\xi) = \frac{1}{\omega_n}\int_{[USp(2n)]}(u(U) - u^*(WU))[\dot{W}]. \tag{12.1.25}$$

§ 12.2 A General Convergence Theorem

$u(U) - s_R^\varphi(U)$ can also be expressed by

$$\frac{(-i)^n\Gamma\left(\frac{n}{2}\right)R^{n^2}}{D(n,\cdots,1)2^n n!2\pi^{\frac{n}{2}}}\int_0^\infty u^{n^2+n-1}\left(\frac{d}{udu}\right)^{n^2}\frac{H_\phi(u)}{u^{n-1}}du$$

$$\cdot\int_{\sigma} s_{\frac{u}{R}\eta}(n,\cdots,1)\phi_v^*\left(\frac{u}{R}\eta\right)Q(\eta_1,\cdots,\eta_n)d\sigma_\eta, \tag{12.2.1}$$

ϕ_v^* being defined by (12.1.25).

Assume that

$$u^{n^2+n-1}\left(\frac{d}{udu}\right)^{n^2}\left(\frac{H_\phi(u)}{u^{n-1}}\right) = O(u^{-n^2-1-p}) \quad (p > 0), \tag{12.2.2}$$

as $R\to\infty$. Divide (12.2.1) into two parts

$$\frac{(-i)^n\Gamma\left(\frac{n}{2}\right)R^{n^2}}{D(n,\cdots,1)2^n n!2\cdot\pi^{\frac{n}{2}}}\left(\int_0^{M(R)}+\int_{M(R)}^{\infty}\right)du=I_1+I_2.$$

Take $M(R)$ such that

$$R^{n^2}/M(R)^{n^2+p}=0(1),\quad M(R)/R=0(1),\tag{12.2.3}$$

and such $M(R)$ can be found. For example, we may take

$$M(R)=R^{1-p/2n^2}.$$

Since

$$\int_\sigma s_{\frac{u}{R}\eta}(n,\cdots,1)\phi_v^*\left(\frac{u}{R}\eta\right)Q(\eta_1,\cdots,\eta_n)d\sigma_\eta=O(1),$$

we have

$$|I_2|=O\left(\int_{M(R)}^{\infty}R^{n^2}u^{-n^2-1-p}du\right)=O(R^{n^2}/(M(R))^{n^2+p}).$$

With the aid of (12.2.3), we get

$$I_2=o(1).$$

By the given conditions for φ,

$$\left(\frac{d}{udu}\right)^{n^2}\left(\frac{H_\varphi(u)}{u^{n-1}}\right)$$

is continuous at $u=0$ and

$$s_{\frac{u}{R}\eta}(n,\cdots,1)=O\left(\frac{u^{n^2}}{R^{n^2}}\right).$$

Since $u(U)$ is continuous, we get

$$\phi_v^*\left(\frac{u}{R}\eta\right)=o(1).$$

This leads to

$$|I_1|=o\left(\int_0^{M(R)}\left|u^{n^2}u^{n^2+n-1}\left(\frac{d}{udu}\right)^{n^2}\frac{H_\varphi(u)}{u^{n-1}}\right|du\right).\tag{12.2.4}$$

By (12.2.2), it yields that

$$\int_0^\infty\left|u^{n^2}u^{n^2+n-1}\left(\frac{d}{udu}\right)^{n^2}\frac{H_\varphi(u)}{u^{n-1}}\right|du$$

is absolutely integrable. Thus, in virtue of (12.2.4),

$$|I_1|=o(1).$$

Consequently we are led to the following

Theorem 12.2.1 *If $u(U)$ is continuous on $US_p(2n)$ and the spherical mean (12.1.2) in a sense of φ of its Fourier series is denoted by $s_R^\varphi(U)$. Then, under the following conditions, $s_R^\varphi(U)$ converges to $u(U)$ as $R\to\infty$:*

(i) *$\varphi(t)$ is absolutely continuous on every finite interval and satisfies*

$$\int_0^\infty |\varphi(t)| t^{\frac{n-1}{2}} dt < \infty; \tag{12.2.5}$$

(ii) *H_φ(12.1.11) determined by $\varphi(t)$ satisfies*

$$\left(\frac{\partial}{\partial \xi_1}\right)^{j_1} \cdots \left(\frac{\partial}{\partial \xi_n}\right)^{j_n} \left(\frac{H_\varphi(|\xi|R)}{|\xi|^{n-1}}\right)\Bigg|_{|\xi|=\infty} = 0 \tag{12.2.6}$$

for $1 \leqslant j_1, \cdots, j_n \leqslant 2n-1$; and when $u\to\infty$, we have

$$\left(\frac{d}{udu}\right)^{n^2} \left(\frac{H_\varphi(u)}{u^{n-1}}\right) = O(u^{-(2n^2+n+p)}) \quad (p>0). \tag{12.2.7}$$

Since, in the proof of the estimate $I_2 = o(1)$, the other properties of $u(U)$ have not been touched upon, the φ-spherical summability of the Fourier series is a local property, too.

§ 12.3 Three Kinds of Spherical Summation and the Proof of Convergence Theorem

As in Chapter 5, we proceed to consider three kinds of summation, namely, the Riesz summation of order δ, the Gauss-Sommerfeld summation and the Abel summation.

1) The Riesz summation of order δ.

In this case we have

$$\varphi(t) = \begin{cases} (1-t^2)^\delta, & \text{if } 0 \leqslant t < 1 \\ 0, & \text{if } 1 \leqslant t. \end{cases}$$

When $R^2 \geqslant m_0$, we have

$$H_\phi = H_R^\varphi(u) = \frac{2^{\delta-\frac{n}{2}+1}\Gamma(\delta+1)}{(1-m_0/R^2)^\delta \Gamma(n/2)} u^{n-1} V_{\delta+\frac{n}{2}}(u),$$

where $V_s(u) = J_s(u)/u^s$ and J_s is the Bessel function of the first kind of order s. Thus

$$\left(\frac{d}{udu}\right)^{n^2} \left(\frac{H_\phi(u)}{u^{n-1}}\right) = \left(\frac{d}{udu}\right)^{n^2} V_{\delta+\frac{n}{2}}(u)$$

$$\cdot \frac{2^{\delta-\frac{n}{2}+1}\Gamma(\delta+1)}{\left(1-\frac{m_0}{R^2}\right)^{\delta}\Gamma\left(\frac{n}{2}\right)} = \frac{(-1)^{n^2}2^{\delta-\frac{n}{2}+1}\Gamma(\delta+1)}{\left(1-\frac{m_0}{R^2}\right)^{\delta}\Gamma\left(\frac{n}{2}\right)}$$

$$\cdot V_{\delta+\frac{n}{2}+n^2}(u) = O(u^{-\delta-\frac{n+1}{2}-n^2}).$$

When $\delta > n^2 + \frac{n-1}{2}$, taking $p = \delta - n^2 - \frac{n-1}{2}$, we get

$$\left(\frac{d}{udu}\right)^{n^2}\left(\frac{H_\phi(u)}{u^{n-1}}\right) = O(u^{-(2n^2+n+p)}).$$

Thus the condition (12.2.7) in Theorem 12.2.1 is satisfied, and both conditions (12.2.5) and (12.2.6) are evidently satisfied.

2) The Gauss-Sommerfeld summation

In this case, we have

$$\varphi(t) = e^{-t^2}.$$

When $R^2 \geqslant m_0$, we have

$$H_\phi = H_R^\varphi(u) = \frac{e^{m_0/R^2}u^{n-1}e^{-\frac{1}{4}u^2}}{2^{n-1}\Gamma\left(\frac{n}{2}\right)}.$$

Besides, for any $p > 0$ we have

$$\left(\frac{d}{udu}\right)^{n^2}\left(\frac{H_\phi(u)}{u^{n-1}}\right) = \frac{e^{m_0/R^2}}{2^{n-1}\Gamma\left(\frac{n}{2}\right)}\left(\frac{d}{udu}\right)^{n^2} e^{-\frac{1}{4}u^2}$$

$$= \frac{\left(-\frac{1}{2}\right)^{n^2} e^{m_0/R^2}}{2^{n-1}\Gamma\left(\frac{n}{2}\right)} e^{-\frac{1}{4}u^2} = O(u^{-(2n^2+n+p)}).$$

Thus the condition (12.2.7) in Theorem 12.2.1 is satisfied. Both (12.2.5) and (12.2.6) are evidently satisfied.

3) The Abel summation

In this case, we have $\varphi(t) = e^{-t}$. When $R^2 \geqslant m_0$,

$$H_\phi = H_R^\varphi(u) = \frac{e^{\frac{\sqrt{m_0}}{R}}(n-1)!}{2^{n-2}\left(\Gamma\left(\frac{n}{2}\right)\right)^2} \cdot \frac{u^{n-1}}{(1+u^2)^{\frac{n+1}{2}}}$$

and

$$\left(\frac{d}{udu}\right)^{n^2}\left(\frac{H_\phi(u)}{u^{n-1}}\right)$$

$$=\frac{e^{\frac{\sqrt{m_0}}{R}}(n-1)!(-1)^{n^2}(n+1)(n+3)\cdots(2n^2+n-1)}{2^{n-2}\left(\Gamma\left(\frac{n}{2}\right)\right)^2}$$

$$\cdot(1+u^2)^{-\left(\frac{n+1}{2}+n^2\right)}=O(u^{-(2n^2+n+1)}).$$

Thus (12.2.7) is satisfied. (12.2.5) and (12.2.6) are obviously satisfied.

Finally we obtain

Theorem 12.3.1 *If $u(U)$ is continuous on $US_p(2n)$, then its Fourier series is Abel summable, Gauss-Sommerfeld summable and Riesz summable of order δ to itself where $\delta > \frac{2n^2+n-1}{2}$.*

Chapter 13. Harmonic Analysis in Classical Domains on Quaternion Field

§ 13.1 Introduction

An important starting point for discussing harmonic analysis on unitary groups in Part I is to regard the unitary group $U_n = U(n, \mathrm{C})$ as the characteristic manifold on the classical domain of the first class of several complex variables.

$$\mathscr{R}_I = \mathscr{R}_I(n, \mathrm{C}) = \{Z \mid I - Z\bar{Z}' > 0,\ Z \text{ is } n \times n \text{ complex matrix}\}.$$

Proceeding from the Poisson-Hua kernel $P_I(Z, U)$ of $\mathscr{R}_I(n, C)$, Hua Luogeng [2] deduced Abel's summation of the Fourier series on $U(n,\mathrm{C})$.

In Part II, the study of harmonic analysis on rotation groups is also based on Hua's idea. Regarding the orthogonal group $O(n)$ as the characteristic manifold on the classical domain of the first class of real variables

$$\mathscr{R}_I(n, R) = \{X \mid I - XX > 0,\ X \text{ is } n \times n \text{ real matrix}\}$$

and starting from the Poisson kernel $P_1(x, \Gamma)$ of $\mathscr{R}_I(n, \mathrm{R})$, we again obtain Abel's summation of the Fourier series on $SO(n)$ (Lu Qikeng [1] and Zhong Jiaqing [2]). This method also becomes one of the starting points of studying harmonic analysis on rotation groups.

In the preceding chapters of Part III, we have already established the theory of harmonic analysis on $USp(2n)$ without using the above-mentioned point of view. There, proceeding from the Cesàro (c, α) kernel corresponding to the Cesàro summability of the Fourier series on $USp(2n)$ and letting $\alpha \to \infty$, we deduced Poisson's kernel corresponding to Abel's summation of the Fourier series on $USp(2n)$.

It is doubtful whether we can, as in the first two parts, find out some bounded domain whose characteristic manifold is just a unitary sym plectic group $USp(2n)$. Long ago, Hua Luogeng pointed out that $USp(2n)$, actually, is a unitary group on the quaternion field and can also be regarded as the characteristic manifold of the classical domain of the first class

$$R_I(Q) = \{Z = (z_{st}) : I - Z\bar{Z}' > 0,\ z_{st} \text{ is a quaternion}\}.$$

From this point of view, harmonic analysis on $R_I(Q)$ can be established and the Poisson kernel of $USp(2n)$ can be deduced. This kernel is in good agreement with that obtained in Chapter 11.

In this chapter, we come to establish the theory of harmonic functions on the quaternion field (see Chen Guangxiao [2]).

In [1], Sun Jiguang discussed the theory of harmonic functions in the classical domain $R_{III}(Q)$ on the quaternion field and proved that the three classes of classical domains, $R_I(Q)$, $R_{II}(Q)$ and $R_{III}(Q)$, belong to the equivalent matrix representations of $s_p(m, n)/s_p(m) \times s_p(n)$, $SU^*(2n)/s_p(n)$ and $s_p(n, \mathrm{C})/s_p(n)$ in the classified list for irreducible global symmetric Riemannian spaces made by E. Cartan (cf. S. Helgason [1]), respectively. Thus $R_I(Q)$ discussed below is irreducible.

§ 13.2 Classical Domain of Square Matrices on Quaternion Field Q

Let Q be the quaternion field on real number field. The element $\bar{q} = a - ib - jx - ky$ is called the conjugate element of $q = a + ib + jx + ky$. $|q| = \sqrt{q\bar{q}}$ is called the absolute value of q; $a = (q + \bar{q})/2$ is the real part of q and denoted by Req. Q^* denotes the multiplicative subgroup of Q. It is easily shown that all quaternions with 1 being the absolute value form the commutator subgroup C of Q^*, $Q^*/C \simeq \mathrm{R}^+$ and $q \to |q|$ is a natural isomorphism. In Hua Luogeng & Wan Zhexian [1] the theory of determinants on general fields was established. Applying the theory to Q, we obtain the following properties of the determinant of any square matrix A of order n on Q:

I. The determinant det A, remains unchanged after exchanging any two rows (column) of A.

II. The determinant becomes $|q|$ det A if a certain row (column) is multiplied on left by an element q inQ^*.

III. The determinant remains unchanged after adding a left (right) multiple of any row (column) to a certain row (column) of A.

From the representation of Q by 2×2 matrices

$$\tau: a + ib + jx + ky \to \begin{pmatrix} a + ib, & x + iy \\ -x + iy, & a - ib \end{pmatrix} \in M(2, \mathrm{C}), \tag{13.2.1}$$

a faithful representation from ring $M(n, Q)$ to $M(2n, \mathrm{C})$

$$\tau: (q_{jk})_{1 \leqslant j,k \leqslant n} \to (\tau(q_{jk}))_{1 \leqslant j,k \leqslant n} \tag{13.2.2}$$

is induced. We are apt to show that

$$\det A = \sqrt{\det \tau(A)}. \tag{13.2.3}$$

Definition. Any $n\times n$ matrix satisfying

$$\overline{U}U' = I \tag{13.2.4}$$

is called *a unitary square matrix* and any $n\times n$ matrix satisfying

$$\overline{H}' = H \tag{13.2.5}$$

is said to be *a Hermite.*

Obviously, τ faithfully represents the unitary group $U(n, Q)$ as the unitary symplectic group $USp(2n)$ of order $2n$ and maps H onto the set consisting of all complex Hermite square matrices of order $2n$ of which each satisfies

$$\tau(H)\cdot J_n = J_n\cdot \overline{\tau(H)}, \left(J_n = I^{(n)} \begin{pmatrix} 0 & 1 \\ -1 & 0 \end{pmatrix}\right).$$

$R_I(n, Q)$ denotes

$$\{Z\in M(n, Q): I - Z\overline{Z}' > 0\}, \tag{13.2.6}$$

i. e. the classical domain of square matrices of the first class on Q. If H is Hermite, "$H > 0$" denotes $\tau(H) > 0$.

Theorem 13.2.1 *Let $Z\in R_I(n, Q)$, and then there exist U and $V\in U(n, Q)$ such that*

$$UZV^{-1} = [\lambda_1, \cdots, \lambda_n], \tag{13.2.7}$$

where $0\leqslant \lambda_1 \leqslant \cdots \lambda_n < 1$.

Proof. Let $Y = Z\overline{Z}$. If Z is nonsingular, then $Y > 0$. Let λ_1^2 be the smallest characteristic root of $\tau(Y)$. By (13.2.3), we know that

$$\det(Y - \lambda_1^2 I) = 0.$$

From §7 in Chapter 3 of Hua Luogeng and Wan Zhexian [1], it follows that the solution of the equation

$$(Y - \lambda_1^2 I)e = 0, \quad |e| = 1$$

exists (we note that the modulus of the column vector $(q_1, \cdots, q_n)$ is $\sqrt{\sum q_i\overline{q}_i}$). Employing Schmidt's orthogonalization on column vectors, we can obtain a unitary square matrix W of which the first column is e and

$$YW = W[\lambda_1^2, Y_1], \quad (\overline{Y}_1' = Y_1, Y_1 > 0).$$

By induction, there exists $U\in U(n, Q)$ such that

$$U'Z\overline{Z}\overline{U}' = [\lambda_1, \cdots, \lambda_n]^2.$$

Thus $V = [\lambda_1^{-1}, \cdots, \lambda_n^{-1}]UZ$ is a unitary square matrix which makes (13.2.7) valid.

If Z is degenerate, then UZ has a zero row vector. Those non-vanishing vectors, having been normalized, form an incomplete system of orthogonal vectors. Again using Schmidt's orthogonalization on the row vectors and then adding some vectors lead to a unitary basis, Thus there exists a unitary square matrix V such that

$$UZV^{-1} = \Lambda.$$

By concerning the differentiation on the field Q, the following principles are employed:

1. Let $\frac{\partial}{\partial q} = \frac{1}{4}\left(\frac{\partial}{\partial a} - i\frac{\partial}{\partial b} - j\frac{\partial}{\partial x} - k\frac{\partial}{\partial y}\right)$ denote the formal differential operator and let

$$\partial_Z = \left(\frac{\partial}{\partial z_{ij}}\right)_{1\leqslant i,j\leqslant n}.$$

Then it is easily checked that, for a real function $u(z)$, there exists

$$du(z) = 4\text{Re tr}(\partial_Z' u \cdot dz), \tag{13.2.8}$$

where $4\text{Re tr}\,(\partial_Z' \cdot Y(Z))$ denotes the formal divergence of the matrix-valued function Y of Q.

2. If $dZ_1 = A^{-1}dZB^{-1}$, then, for a real function u, a real matrix-valued function R and a matrix-valued function Y on Q, we have

$$\partial_{Z_1}' u = B \cdot \partial_Z' u \cdot A,$$

$$\text{tr}(\partial_Z' \cdot R) = \text{tr}(B \cdot \partial_Z' \cdot A \cdot R), \tag{13.2.9}$$

$$\text{Re tr}\,(\partial_{Z_1}' \cdot Y) = \text{Re tr}(B \cdot \partial_Z' \cdot A \cdot Y) \tag{13.2.10}$$

respectively.

The proof of (13.2.9): Let $I_{\alpha\beta}$ be the matrix obtained by changing the element of $O^{(n)}$ lying in the αth row and βth column into 1. The statement that the (β,α) elements in two matrices are equal can be expressed as

$$\text{tr}(I_{\alpha\beta}C) = \text{tr}(I_{\alpha\beta}D) \text{ or } \text{tr}(C\ I_{\alpha\beta}) = \text{tr}(DI_{\alpha\beta}).$$

By the formal matrix identity $\partial_{Z_1}' = B \cdot \partial_Z' \cdot A$, we have that (13.2.9) holds. By means of $iI_{\alpha\beta}, jI_{\alpha\beta}$ and $kI_{\alpha\beta}$, (13.2.10) can also be proved.

3. Let A be a constant matrix on Q. From the formal operations

$$\frac{\partial}{\partial z} \cdot q \cdot dz \left(= dz \cdot q \cdot \frac{\partial}{\partial z}\right) = \text{Re}q,$$

$$\frac{\partial}{\partial z} \cdot q \cdot \bar{d}z = -\frac{1}{2}\bar{q},$$

we get

$$\partial'_Z \cdot A \cdot Z = (\mathrm{Re\ tr}\ A) \cdot I, \quad \partial'_Z \cdot A \cdot \bar{Z} = -\frac{1}{2}(\mathrm{tr}\,\bar{A})I, \tag{13.2.11}$$

$$\partial'_Z \cdot A \cdot Z' = \mathrm{Re}A', \quad \partial'_Z \cdot A \cdot \bar{Z}' = -\frac{1}{2}\bar{A}'.$$

4. For real functions u and v we have

$$\partial'_Z(uv) = u\partial'_Z v + v\partial'_Z u,$$
$$\partial'_Z f(u) = f'(u)\partial'_Z u.$$

By (13.2.3) we get

$$\begin{aligned}&(\det Z)^{-1}d(\det Z) = \mathrm{Re}\ (\mathrm{tr}(Z^{-1}dZ)),\\ &\partial'_Z(\det Z) = (\det Z)Z^{-1}.\end{aligned} \tag{13.2.12}$$

§ 13.3 Continuous Group of Motions of $\mathscr{R}_I(n,Q)$, Laplacian

Let

$$G = \{x \in M(2n, Q),\ \bar{x}Kx' = K,\ \det x = 1,\ K = [I^{(n)}, -I^{(n)}]\}, \tag{13.3.1}$$

and

$$N = \{[U, V], U, V \in U(n, Q)\}. \tag{13.3.2}$$

Then N is a compact subgroup of G. Write x as

$$\begin{pmatrix} A & B \\ C & D \end{pmatrix}, A = A^{(n)},\ B = B^{(n)},\ C = C^{(n)},\ D = D^{(n)}. \tag{13.3.3}$$

It is easy to show that

$$x^{-1} = \begin{pmatrix} \bar{A}', & -\bar{C}' \\ -\bar{B}', & \bar{D}' \end{pmatrix}, \tag{13.3.4}$$

$$BD^{-1} = \overline{CA^{-1}}' = \bar{A}'^{-1}C^{-1},\quad \overline{A^{-1}B}' = D^{-1}C. \tag{13.3.5}$$

Consider the mapping σ: $G \ni x \to BD^{-1} \in \mathscr{R}_I(n, Q)$. From (13.3.4) and (13.3.5), we yield the following principal condition for $\sigma(x) = \sigma(x_1)$:

$$\sigma(x^{-1}x_1) = 0 \text{ or } x^{-1}x_1 \in N.$$

Thus it can be easily shown that σ induces a one-to-one surjective mapping σ' from the space G/N of left cosets onto $R_I(n, Q)$. Therefore, under the following meaning, the group G is transitive over $R_I(n, Q)$:

$$f_x: R_I(n, Q) \ni Z \to f_x(z) = (AZ + B)(CZ + D)^{-1} \in R_I(n, Q).$$

This transformation maps $Z_0 = -A^{-1}B$ onto the origin. We can easily find out that the transformation which maps $Z_0 \in R_I(n, Q)$ into O must be

$$Z_1 = P^{-1}(Z - Z_0)(I - \bar{Z}_0'Z)^{-1}R, \tag{13.3.6}$$

$$P\bar{P}' = I - Z_0\bar{Z}_0', \quad R\bar{R}' = I - \bar{Z}_0'Z_0. \tag{13.3.7}$$

The transformation

$$Z_1 = UZV(U, V \in U(n, Q)), \tag{13.3.8}$$

which keeps the origin unchanged, is transitive over $U(n, Q)$. (13.3.6) keeps $U(n, Q)$ unchanged. Thus the subgroup which keeps Z_0 unchanged is also transitive over $U(n,Q)$. By Theorem 13.2.1 we obtain

Theorem 13.3.1 *The group of motions of $R_I(n, Q)$ is generated by the transformations* (13.3.8) *and*

$$Z_1 = M^{-1}(Z - \Lambda)(I - \Lambda Z)^{-1}M, \tag{13.3.9}$$

where

$$\Lambda = [\lambda_1, \cdots, \lambda_n], \ M = [m_1, \cdots, m_n],$$

$$0 \leqslant \lambda_1 \leqslant \lambda_2 \leqslant \cdots \leqslant \lambda_n < 1, \ m_\alpha = \sqrt{1 - \lambda_\alpha^2}.$$

Next we proceed to calculate the Laplacian in $R_I(n, Q)$. Differentiating (13.3.6) yields

$$dZ_1 = P^{-1}(I - Z\bar{Z}_0')^{-1}dZ(I - \bar{Z}_0'Z)^{-1}R.$$

So we have

$$\partial_{Z_1}' = R^{-1}(I - \bar{Z}_0'Z)\partial_Z'(I - Z\bar{Z}_0')\bar{P}'^{-1}, \tag{13.3.10}$$

$$\bar{\partial}_{Z_1} = P^{-1}(I - Z_0\bar{Z}')\bar{\partial}_Z(I - \bar{Z}'Z_0)\bar{R}'^{-1}, \tag{13.3.10'}$$

where the matrices ∂_Z' and $\bar{\partial}_Z$ do not execute differential operation on the matrices behind them. Noticing that the Laplacian becomes $16\mathrm{Re}\ \mathrm{tr}(\partial_{Z_1}'\bar{\partial}_{Z_1})|_{Z_1=0}$ at the origin, we obtain

Theorem 13.3.2 *The value at $Z = Z_0$ of the Laplacian in $R_I(n, Q)$ is equal to*

$$16\mathrm{Re}\ \mathrm{tr}A_Z u(z)|_{Z=Z_0} = 16\mathrm{Re}\ \mathrm{tr}\{\partial_z'(I - Z_0\overset{\downarrow}{\bar{Z}'})(\bar{\partial}_z^{\downarrow} u)\cdot(I - \overset{\downarrow}{\bar{Z}'}Z_0)\}|_{Z=Z_0}, \tag{13.3.11}$$

where ∂_L' exectes the operation on the variable indicated by the arrow.

Proof. Applying the properties of differentiation on Q given in § 13.2, by (13.3.7) and

$$\mathrm{Re}\,\mathrm{tr}\,(AB) = \mathrm{Re}\,\mathrm{tr}(BA)$$

we obtain (13.3.11).

Example. Set $n=1$, $R_I(1,Q)=\{q\in Q,\ |q|<1\}$ and $U(1,Q)=\{|q|=1\}$. Then its Laplacian is expressed as

$$16(1-q\bar{q})\left\{(1-q\bar{q})\frac{\partial^2 u}{\partial q\partial\bar{q}}+\frac{1}{2}\left(\bar{q}\frac{\partial u}{\partial\bar{q}}+\frac{\partial u}{\partial q}q\right)\right\}$$
$$=(1-a^2-b^2-x^2-y^2)$$
$$\cdot\left\{(1-a^2-b^2-x^2-y^2)\left(\frac{\partial^2 u}{\partial a^2}+\frac{\partial^2 u}{\partial b^2}+\frac{\partial^2 u}{\partial x^2}+\frac{\partial^2 u}{\partial y^2}\right)\right.$$
$$\left.+4\left(a\frac{\partial u}{\partial a}+b\frac{\partial u}{\partial b}+x\frac{\partial u}{\partial x}+y\frac{\partial u}{\partial y}\right)\right\}.$$

Laplacian (13.3.11) can also be written as

$$16\mathrm{Re}\,\mathrm{tr}\{(I-\bar{Z}'Z)\partial_Z'(I-Z\bar{Z}')\overset{\downarrow}{\bar{\partial}}_Z u\}$$
$$+8\mathrm{Re}\,\mathrm{tr}\{(I-\bar{Z}'Z)(\bar{Z}'\bar{\partial}_Z u+\partial_Z' u\cdot Z)\},\qquad(13.3.12)$$

where the arrow indicates that ∂_z' executes differential operation only on $\bar{\partial}_Z u$.

Definition 13.3.1 Any real function in $R_I(n,Q)$ which satisfies

$$\mathrm{Re}\ \mathrm{tr}\Delta_Z u=0$$

is referred to as *a harmonic function in* $R_I(n,Q)$.

§ 13.4 Extremum Principle for Harmonic Functions of Class ξ

Let $\partial R_I(n,Q)$ denote the boundary of $R_I(n,Q)$ and $L^{(r)}$ be the set of those points where the rank of $I-Z\bar{Z}'$ is r. Then $\partial R_I(n,Q)$ is the union of the disjoint sets

$$L^{(0)}=U(n,Q),\ L^{(1)},\cdots,L^{(n-1)}.$$

Imitating the idea in § 5.8 of Hua [1], we introduce the concept r-cover as follows: For any two given unitary square matrices U and V, the set of all those points of form

$$U[I^{(n-r)},Z_1]V,\quad Z_1\in R_I(r,Q)\qquad(13.4.1)$$

is called a r-cover. Thus each point in $L^{(r)}$ belongs to some r-cover. However, any two r-cover may have common points. Let real function $u(Z)$ be defined on $\overline{R_I(n,Q)}\backslash U(n,Q)$ and, for any r-cover of any given point $Z_1(r>0)$, $u(Z)$ has continuous partial derivatives of order 2. We now study the values of $\mathrm{Re}\,\mathrm{tr}\Delta_Z u(Z)$ at the point (13.4.1). As (13.3.11), under the transformation $W=UZV$, keeps unchanged, so what we need to study is the values of $\mathrm{Re}\,\mathrm{tr}\Delta_Z u(Z)$ at the points of form

$$[I^{(n-r)},Z_1],\quad Z_1\in R_I(r,Q).\qquad(13.4.2)$$

We are apt to show that, at those points of form (13.4.2), $\operatorname{Re}\operatorname{tr}\Delta_Z u(Z)$ can be simplified as

$$\operatorname{Re}\operatorname{tr}\{\partial'_{z_1}(I^{(r)}Z_1\overset{\downarrow}{\bar{Z}'_1})\bar{\partial}_{z_1}\overset{\downarrow}{u}([I^{(n-r)}, Z_1])(I^{(r)}-\overset{\downarrow}{\bar{Z}'_1}Z_1) \tag{13.4.3}$$

or as

$$\operatorname{Re}\operatorname{tr}\Delta_{Z_1}u([I^{(n-r)}, Z_1]).$$

Definition 13.4.1 Any real function $u(Z)$ with the above differentiation properties satisfies

$$\operatorname{Re}\operatorname{tr}\Delta_Z u(Z) = 0 \tag{13.4.4}$$

on $\overline{R_I(n, Q)}\backslash U(n, Q)$ and is continuous on $\overline{R_I(n, Q)}$, is called *a harmonic function of* Class ξ.

Theorem 13.4.1 *Every harmonic function of Class ξ achieves its maximum (minimum) on $U(n, Q)$.*

It suffices to prove that any harmonic function attains its extremum on the boundary, since any problem on one point of $L^{(r)}(0 < r < n)$ is equivalent to that on $R_I(r, Q)$. First we show

Theorem 13.4.2 *If real function $u(Z)$ has continuous partial derivatives of order 2 and*

$$\operatorname{Re}\operatorname{tr}\Delta_Z u(Z) > 0 \tag{13.4.5}$$

in $R_I(n,Q)$, then $u(Z)$ can not attain its maximum inside $R_I(n, Q)$.

Proof. Suppose $u(Z)$ attains its maximum at some interior point Z_0. By changing variables, we could assume that $Z_0 = 0$. By (13.4.5) we get

$$\sum_{\alpha,\beta=1}^{N} \frac{\partial^2}{\partial z_{\alpha\beta}\partial \bar{z}_{\alpha\beta}} u(Z)\bigg|_{Z=0} > 0. \tag{13.4.6}$$

Since $u(Z)$ attains its maximum at $Z = 0$, we have

$$\frac{\partial^2}{\partial z_{\alpha\beta}\partial \bar{z}_{\alpha\beta}} u(Z)\bigg|_{Z=0} \leqslant 0,$$

which contradicts (13.4.6).

The proof of Theorem 13.4.1. Let

$$M = \max_{Z\in\partial R_I(n,Q)} u(Z).$$

and, at the interior point Z_0,

$$u(Z_0) > M + \varepsilon. \tag{13.4.7}$$

Put $\quad u_\eta(Z) = u(Z) + \eta \mathrm{tr}((Z - Z_0)(\overline{Z - Z_0})'), \ (\eta > 0).$

Then, with the help of the formula in §13.2, Re tr $\Delta_Z u(Z)$ is equal to

$$\eta \operatorname{Re} \mathrm{tr}\Delta_Z \mathrm{tr}((Z-Z_0)(\overline{Z - Z_0})') = 2\eta \operatorname{Re} \mathrm{tr}\ (I - Z\bar{Z}') \operatorname{Re} \mathrm{tr}(I - \bar{Z}'Z)$$
$$+ 2\eta \operatorname{Re} \mathrm{tr}\bar{Z}'Z \operatorname{Re} \mathrm{tr}\ (I - \bar{Z}'Z) = 2n \cdot \eta\, \mathrm{tr}(I - \bar{Z}'Z) > 0.$$

In virtue of Theorem 13.4.2, $u_\eta(z)$ cannot attain its maximum at any interior point. However,

$$\max_{z \in \partial R_I(n,Q)} u_\eta(Z) \leqslant M + \eta \max_{z \in \partial R_I(n,Q)} \mathrm{tr}((Z-Z_0)(\overline{Z-Z_0})');$$

$$u_\eta(Z_0) = u(Z_0) > M + \varepsilon.$$

Thus, for sufficiently small η,

$$u_\eta(Z_0) > \max_{z \in \partial R_I(n,Q)} u_\eta(Z) + \frac{\varepsilon}{2},$$

so $u(Z_0)$ must attain its maximum at some interior point of $R_I(n, Q)$. This leads to a contradiction. Therefore (13.4.7) does not hold, i.e $u(Z) \leqslant M$ inside $R_I(n, Q)$.

We give the following consequences.

Theorem 13.4.3 *If two harmonic functions of Class ξ are equal on $U(n, Q)$, then they are identical on $\overline{R_I(n, Q)}$.*

Theorem 13.4.4 *If a given harmonic function of Class ξ attains its maximum (or minimum) at some interior point of $R_I(n, Q)$, then it must be a constant.*

In the next section, we shall find out the Poisson integral formula. Consequently, the assertion that $U(n,Q)$ is the characteristic manifold of $R_I(n, Q)$ would be proved.

§ 13.5 Poisson Kernel and Poisson Formula

The transformation (13.3.6) induces the following transformation

$$U_1 = P^{-1}(U - z_0)(I - \bar{z}_0'U)^{-1}R = \bar{P}'(I - U\bar{z}_0')^{-1}(U - z_0)\bar{R}'^{-1} \qquad (13.5.1)$$

on $U(n, Q)$, whose Jacobian $\dot{U}_1/\dot{U}$ is called the poisson kernel of $R_I(n, Q)$ (cf. Hua [4]) and is denoted by $P(z_0, U)$. We have

Theorem 13.5.1 *The Poisson kernel $P(Z, U)$ of $R_I(n, Q)$ can be written as*

$$\left(\frac{\det (I-Z\bar{Z}')}{\det (I - \bar{Z}'U)\det(I-\bar{U}'Z)}\right)^{2n+1}. \qquad (13.5.2)$$

Proof. Differentiating (13.5.1) gives us

$$dU_1 = \bar{P}'(1 - U\bar{Z}_0')^{-1}dU(I - \bar{Z}_0'U)^{-1}R,$$

so

$$U_1^{-1}dU_1 = \bar{R}'(U - Z_0)^{-1}dU(I - \bar{Z}_0'U)^{-1}R,$$

namely

$$\delta U_1 = \bar{R}'(I - \bar{U}Z_0)^{-1}\delta U(I - \bar{Z}_0'U)^{-1}R,$$

where

$$\delta U_1 = U_1^{-1}dU_1, \quad \delta U = UdU.$$

By Theorem 13.2.1, there exist W_1 and W_2. Both of them belong to $U(n, Q)$ such that

$$(I - \bar{Z}_0'U)^{-1}R = W_1 L W_2^{-1}, \tag{13.5.3}$$

where

$$L = [l_1, \cdots, l_n], \; 0 < l_1 \leqslant l_2 \leqslant \cdots \leqslant l_n < 1.$$

Therefore

$$\bar{W}_2'\delta U_1 W_2 = L(\bar{W}_1'\delta U W_1)L.$$

Using the results in Chapter 10, we get

$$\dot{U}_1 = \dot{U}(\det L)^{4n+2}.$$

From (13.5.3), we obtain

$$P(Z_0, U) = (\det L)^{4n+2} = \frac{(\det R)^{4n+2}}{\det (I - \bar{Z}_0'U)^{4n+2}}.$$

In virtue of (13.3.7), we have

$$(\det R)^2 = \det(I - \bar{Z}_0'Z_0) = \det(I - Z_0\bar{Z}_0'),$$
$$\det (I - \bar{Z}_0'U) = \det(I - \bar{U}'Z_0).$$

Thus the desired conclusion is proved.

Theorem 13.5.2 *Under the assumption that the transformation* (13.3.6) *maps Z_1 and U_1 onto Z_2 and U_2 respectively, we have*

$$P(Z_2, U_2) = P(Z_1, U_1)P(Z_0, U_1)^{-1}. \tag{13.5.4}$$

Proof. By the definition of the Poisson kernel, we have

$$\dot{U} = P(Z_1, U_1)\dot{U} = P(Z_2, U_2)\dot{U}_2,$$
$$\dot{U}_2 = P(Z_0, U_1)\dot{U}_1.$$

Thus (13.5.4) follows.

By the definition, we obtain that the Poisson kernel is harmonic. Set

$$Q(Z, U) = \operatorname{Re} \operatorname{tr} \Delta_Z P(Z, U).$$

By the invariance of Laplacian and (13.5.4), we easily get

$$\operatorname{Re} \operatorname{tr} \Delta_Z P(Z_1, U_1) = (\operatorname{Re} \operatorname{tr}\Delta_{Z_2} P(Z_2, U_2)) \cdot P(Z_0, U_1).$$

When $Z_2 = 0$ in particular, we have $Z_1 = Z_0$. Thus

$$Q(Z_1, U_1) = Q(0, U_2)P(Z_1, U_1).$$

It is easily to show that $Q(0, U_2)$ is independent of U_2, and is denoted simply by e. Differentiating the identity

$$\int_{U(n,Q)} P(Z,U)\dot{U} = C_n$$

under the integral sign, where C_n is the volume of $U(n, Q)$, we get

$$\int_{U(n,Q)} Q(Z, U)\dot{U} = 0.$$

Consequently,

$$0 = \int_{U(n,Q)} e\, P(Z, U)\dot{U} = eC_n.$$

Thus $e = 0$, namely

$$\text{Re tr}\, \Delta_Z P(Z, U) = 0. \tag{13.5.5}$$

Therefore, without using the explicit expression of $P(Z, U)$, we obtain only by the differentiability of $P(Z, U)$

Theorem 13.5.3 *The Poisson kernel $P(Z, U)$ is a harmonic function of Z.*

By differentiation under the integral sign, we know that the Poisson integral

$$u(Z) = \frac{1}{C_n}\int_{U(n,Q)} \varphi(U)P(Z, U)\dot{U} \tag{13.5.6}$$

of any continuous function $\varphi(U)$ on $U(n, Q)$ is a harmonic function. Moreover, we shall prove

Theorem 13.5.4 *$u(Z)$ defined by the Poisson integral* (13.5.6) *is a harmonic function of Class ξ.*

Proof. We only need to prove the following two formulae

$$\lim_{Z\to V} u(Z) = \varphi(V), \tag{13.5.7}$$

and

$$\lim_{Z\to[1,z_1]} u(z) = \frac{1}{C_{n-1}}\int_{U(n-1,Q)} \varphi([1, U_1])P_{n-1}(Z, U_1)\dot{U}, \tag{13.5.8}$$

$P_{n-1}(Z_1, U_1)$ being the Poisson kernel of $R_I(n-1, Q)$.

First we show (13.5.7) for $n = 1$, i.e. show

$$\lim_{\rho\to 1} \frac{1}{C_1}\int_{|q|=1} \varphi(q)\,\frac{(1-\rho^2)^3}{|1-\rho q|^6}\,\dot{q} = \varphi(1), \tag{13.5.9}$$

where $\dot{q}$ is the volume element of $U(1, Q)$. Let

$$q = (1 + h)(1 - h)^{-1}, \quad (h = (q - 1)(q + 1)^{-1}).$$

It can be easily shown that, when q runs over $U(1,Q)-\{-1\}$, we have $\operatorname{Re} h = 0$, h running over all "pure imaginary numbers" $ib + jx + ky$ and

$$q^{-1}dq = \delta q = 2(1 + h)^{-1}dh(1 + h)^{-1}, \quad \dot{q} = 2^3(1 - h^2)^{-3}\dot{h},$$

$$C_1 = 2^3 \iiint_{-\infty}^{\infty} \frac{dbdxdy}{[1 + b^2 + x^2 + y^2]^3} = 2^3 \cdot 4\pi \int_0^{\infty} \frac{R^2dR}{(1 + R^2)^3} = 2\pi^2. \tag{13.5.10}$$

$q_1 = (q - \rho)(1 - \rho q)^{-1}$ and $h_1 = \dfrac{1 + \rho}{1 - \rho}h$ being set, (13.5.9) can be re-written as

$$\lim_{\rho\to 1} \frac{1}{C_1} \int_{|q|=1} \varphi(h)2^3(1 - h_1^2)^{-3}\dot{h}_1 = \psi(0), \ (\psi(h) = \varphi(q))$$

or

$$\lim_{\rho\to 1} \frac{1}{C_1} \int_{(h_1)} \psi\left(\frac{1 - \rho}{1 + \rho} h_1\right) 2^3(1 - h_1^2)^{-3}\dot{h}_1 = \psi(0).$$

Thus, from the absolute convergence of the integral (13.5.10) and the continuity of $\psi(h)$, the above desired formula follows.

The case $n > 1$ of (13.5.7) can be treated by the same method (cf. Chapter 11).

For proving (13.5.8), we first show

$$\lim_{\rho\to 1} u([\rho, 0, \cdots, 0]) = \frac{1}{C_{n-1}} \int_{U(n-1,Q)} \varphi([1, U_1])\dot{U}_1. \tag{13.5.11}$$

Setting

$$\begin{aligned} H &= (U - I)(U + I)^{-1}, \\ U_1 &= M^{-1}(U - [\rho, 0^{(n-1)}])(I - [\rho, 0^{(n-1)}]U)^{-1}M, \\ H_1 &= LHL, \end{aligned} \tag{13.5.12}$$

where

$$M = [\sqrt{1 - \rho^2}, I^{(n-1)}], \quad L = \left[\sqrt{\frac{1 + \rho}{1 - \rho}}, I^{(n-1)}\right],$$

we have

$$\dot{U}_1 = P(\Lambda, U)\dot{U}, \quad \dot{H}_1 = \dot{H}(\det L)^{4n+2}, \tag{13.5.13}$$

where $\Lambda = [\rho, 0^{(n-1)}]$. However (cf. Chapter 10)

$$\dot{U} = 2^N\det(I - H^2)^{-\frac{N}{n}}\dot{H},$$

$$\dot{U}_1 = 2^N\det(I - H_1^2)^{-\frac{N}{n}}\dot{H}_1,$$

where $N = \dim U(n, Q) = \dim USp(2n) = n(2n+1)$; and, in order to prove (13.5.11), we should show

$$\lim_{\rho\to 1}\frac{1}{C_n}\int_{(H_1)}\phi(L^{-1}H_1L^{-1})2^N\det(I-H_1^2)^{-(2n+1)}\dot{H}_1$$

$$=\frac{1}{C_n}\int_{(H_1)}\phi([0,\tilde{H}])2^N\det(I-H_1^2)^{-(2n+1)}\dot{H}_1$$

$$=\frac{1}{C_{n-1}}\int_{(\tilde{H})}\phi([0,\tilde{H}])2^{(n-1)(2n-1)}\det(I-\tilde{H}^2)^{-(2n-1)}\dot{\tilde{H}}. \qquad (13.5.14)$$

The first equality follows from the continuity of $\phi(H_1)$ and the absolute convergence of the integral

$$\frac{1}{C_n}\int_{(H_1)}2^N\det(I-H_1^2)^{-(2n+1)}\dot{H}_1 = 1.$$

The second one can be deduced from the recursion formula used in calculating C_n in Chapter 10.

Thus we have proved that, along the special path, (13.5.7) and (13.5.8) are valid. By Theorem 13.5.2 and the uniform continuity, we know that (13.5.7) and (13.5.8) hold valid along any path.

From § 13.4 and Theorem 13.5.4, it follows that the solution for the Dirichlet problem in $R_I(n,Q)$ is uniquely given by

$$u(Z)=\frac{1}{C_n}\int_{U(n,Q)}\varphi(U)P(Z,U)\dot{U}.$$

Taking $Z=\rho I\,(0<\rho<1)$ in particular, we obtain a convergence theorem of the Abel summation on unitary symplectic group $USp(2n)$ when $\rho\to 1$. And, if $\varphi(U)$ is continuous on $U(n, Q)$, the corresponding kernel is the same as the Poisson kernel obtained here.

Epilogue

In this book, we deal with harmonic analysis on the three most important compact Lie groups, namely, unitary, rotation and unitary symplectic groups.

Let G denote any one of these three groups. The main points are sketched out as follows.

Since all irreducible representations of any compact Lie group form a complete orthogonal system with regard to continuous functions, the continuous function $u(g)$ on G can be expanded to the Fourier series.

1. Regard unitary groups as characteristic manifolds of classical domains for the first class of several complex variables $I - Z\bar{Z}' > 0$. Since the classical domain has the Poisson kernel, the Poisson kernel on unitary groups can be deduced. Having expanded the Poisson kernel in characters, we obtain the Abel summation of the Fourier series on unitary groups.

Similarly, consider rotation groups as characteristic manifolds of real classical domains for the first class of several real variables $I - XX' > 0$. Since the real classical domain has the Poisson kernel, the Poisson kernel on rotation groups can be deduced. After the Poisson kernel being expanded in characters, Abel's summation of the Fourier series on rotation groups is produced.

In the same way, regard unitary symplectic groups as characteristic manifolds of the classical domains of the first class on the quaternion field $I - Z\bar{Z}' > 0$. Then,as the class of unitary groups and rotation groups, the Poisson kernel and the Abel's summation of the Fourier series on unitary symplectic groups are established.

All the Poisson kernels thus obtained are positive and have the properties of the δ-function. By the aid of the skills in group representations and matrix integrals, the explicit expressions for the coefficients in the expansion of the Poisson kernel can be fully determined. The following Theorem is formulated.

Theorem 1 *If $u(g)$ is a continuous function on G, and $g \in G$, then its Fourier series is Abel summable to itself.*

2. On these three groups, the Cesàro (c, α)-sum can be established. This (c,α)-sum, different from the Abel sum, is finite, its corresponding Cesàro (c, α)-kernel is semi-positive and has quite a simple form. When α tends to infinity, the above Cesàro kernel is just the preceding Poisson kernel, and we propose

Theorem 2 *If $u(g)$ is a continuous function on G and $g \in G$, then its Fourier series, under a proper restriction on α, is Cesàro (c, α) summable to itself.*

If $u(g) \in \mathrm{Lip}\, p$, then the estimate of the error between $u(g)$ and the Cesàro (c, α) sum of its Fourier series can be obtained.

The coefficients of Cesàro's summation thus defined can be calculated explicitly. In particular, when $\alpha = 1$, the Fejér summation is obtained with its Fejér kernel being positively definite.

3. On these three groups, the partial sum of the Fourier series and the corresponding Dirichlet kernel can be determined and, thus, the corresponding convergence criterion can be given out.

Theorem 3 *If $u(g)$ is a continuous function on G and $g \in G$, then, when $u(g) \in C^{p+N/2}(0 < p \leqslant 1)$, the partial sum of its Fourier series converges to itself, where N is equal to the difference between the dimension of G and the number of the independently variable eigenvalues of G.*

In addition, some criteria for absolute convergence of the Fourier series can also be deduced.

Moreover, the difference between the partial sum and $u(g)$ is given by

$$O\left(\max\left(\left(\frac{\ln^{\beta^2} N}{N}\right)^{\frac{1}{\beta+1}}, \frac{\ln^{\beta-1} N}{NP}\right)\right),$$

where β refers to the number of the independently variable eigenvalues of G.

4. On these three groups, the "spherical summation" of the Fourier series can be well defined and the corresponding integral expression can be deduced. Furthermore, we have

Theorem 4 *If $u(g)$ is a continuous function on G, and $g \in G$, then, when $\delta > (\dim_{\mathbb{R}} G - 1)/2$, its Fourier series is Abel-, Gauss-, Sommerfeld- and Riesz- summable of order δ to itself.*

Our problem involved is whether the results mentioned above hold true for the general compact Lie group. Concretely speaking, it covers how to establish the Poisson kernel, Abel's sum and the corresponding summability theorem, how to establish the Cesàro (c, α)-kernel, the Cesàro (c,α)-sum and the corresponding summability theorem, how to determine the corresponding Dirichlet kernel and how to deduce the convergence criterion for the partial sum of the Fourier series, and how to establish the integral expression and the corresponding convergence theorem for "spherical summation" on general compact Lie groups, etc. All these problems demand much further research effort.

References

S. Bochner

[1] Summation of multiple Fourier series by spherical means, *Tran. Am. Math. Soc.*, **46**(1936), 175—207.

H. Boerner

[1] Representations of groups, North-Holland Publishing Co., Amsterdam, 1963.

E. Cartan

[1] Sur les domaines bornés homogènes de l'espace de n variables complexes, *Abh. Math. Semin. Univ. Hamb.* **11**(1935), 116—162.

K. Chandrasekharan and S. Minakshisundaran

[1] Typical means, Oxford University Press, 1952.

Chen Guangxiao

[1] The volume of unitary-symplectic groups (Chinese), *Kexue Tongbao,* **19** (1981), 1212.

[2] Harmonic analysis on bounded domain that has $USp(2n)$ as its characteristic manifold (Chinese), *J. Math. Res. and Expo.*, **1**(1984), 14—16.

Chen Guangxian and He Zuqi

[1] Harmonic analysis on unitary-symplectic groups I: convergence criterion of Fourier series (Chinese), *J. Math. Res. Expo.*, **1**(1981), 29—42.

[2] Harmonic analysis on unitary-symplectic groups II: Cesàro summability of Fourier series (Chinese), *J. Math. Res. Expo.*, **1**(1983), 97—100.

[3] Harmonic analysis on unitary-symplectic groups III:Abelian summability of Fourier series (Chinese), *J. Math. Res. Expo.*, **2**(1983), 23—26.

[4] Harmonic analysis on unitary-symplectic groups IV: spherical summability of Fourier series (Chinese), *J. Math. Res. Expo.*, **3**(1983), 51—54.

Chen Jiangong (K. K. Chen)

[1] Trigonometric series, Sci. and Tech. Press of Shanghai, 1976.

C. Chevalley

[1] Theory of Lie groups I, Princeton University Press, 1964.

Dong Daozhen and Wang Shikun

[1] Harmonic analysis on rotation groups I: convergence criterion of Fourier series (Chinese), *Chin. Ann. Math.*, Ser. 4A, **2**(1983), 195—206.

[2] Harmonic analysis on rotation groups II: Cesàro summability of Fourier series (Chinese), *Chin. Ann. Math. Ser.*, 4A, **3**(1983), 369—378.

[3] Harmonic analysis on rotation groups III: spherical summability of Fourier series (Chinese), *Chin. Ann. Math.* Ser., 4A, **5**(1983), 547—556.

A. Erdélyi, W. Magnus, E. Oberhettinger and F. G. Tricomi

[1] Higher transcendental functions, McGraw-Hill, New York, 1953.

E. M. Gelfand and M. A. Naimark

[1] Unitary representations of the classical groups. Trudy Mat. Inst. Steklov, V. **36.** Moscow, 1950. *(Russian)*

E. M. Gelfand, L. A. Minloe and E. A. Shapiro

[1] Representations of rotation groups and Lorentz groups, Moscow, 1958. (Russian)

Gong Sheng (Sheng Kung)

[1] Fourier analysis on unitary groups I, *Acta Math. Sin.*, **10**(1960), 239—261.

[2] Fourier analysis on unitary groups II, *Acta Math. Sin.*, **12**(1962), 17—31.

[3] Fourier analysis on unitary groups III, *Acta Math. Sin.*, **13**(1963), 152—161.

[4] Fourier analysis on unitary groups IV, *Acta Math. Sin.*, **13**(1963), 323—331.

[5] Fourier analysis on unitary groups V, *Acta Math. Sin.*, **15**(1965), 305—325.

[6] Partial sum of Fourier series on rotation groups, *J. Univ. of Sci. and Tech. of China*, **9**(1979), 25—30.

S. Helgason

[1] Differential geometry, Lie groups and symmetric spaces, Academic Press, New York San Francisco London, 1978.

Hua Luogeng (Hua Loo-Keng)

[1] Harmonic analysis of functions of several complex variables in the classical domains (in Chinese), Science Press, 1958.

[2] A convergence theorem in the space of continuous functions on compact group, *Science Record*, **2**(1958), 341—344.

[3] A system of partial differential equation, *Science Record*, **1**(1957), 7—8.

[4] Starting with the unit circle, Springer-Verlag, Berlin Heidelberg New York, 1981.

Hua Luogeng and Lu Qikeng

[1] Theory of harmonic functions in the classical domains, *Sci. Sin.*, Ser, A **8** (1959), 1031—1094.

Hua Luogeng and Wan Zhexian

[1] The classical groups, Sci. and Tech. Press of Shanghai, 1963.

D. Jackson

[1] The theory of approximation, *Amer. Math. Soc.* Colloquium Publications, Vol. 9, New York, 1930.

B. M. Levitan

[1] Almost periodic functions, 1953. (Russian)

L. H. Loomis

[1] An introduction to abstract harmonic analysis, D. Van Nestrand Co., New York, 1953.

Lu Qikeng

[1] The classical manifolds and the classical domains, Sci. and Tech. Press of Shanghai, 1963.

J. Masielak

[1] On absolute convergence of multiple Fourier series, *Ann. Pol. Math.*, **3**(1958—1959), 107—120.

F. D. Murnaghan

[1] The theory of group representations, The Johns Hopkins Press, Baltimore, 1938.

[2] The unitary and rotation groups, Lectures on applied mathematics, Vol 3, 1962.

M. A. Naimark

[1] Linear representations of Lorentz groups, Moscow, 1958. (Russian)

E. N. Natanson

[1] Construction theory of functions, 1949. (Russian)

F. Peter and H. Weyl

[1] Die Vollständigkeit der primitiven Darstellungen einer geschlossen kontinuierlichen Gruppe, *Math. Ann.*, **97** (1927), 735—755.

L. S. Pontryagin

[1] Continuous groups, Moscow, 1954. (Russian)

Sun Jiguang

[1] Harmonic functions of a class of symmetric bounded domains, *J. Uni of Sci. and Tech. of China*, **2**(1973), 43—55.

E. C. Titchmarch

[1] An introduction to the theory of Fourier integral, Oxford University Press, 1948.

A. Weil

[1] L'integration dans les Groups Topologiques et ses applications, 1953.

H. Weyl

[1] Harmonic analysis on homogeneous manifold, *Ann. Math.*, **35**(1934), 486—499.

[2] The classical groups, Princeton Univ. Press, 1946.

W. H. Young

[1] On multiple Fourier series, *Proc. Lond. Math. Soc.*, II. Ser., **11**(1911—1912), 133—184.

Zhong Jiaqing

[1] A class of integral determinants and its applications to the theory of group representations, *Acta Math. Sin.* **19**(1976), 88—106.

[2] Harmonic analysis on rotation groups——Abelian summability. *J, Uni. of Sci. and Tech. of China*, **9**(1979), 31—43.

[3] A note on Schubet calculus, Proc. of the Symposium on Differential Geometry and Differential Equations, 1679—1708, 1980, Beijing.

A. Zygmund

[1] Trigonometric series, vols I and II, Cambridge University Press, 1959.

[2] On the differentiability of multiple integrals, *Fundam. Math.*, **23**(1934), 143—149.

Index

The manufacturer's authorised representative in the EU is Springer Nature Customer Service Centre GmbH, Europaplatz 3, 69115 Heidelberg, Germany. If you have any concerns regarding our products, please contact ProductSafety@springernature.com

Printed and bound by CPI Group (UK) Ltd, Croydon, CR0 4YY
28/04/2026
02098503-0005